Building in the 21st Century

D0736246

Building in the 21st Century

Robert Cooke

Blackwell Publishing editorial offices:
Blackwell Publishing Ltd, 9600 Garsington Road, Oxford OX4 2DQ, UK
  Tel: +44 (0)1865 776868
Blackwell Publishing Inc., 350 Main Street, Malden, MA 02148-5020, USA
  Tel: +1 781 388 8250
Blackwell Publishing Asia Pty Ltd, 550 Swanston Street, Carlton, Victoria 3053, Australia
  Tel: +61 (0)3 8359 1011

First published 2007 by Blackwell Publishing Ltd

ISBN: 978-14051-5655-4

Library of Congress Cataloging-in-Publication Data

Cooke, Robert.
    Building in the 21st century / Robert Cooke.
        p. cm.
    Includes index.
    ISBN-13: 978-1-4051-5655-4 (pbk. : alk. paper)
    ISBN-10: 1-4051-5655-4 (pbk. : alk. paper) 1. Building–Textbooks. I. Title. II.
Title: Building in the twenty first century.
    TH146 C66 2007
    690–dc22   2007017322

A catalogue record for this title is available from the British Library

Set in 10/12 Palatino by Aptara Inc., New Delhi, India
Printed and bound in Singapore by C.O.S. Printers Pte Ltd

For further information on Blackwell Publishing, visit our website:
www.blackwellpublishing.com/construction

Contents

Preface

Some may ask why there is a need for another book on construction. The answer is that the industry has changed, there are new challenges such as global warming and climate change, and perhaps more importantly for students and lecturers, so have the BTEC/Edexcel units. This volume is completely new and has been written specifically for students of the 21st century studying for a National Award or National Certificate in construction.

With hundreds of full colour graphics and full explanations, this textbook will be useful for students up to first year degree level, especially for those who have entered university without taking the BTEC/Edexcel route; students taking the new GCSE in construction theory units will also find the book useful. The information provided by this book will be invaluable to the construction student for many years.

Starting with the topic 'surveying' the book can be used for quick reference on how to achieve a specific outcome or to introduce the reader to the many facets of the professional surveyor. Health, safety and welfare together with risk assessment have been integrated throughout the book, emphasising that they are not a 'bolt on' unit but part of the everyday routine. The approach has been styled to help the reader to understand why and how procedures work. Surveying can require calculations; therefore formulae and worked examples have been included as required to help with analytical methods. Chapters 1–3 should be very useful for those studying any of the surveying units.

A feature of this book has been to address construction using an elemental approach, working through the building in Chapters 4–7 and then looking at construction techniques and methods in Chapter 8, using full colour graphics and detailed explanations of each element, showing how they connect. The science and reasons behind common failures and specific design issues are explained. Chapters 4–8 cover 'building technology' and 'construction technology and design'.

Chapter 9 on 'site issues' explains the problems and requirements for a safe site. Full colour photographs of actual site issues, including modern plant at work, enable the reader to gain an insight to life on site.

Chapter 10 on 'construction and the environment' follows the theme of the unit with the same title. Pollution and climate change are major issues in the world around us and constantly in the news. This chapter looks at the subject beginning with the fundamental causes, through to a more in-depth approach leading onto a dedicated chapter on energy.

Global energy needs are increasing at a faster rate than the population explosion. Chapter 11 presents a full range of methods for generating electricity and sourcing alternative energy and fuels. The whole spectrum of energy supply is addressed, from atomic sources through to the latest 'green' alternatives. Simple calculations will enable the reader to make comparisons of the common emotive issues on a technical basis.

The unique approach of this book integrates a range of skills enabling the reader to use analytical methods to solve practical problems, such as those of foundation design and surveying. The selection of materials and design of the elements has been approached incorporating the 'science of materials' unit, emphasising the importance of a holistic approach to learning.

Why not try the free *Building in the 21st Century* website (www.blackwellpublishing.com/cooke) where there are resources for students, teachers, and lecturers, and at all levels of study?

Good luck with your studies!

We should never stop learning.

Robert Cooke

Acknowledgements

I would like to thank Julia Burden for taking the risk on such a new style of textbook and her team for their advice over the past year. Thanks to Keith Jenner, ex-colleague who proofread the book – sorry if some of the text goes against the grain – and to the following people and organisations for their help in ensuring the latest information for this book:

David Langston – Wavegen
Keith Lovell – BovisLendLease
Leonid Ragozin – BBC Russian Correspondent
Roger Johnson – Avongard
Carey's
CDC Demolition Ltd
Drax Power Ltd
Geofirma
Leica Geosystems
Marine Current Turbines TM Ltd
Morrisons
The Open University
Three Cross Demolition Contractors
Whitnell Contracts

Websites and further reading

You may find the following Internet websites of further interest:
http://news.bbc.co.uk
http://www.coal.gov.uk
http://www.crossriverpartnership.org
http://www.dti.gov.uk/energy
http://www.eon-uk.com
http://www.ecn.nl/en
http://www.greenspec.co.uk
http://www.geographyinaction.co.uk
http://www.halfen.co.uk
http://www.homeoffice.gov.uk/rds/prgpdfs/brf117.pdf
http://www.hsedirect.com/search/SQL/dataitem.asp?id=189421
http://www.hse.gov.uk/aboutus/hsc/iacs/coniac/201103/201110.pdf
http://www.hse.gov.uk/LAU/lacs/90-3.htm#appendixi
http://www.landreg.gov.uk/legislation/
http://www.letchworthgardencity.net/heritage/index-3.htm
http://www.marineturbines.com
http://www.metoffice.gov.uk
http://www.nordex-online.com
http://www.opsi.gov
http://www.publications.parliament.uk
http://www.planningportal.gov.uk
http://www.priweb.org
http://www.rise.org.au/info/Tech/hightemp/index.html
http://www.tfl.gov.uk
http://www.tve.org
http://www.wavegen.com
http://www.world-nuclear.org/info/chernobyl/info07.htm

Further reading

Stephen Emmitt & Christopher Gorse (2005) *Barry's Introduction to Construction of Buildings*. Blackwell Publishing, Oxford.

Stephen Emmitt & Christopher Gorse (2006) *Barry's Advanced Construction of Buildings*. Blackwell Publishing, Oxford.

Al Gore (2000) *Earth in the Balance*. Earthscan Publications, London.

Stephen Pople (1995) *Explaining Physics – GCSE Edition*, 2nd edn. Oxford University Press, Oxford.

Stephen Halliday (2004) *Water – A Turbulent History*. Sutton Publishing, Stroud.

Chapter 1

Surveying processes

This chapter has been divided into two parts: Part A concentrates on **lineal methods of measurement**, including the equipment required and practical techniques used in the field and drawing office; Part B concentrates on the techniques used for **levelling**. Health and safety, including risk assessment, are included as an integral part of the text, emphasising that these subjects are part of everyday routine.

PART A

1.1 The role of the surveyor

The role of the surveyor is taken back to the fundamentals to introduce the many facets of their job. The surveyor's role includes:

- Measuring and recording what exists or setting out for what will be built. For example, a land surveyor may be asked to measure a large area of land. In contrast, a building surveyor would measure structures. Both are measuring and recording what actually exists.
- If the plot of land is to be developed into, say, a housing estate, the surveyor will set out the boundaries of each plot, the positions of the new roads, services, positions of the new structures, etc. – 'setting out for what will be'. The building surveyor may be assigned to provide the builders' levels and setting out points for new buildings.

1.2 Land surveying

A surveyor would normally carry out the measurement and recording of land and anything upon it. Depending on the size of the task and the existing land use, various methods would be used.

Let's start with the terminology. **Green field sites** are land that has not been previously built on. Examples include farmland that has permission for **change of use** to building/development land, and school playing fields that have been sold off for housing development.

Brown field sites are land that has previously been developed but where the buildings or structures are no longer required. For example; a site contained an old factory unit surrounded by housing, but the local planning authority would not allow the owners to demolish the old factory and build a new one. Therefore the owners applied for a change of use and sold the land for redevelopment. The local authority planning department restricted expansion and the factory closed down. The land was re-designated as **residential land**; therefore it increased in value and a new block of flats was built on the site. The Environment Agency has estimated that there are 66 000 hectares of brown field sites in England.

Brown field sites should not be confused with **green belt** land or **white land** which are areas defined by local authorities and national government (the Crown).

All survey procedures should start with planning:

- **Location** – site address or map reference as with farmland
- **Reason for survey** – valuation for asset or tax purposes, boundary disputes, future development, land registration requirement for sale of land
- **Equipment** – the type of survey will determine equipment requirement
- **Health and safety issues** – personal protective equipment (PPE). To comply with the Health and Safety at Work Act 1974, the surveyor should have a risk assessment in place before any survey is carried out.

Location

A postal address including the post code should be used if available. Where land has not been previously built on, map references will be required. The Ordnance Survey (OS) produces both paper and digital map data files and charts for use with computer software. They range from regional mapping with a scale of 1:50 000 down to detailed 1:9999 in suburbs and villages. For built-up areas such as cities and small counties the scale ranges from 1:10 000–1:49 999. The utility companies and local authorities record their information digitally using OS mapping.

The OS maps cover the whole of Great Britain with a point of origin southwest of Land's End in Cornwall, meaning that all map references have positive numbering to the east and north of the point of origin. A grid system has been used to divide Great Britain into manageable maps. Each grid has a prefix letter from the words 'St John'.

The Ordnance Survey dates back to the mid 1700s when King George II, fearing the possibility of invasion, commissioned the army to carry out a military survey of the highlands of Scotland. At the end of the eighteenth century the threat of invasion of Britain by French troops under Napoleon increased. The army required easily referenced and accurate maps of the land, especially of the south coast, so the Royal Ordnance Corps started mapping Kent and then Essex, and continued with the rest of Great Britain. Previously, maps had been drawn up by various people using their own scales or no scale at all; at best these gave general information about an area. Today the OS has an excellent website at http://www.ordnancesurvey.co.uk/oswebsite/ which is worth a visit. In particular, although designed for children, a leaflet showing how to use an OS map can be downloaded from http://www.ordnancesurvey.co.uk/oswebsite/freefun/nationalgrid/nghelp1.html. Further and higher education students can visit http://edina.ac.uk/digimap/.

The map of the United Kingdom is divided into 100 km squares, each referenced using a prefix letter from the words 'St John' and a second letter of the alphabet. Each 100 km square is then divided into 10 km spacings, starting at the bottom left hand corner. Each 10 km square can be divided by 10 vertically and horizontally, enabling a grid reference to the nearest 1000 m. When reading a grid reference, always start with the longitude (west to east) followed by the latitude (south to north). One way to remember which comes first is *'along the*

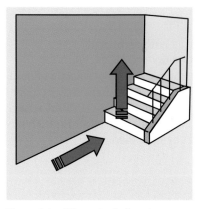

Figure 1.1 'Along the hall and up the stairs'.

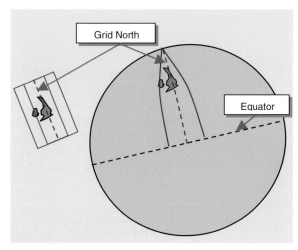

Figure 1.2 Grid North.

hall and up the stairs' (Figure 1.1). If a more accurate location is required, a larger scale map should be used. Each map will indicate the true north, grid north and magnetic north. Magnetic north moves as the tilt of the planet changes; therefore the date of the map must be taken into account. Grid north is only correct on maps 2° west of the central meridian. All other maps either side will run parallel and therefore will not be true north. This is because the Earth is widest at the equator and becomes narrower as it reaches the poles. In contrast, the grid lines on the maps are the same width throughout (Figure 1.2).

Reasons for land surveys

Land surveys can be used for **cadastral** purposes, which enables the taxation department to apply tax appropriate to the size, use and location of the land. A hectare of land in the West End of London, for example, will be more valuable than, say, a hectare on a mountain in North Wales.

The Land Registration Act 2002 requires all land to be registered. When a plot of land or property is sold, the Act requires the new owner to make the declaration. Unregistered land can be voluntarily registered if there are major changes to interests in land possession without the need for sale. For more detail visit the Land Registry website at http://www.landreg.gov. uk/legislation/.

The Act can be used to trace the owners of land that has become a health hazard by illegal dumping or storage of hazardous materials. Previously it had been impossible to serve a summons if the owner could not be traced. Another reason for land owners not wanting their interests known was to avoid tax. It was not possible to levy tax where the owner of the land could not be identified.

Equipment

Health and safety equipment

- Warning triangles for use when surveying near roads

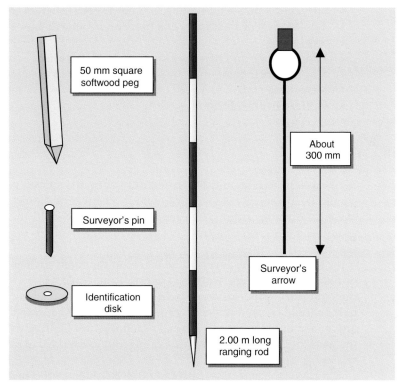

Figure 1.3 General surveying equipment.

- High visibility (known as hi-viz) vests or jackets – a legal requirement when surveying on a public footpath or near a road
- Stout footwear – ideally waterproof with steel toe protection and ankle support such as leather boots
- Hard hat – although it is not essential unless there is a risk from overhead such as rock fall or parts from buildings etc., a hard hat also increases your visibility making you more easily seen
- Gas monitor and respirator – if working below ground level
- Mobile phone for emergency help – land surveyors may be working in remote areas or working on their own
- Warm clothing – the weather can change significantly whilst out surveying especially in remote areas.

Surveying equipment (Figure 1.3)

Not all of the following equipment is required for every survey. The list includes optional equipment used for different techniques.

- Timber pegs, mallet or club hammer (and metal spike for hard ground), coloured paints, chalks

- Surveyor's pins – for nailing into hard surfaces when setting up stations
- Total station or theodolite and electromagnetic distance measure (EDM)
- Steel tapes
- Ranging rods and tripods for use on hard surfaces
- Magnetic compass
- Measuring staff
- Optical level and tripod.

Health and safety issues

A risk assessment should be carried out before the survey takes place. There should be a general assessment to which all surveys should adhere, and any specific issues regarding each individual survey should then be added. For example, if the survey requires measuring near a road it is important to ensure that the surveyors are seen, so high visibility vests or jackets should be worn. If the survey requires levels to be taken on the road, such as manhole covers or road gully gratings, warning triangles and traffic cones will be needed. If the light during the survey is poor, as at dusk, then flashing yellow warning lights should be used or floodlighting from a lighting mast. Where road works are carried out at night it is normal practice to use lighting masts, not only for the workforce to see what they are doing, but to ensure that drivers can see the works for some distance.

A copy of the risk assessment should be given to the main contractor for inclusion in their overall risk assessment for the project. There will be occasions when a risk may not have been foreseen: for example, when using a total station and reflectors the tripod legs may obstruct the footpath. If then someone in a wheelchair, or with a pram or buggy cannot get past, there is a potential risk. Do they ask you to move (they have every right to unless you have a permit to temporarily close the footpath), knock the tripod as they go past or be put at risk by going onto the road to get past? Identifying those types of risk comes with experience, and hopefully the surveyor will consider them when setting up the equipment.

1.3 Land survey methods

There are two main methods of surveying land:

- optical method
- lineal method.

The **optical method** is based on triangulation. An optical instrument such as a theodolite with an EDM (or a total station) is used to take angular readings and distances radiating from specific points termed **stations**. Some optical levels can read horizontal angles using the horizontal circle similar to a 360° protractor.

The word **lineal** originates from the ancient Egyptian word for linen. Surveyors used strands of linen to show the shortest distance between two points. If the linen (line) was pulled taut then the line must be straight. Many, if not most, of the surveying methods originated with Egyptian surveyors. The **lineal method** is based on use of the old metal chains and measuring **offset**; however, except for some forestry surveying, chains are rarely used today, although the principles are still very useful and the chains have been replaced by tapes.

Optical method

1. Decide the area to be surveyed and sketch out a plan showing any useful reference points such as roads and buildings, or the orientation (which way is magnetic north?). The drawing will be used for reference to the more detailed survey sheets (similar to a route map of, say, the whole of Wales as a reference to the more detailed maps of roads and towns, etc.).

2. Set up a reference point (station) to which the survey can be referenced. It could be a surveyor's pin driven into a flat hard surface (Figure 1.4) or a temporary bench mark (TBM). On very large sites one method is to dig a hole about the size of a bucket and fill it with a bag of ready mixed dry concrete. Next drive a steel rod (commonly an offcut of steel reinforcement termed **rebar**) into the dry mix and *slowly* pour about 0.5 litre of clean water over to activate the cement and then leave it to set. Position the TBM so that it will not be removed or knocked (Figure 1.5; also see Chapter 9).

3. As the area may be too large to record on a single A4 sheet, a series of sheets will be required. On each sheet sketch the plan area to be surveyed and mark up with orientation or identification that relates to the main sketch plan. For example, one edge of a sheet could be labelled 'AA' and the adjoining sheet would also have 'AA' to show where the information follows on. It is easy on site to see which sheet follows on; however, back in the office the codes are of great help in sorting them out (Figure 1.6).

4. Write your name and the date the survey has been carried out. It is good practice also to write a job number or location on every sheet used, as there may be more than one project for a client, or the area may have been

Figure 1.4 Surveyor's pin.

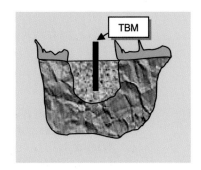

Figure 1.5 Temporary bench mark pin.

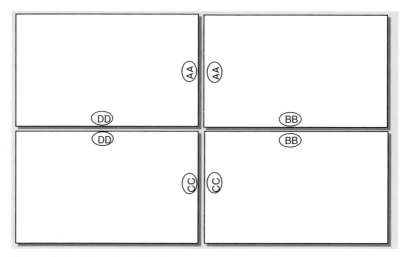

Figure 1.6 Booking sheet identification.

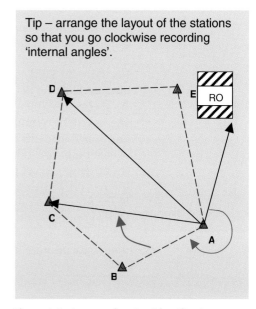

Tip – arrange the layout of the stations so that you go clockwise recording 'internal angles'.

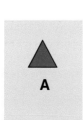

Figure 1.7 Station identification.

Figure 1.8 Layout of station identification.

surveyed previously. Back in the office it may be hard to know which sheet is the latest.

5 Decide where the other stations/points are to be located and note these on your hand drawn plan using upper case lettering and a triangle (Figure 1.7).

6 Using a total station (TS) and reflecting prism (RP) set up over the TBM or surveyor's pin, take a reading either from north (in this case you will

Station	Target	Recorded Angle	Distance	Comments
Name: *A. Higgins*			**Date: *25ᵗʰ Aug 06***	
Location: *Two Tree Farm Chelmsford Essex CM3 1OU.*				
			Job No: 032	
Station	Target	Recorded Angle	Distance	Comments
A	RO	000° 00' 00"	18.356 m	Reference Object
	B	225° 15' 05"	19.250 m	
A	B	000° 00' 00"		
	C	43° 27' 30"	27.358 m	
A	C	000° 00' 00"		
	D	37° 25' 29"	33.504 m	
A	D	000° 00' 00"		
	E	23° 17' 26"	26.425 m	

Table 1

Station	Target	Recorded Angle	Distance	Comments
Name: *A. Higgins*			**Date: *25ᵗʰ Aug 06***	
Location: *Two Tree Farm Chelmsford Essex CM3 1OU.*				
			Job No: 032	
Station	Target	Recorded Angle	Distance	Comments
A	RO	000° 00' 00"	18.356 m	Reference Object
	B	225° 15' 05"	19.250 m	
	C	268° 42' 35"	27.358 m	
	D	306° 08' 04"	33.504 m	
	E	329° 25' 30"	26.425 m	

Table 2

Figure 1.9   Examples of booking sheets.

need a magnetic compass) or from a reference object (RO) which can be any immoveable object or building (Figure 1.8). Magnetic north or the reference station will be horizontal 'zero'. Rotate the TS clockwise and book the horizontal angle and the distance using the EDM. In Figure 1.8 the angle from the RO to station B will be greater than 180°. For ease of later calculations it may be easier to book reading RO–station B, then zero the instrument and take all the other readings from station B (see Table 1 in Figure 1.9). The more common approach is to take all of the readings from the RO as shown above (see Table 2 in Figure 1.9). This method is quicker.

With modern electronic equipment accuracy is more easily obtained than with the older optical theodolites using vernier scales. However, a good surveyor should be able to achieve the same degree of accuracy with either piece of equipment. The difference is speed and ease of operation.

Lineal method

Figure 1.10 Ranging rod with supporting tripod.

Alternatively, the land can be measured using steel tapes and pegs by the **lineal survey** method. Land surveys commonly require the use of ranging rods which are 2.00 m long red and white banded poles used as markers (Figure 1.3). When they are not being used for a survey, they should be pushed into the ground at an angle so they cannot be mistaken for those being used. Do not lay them flat on the ground, especially when working in long grass, as they are difficult to find again, or vehicles operating in the area may run over them. When working on a hard surface such as a road, metal tripods should be used to hold the ranging rods vertical. Some are available with one adjustable leg for use on inclines (Figure 1.10).

The lineal method of surveying follows a number of the features of the optical survey method:

1 Sketch area to be surveyed.
2 Set up a reference point – this time, however, **witness** from static points (Figure 1.11).
3 Sketch the plan area to be surveyed – include name, site, date, etc.
4 Decide where the other stations/points will be and hammer in wooden pegs as markers or use ranging rods.
5 Measure the lineal dimension using a tape or band between the ranging rods and note it on the drawing (remember only two decimal places are needed for land surveys).
6 Measure the diagonals between the stations termed **tie lines** or **check lines**.
7 Check that all the dimensions have been recorded before removing any of the ranging rods.
8 To ensure the accuracy of the survey, a procedure termed **working from the whole to the part** is used. This means that the surveyor should measure the periphery (the outside edge) of the plot first. When the whole plot has been recorded everything is referenced back to those lines. Figure 1.12 shows a plot referenced and measured using a series of straight lines

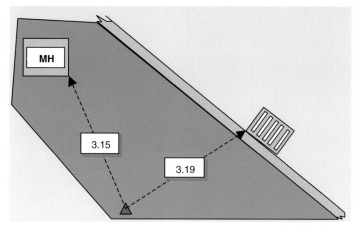

Figure 1.11 Station witnessing.

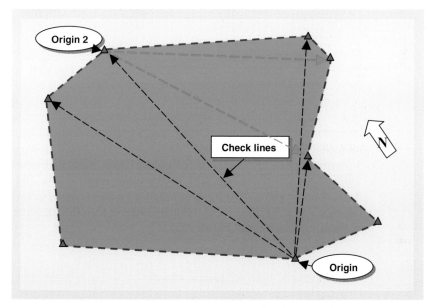

Figure 1.12 'Whole to the part' check lines.

forming triangles. From those lines the objects can then be referenced. This ensures that if an error is made recording one object it should not become compounded when measuring subsequent objects.

9   If the land is hilly then step chaining will be required (see Section 1.6).

1.4   Measuring irregular edges

Often land will have irregular shapes such as around ponds, lakes, etc., or along the boundaries. Using techniques similar to those previously described

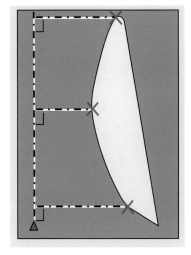

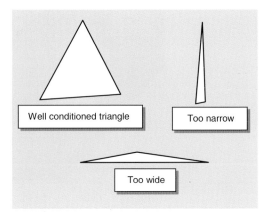

Figure 1.13 Rectangular offsets.　　**Figure 1.14** Triangle shapes.

to set up stations for the lineal survey, lay out tapes (base tapes) on the ground between the stations. If the distance between the stations is longer than the tape, then **sight through** using ranging rods (again, see Section 1.6).

Using rectangular offsets (also known as **perpendicular offsets**) measure and book the irregular shapes from the base tapes. For instance, to book an irregular shape such as a mound or a pond **arrows** could be used to mark the edge at ground level (Figure 1.13).

Arrows are metal wire rods with a loop head, some with material flashes that make them stand out from the background (Figure 1.3). The surveyor places the arrows at regular points or points that need noting. Now the distance from the tape to each arrow is measured, ensuring that the measurements are made at 90° to the base tape (Figure 1.13).

Method

1　Draw a plan of the area to be surveyed.
2　Note the location, date and orientation.
3　Position stations A and B and witness as required.
4　Position arrows at the points to be booked.
5　Note the position in relation to the chain line when at right angles to the line (see notes on the limits of rectangular offsets).
6　Book readings.
7　Position station C and measure the tie line between tapes. Tie lines are required to relate the positions of the two tapes. Ensure the triangles are **well conditioned** (proportioned) as in Figure 1.14.

When surveying a large irregular area such as a whole pond, mound or depression, a series of chain lines/base lines should be referenced together. Figure 1.15 shows four stations located around the pond each referenced by

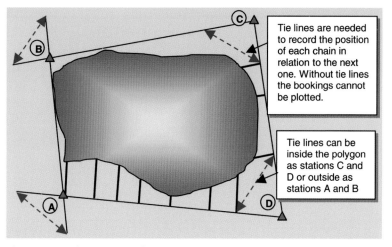

Tie lines are needed to record the position of each chain in relation to the next one. Without tie lines the bookings cannot be plotted.

Tie lines can be inside the polygon as stations C and D or outside as stations A and B

Figure 1.15 Alternative 'tie lines'.

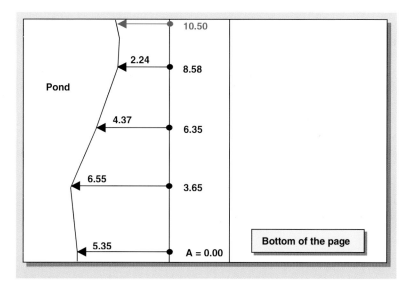

Figure 1.16 Example of a section of a booking sheet.

means of tie lines. The tie lines can be either inside the chain lines as stations C and D, or outside the chains as with stations A and B. It is essential to reference the tie lines, as without the measurements the survey cannot be drawn up.

Booking

The booking sheets comprise two parallel lines down the centre of the page as shown in Figure 1.16. They represent the tape (chainage). Only lineal measurements are shown between the lines. The station is indicated at the bottom of the page as 'A = 0.00 m'. At the top of the page 'B' would be shown equalling the length of the tape – say 23.00 m (Figure 1.17). The rectangular offsets are

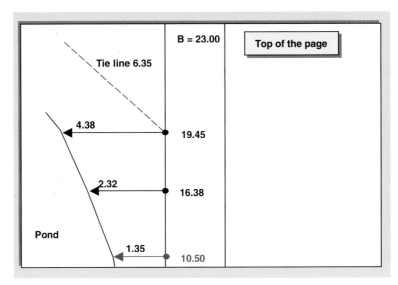

Figure 1.17 Example of a section of a booking sheet.

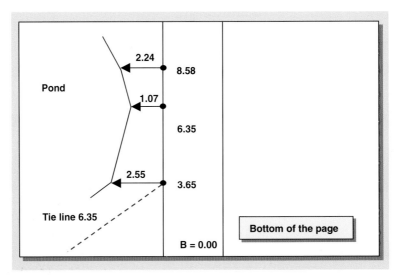

Figure 1.18 Example of a section of a booking sheet.

then shown by a ball to start and arrowhead to finish with the length of the offset written in *above* the line. In this example the pond is to the left of the tape; therefore all booking will be shown to the left of the parallel lines.

At the top of the sheet the last rectangular offset is shown as being 4.38 m long and 19.45 m from station A. It is convenient also to take the **tie line** from the same point, providing a well conditioned triangle.

A new booking sheet from station B to C would then be drawn up, noting where the tie line connected (Figure 1.18).

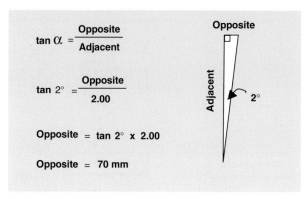

Figure 1.19 Calculating errors.

Back in the office the bookings can be plotted either manually or using computer software such as AutoCAD. Both methods require the main framework of the tapes (chain lines) to be plotted and the orientation noted. From the framework of chain lines the offsets can be plotted. All the points or nodes can be joined up to show the shape of the surveyed pond.

The maximum length of a **rectangular offset** no longer relies on the accuracy of the draughtsman as computer software is wholly accurate. The weak link is the surveyor's eyesight. For example, a surveyor should be able to achieve a right angle to about 2° accuracy by eyesight alone. If the offset is, say, 2.00 m and the possible optical error is 2° then the object could be 70 mm either side of the true position (Figure 1.19). It depends on what is being booked and for what purpose. If it is a tree, then 70 mm either way is unlikely to make much difference. However, if it is the corner of a building then perhaps triangulation would be more appropriate for greater accuracy. If the offset is 5.00 m, the possible error increases to 175 mm either side of the true point, so it is of limited use (Figure 1.20). In general, it is best to keep offsets to less than 3.00 m. If the distance is greater than 3.00 m then either **trilateration** or **triangulation** can be used.

$$\tan \alpha = \frac{\text{Opposite}}{\text{Adjacent}}$$

$$\tan 2° = \frac{\text{Opposite}}{5.00}$$

$$\text{Opposite} = \tan 2° \times 5.00$$

$$\text{Opposite} = 175 \text{ mm}$$

Figure 1.20 Comparative example.

Trilateration is a method using two lineal measurements from two known points forming a triangle. **Triangulation** requires two angles from known points (Figure 1.21).

1.5 Drawing office procedure

1 Draw line A–B 23.00 m long to an appropriate scale (Drg 1: Figures 1.22 and 1.17).

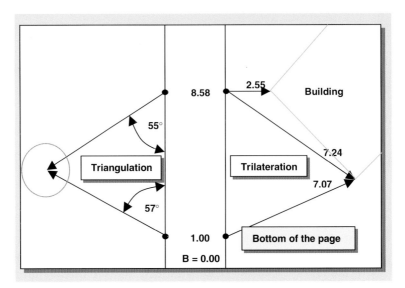

Figure 1.21 Example of a section of a booking sheet.

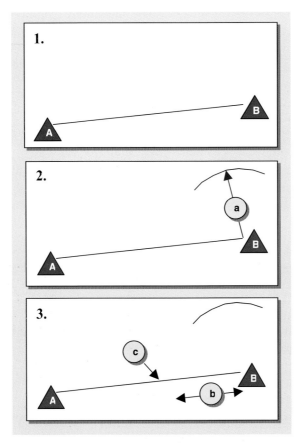

Figure 1.22 Drawing office procedure.

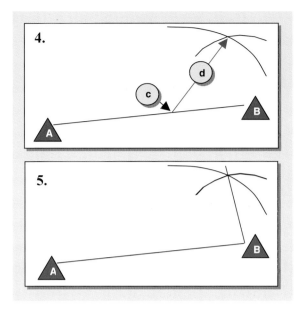

Figure 1.23 Drawing office procedure.

2 From station B strike an arc the length (in this example 3.65m) where the tie line meets line B–C (Drg 2: Figures 1.22 and 1.18).

3 Again from station B measure back towards station A to the point c from which the tie line started (in this case b = 23.00−19.45 m = 3.55 m) (Drg 3: Figures 1.22 and 1.17).

4 Now strike an arc the length of the tie line (in this example 6.35 m) d from point c (Drg 4: Figures 1.23 and 1.17).

5 Finally draw a line from station B through the point that the two arcs cross. This will be tape B–C (Drg 5: Figure 1.23).

1.6 Measuring over sloping distances

If the terrain is on an incline or contains mounds or depressions, additional techniques will be required for measuring over sloping distances. When chaining over hills, dips or smaller gradients the plan length is not the chain length (Figure 1.24). To record the information on a map or chart we must convert the third dimension (height) into a plan dimension. If compared to a right angled triangle the slope will be the hypotenuse and the plan will be the adjacent: the hypotenuse is always longer than the adjacent. Site measurement can be achieved by angular measurement of the slope, or stepping the chain. Angular measurement requires the angle of the slope from the horizontal using an **Abney level** or a **clinometer**.

An Abney level is a small hand held instrument that enables the target to be sighted whilst the angle from the horizontal is being taken. The horizontal is found using the built in spirit level which can be seen by the observer via a 45°

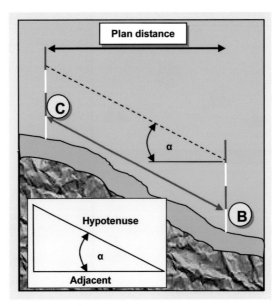

Figure 1.24 Sloping measurement.

Figure 1.25 An Abney level.

polished metal prism. The observer will be able to line up the target using the sighting wire. By using the other hand to turn the adjusting wheel, the spirit bubble in the level can be adjusted until it is between the horizontal markers. The surveyor then reads off the angle from the protractor (Figure 1.25).

Method for using an Abney level

1 Set up ranging rods in the direction you wish to follow. For example, ranging rod A in Figure 1.26 may be the last in a line before the hill. Position a

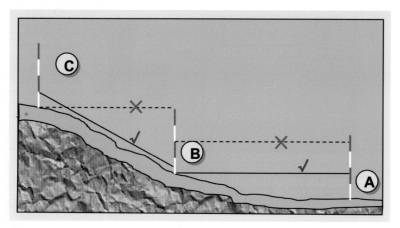

Figure 1.26 Measuring over an incline.

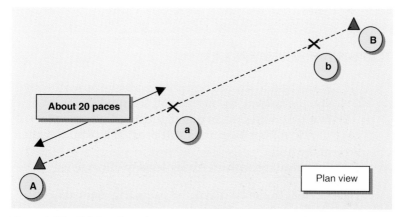

Figure 1.27 Sighting through.

ranging rod in vision with the direction you want to proceed: that will be point C. Then sight an intermediate ranging rod at point B. This method will help keep the line of ranging rods straight. Sight through two ranging rods to position the third. If practical, you may be able to sight through more ranging rods as shown in Figure 1.27. Note the stations are A and B denoted by the red triangles. Points a and b are intermediates purely to enable shorter runs for step chaining. In Figure 1.26 step chaining between station B and C would be impractical as the tape would be too high at station B to be accurate. The runnage measured between A and B should be as low as practical to ensure accuracy.

2 The two methods of finding the plan lengths are:
   - angular measurement of the slope
   - stepping the chain.

To measure the **plan length** between ranging rods B and C the angle of incline must be measured (Figure 1.26). For land measurement using tapes, the accuracy of the incline to the nearest degree is adequate.

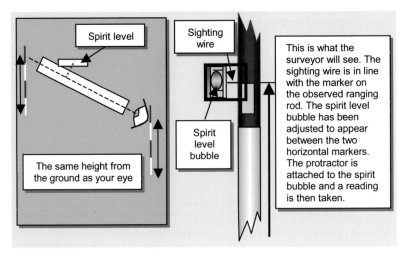

Figure 1.28 Measuring a regular incline angle.

Method for measuring plan length

1 Stand next to the ranging rod B and note how far above the ground level you are holding the instrument (Figure 1.28).
2 Mark the same height on ranging rod C using a piece of masking tape or a wide rubber band (this will provide a parallel sight line to the incline of the hill).
3 Take reading with an Abney level or clinometer.

The plan length can be calculated using trigonometry.

Example

If the measured length between station B and C is, say, 8.25 m and the angle from the horizontal is 32°, what will the plan length be?

Method

1 Sketch the problem to be solved
2 Write the appropriate formula to solve the problem
3 Transpose the formula so that the unknown (in this case the adjacent) is on the left of the equals sign
4 Complete the calculation
5 Show the answer clearly.

The formula

$$\text{cosine } \alpha = \frac{\text{Adjacent}}{\text{Hypotenuse}}$$

$$\cosine 32^\circ = \frac{\text{Adjacent}}{8.25}$$
$$\text{Adjacent} = \cosine 32^\circ \times 8.25$$
$$\text{Adjacent} = 7.00\,\text{m}$$

Therefore the plan measurement is 1.25 m less than the measured length.

1.7 Levelling surveys

Height is the measurement of the difference between points in a vertical plane.

Levelling is the process of measuring **differences in height** on the Earth's surface. There are no units for height; therefore lineal units are used such as millimetres and metres. A height can be said to be so many metres above another. To enable comparison anywhere in the UK, a static point has been set at Newlyn in Cornwall. The sea levels have been averaged out and the height is the origin for all the OS points throughout Great Britain. Figure 1.29 shows a bracketed Ordnance Survey bench mark (OSBM) on a wall in Romford, Essex. A brass bracket can be hooked into the slots providing a shelf at the precise height of the bench mark. The surveyor's assistant stands a staff on the shelf. Other OSBMs are carved into walls of buildings, and gateways. Figure 1.30 shows an OSBM carved in the stone plinth of a gate support.

Temporary bench marks (TBM) are used to relate all other levels to one point (see Section 1.11, Figures 1.48 and 1.49). When setting out levels for a new building it is commonplace to find the depth of the sewer **invert** level

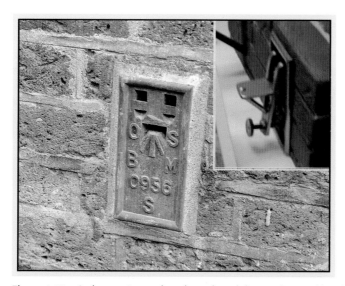

Figure 1.29 Ordnance Survey bench mark and datum plate and bracket (inset).

Figure 1.30 A surveying staff held against an Ordnance Survey bench mark.

and relate it to the TBM (Figure 1.31). All other levels, such as the depth of the foundations, height of the finished floor, etc., are then taken from the TBM. However, where major civil engineering works are to be carried out OSBMs are used. The whole of the British Isles has been surveyed and spot heights set up at set points. Each OSBM can be found on the OS maps either on paper, or more recently electronically for use with many types of computer software. In practice, the heights on the maps are used to produce calculations on falls for main sewers, and heights for inclines and declines of roads. This is especially important when new roads are being laid out from many different starting points. However, where possible, checks should be made on site.

The OSBMs found on buildings, walls, curbs, etc., are accessible to the surveyor; however, at set intervals other survey stations have been set up that have very restricted access. They are used if the other OSBMs are damaged or new points have to be set due to demolition works.

A **level plane (level surface)** is a line that runs at right angles to the Earth's gravity. It can be found using a plumb bob and measuring 90 degrees to the line. It is not a straight line as it runs parallel to the Earth's surface. (The horizon is curved so the level line will follow that same curve.) Figure 1.32 shows how a series of level lines termed **height of collimation** run at a tangent to

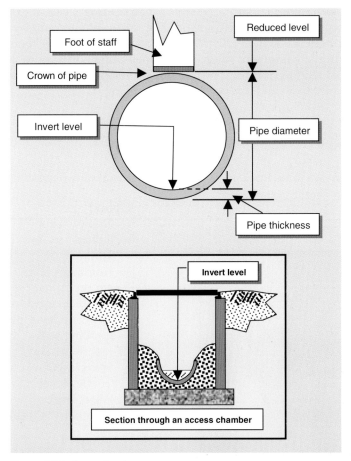

Figure 1.31 Invert levels.

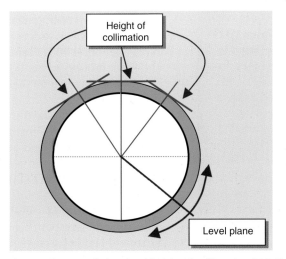

Figure 1.32 Level plane and height of collimation (HOC).

the surface of the Earth; therefore when reading levels at extended distances the curvature of the Earth should be taken into account by calculation. For distances of less than 120 m the difference is negligible at about 1 mm. Modern electronic equipment often has compensating software to provide both the actual reading and the true reading. Generally it is good practice to keep the distance to about 50 m or less when levelling.

Levelling equipment

There are three main groups of instruments used for levelling:

Figure 1.33 Levelling staff being used to take a spot height of the footpath.

1 Optical
 • Dumpy level
 • Ball and socket – quick-set
 • Tilting level
 • Automatic level
2 Electronic
 • Rotating laser level
 • Bar code level plus a tripod and levelling staffs.
3 Others
 • Spirit level
 • Cowley level
 • Water level.

(See Section 1.14.)

Figure 1.33 shows the staff being held correctly, exposing the entire staff face to the observer. Note the high-viz vest required when working near roadways. Figure 1.34 shows an automatic optical level. Note the larger focus knob top right with the slow motion screw beneath it. Two of the foot-screws can be seen below the **cat's eye** spirit level.

Figure 1.35 shows the latest bar code level. The surveyor aims the instrument at the bar coded staff and a digital numeric readout appears on the screen. This instrument incorporates an EDM that records the distance between the instrument and the staff using infrared pulses. If the battery runs out of energy during a survey, this instrument can be used in the conventional way. Levelling staffs should be held with both hands as shown in Figure 1.36.

Figure 1.34 A simple automatic optical level.

Setting up an optical level

1. Place the tripod under your arm, release the thumbscrews and extend the tripod legs to about 100 mm from full extension (this is to allow for later adjustment). Tighten thumbscrews and erect the tripod.
2. Set up tripod with the head as near to level as possible by extending or shortening the legs (Figure 1.37).
3. Attach the levelling instrument using the captivated screw into the trivet.
4. Now level up the **tribrach** as follows:
   a. Position the instrument body across two of the foot screws (Figure 1.37, Drg A). Adjust by rotating both screws inward or outward, termed **thumbs in** or **thumbs out**. Do *not* use one screw at a time (the bubble will follow the direction of your left thumb).
   b. When level, rotate the instrument body 90° and check the position of the bubble (Figure 1.37, Drg B). Adjustment is made *only* on the third screw, otherwise the level in the first plane will be changed.
   c. Rotate the instrument another 90° and re-check for level. If it is slightly out of level adjust as previously described for half the difference. Re-check after rotating 180°. Repeat for the other plane.

Other instruments have different methods of set up. Tilting instruments are easier to set up initially, but before each sighting the tilting adjustment must be made. Automatic levelling instruments are the most popular. They rely on rough levelling using a cat's eye level. When the bubble is within the circle the instrument is able to function. Fine adjustment takes place by a suspended prism that by gravity ensures a horizontal line of sight.

Figure 1.36 A bar coded staff being used to take a spot height of a curb stone.

Figure 1.35 A bar code reading digital level.

Some instruments have a spirit levelling device. Adjusting the foot screws as previously described will enable the bubble to be positioned in between the two marks. Drg 1 in Figure 1.37 indicates the instrument is too high to the left; Drg 2 indicates it is too high to the right and Drg 3 shows the instrument is level.

Now adjustments to the optics should be made. Basically the instrument comprises a telescope that can be rotated 360° through the horizontal/level plane. To enable heights to be compared **stadia** lines are placed within the telescope on a clear disc termed a **reticule** (Figures 1.38 and 1.39). As surveyors' eyesight differs, adjustment of the stadia lines must be made; this is known as **adjusting for parallax**.

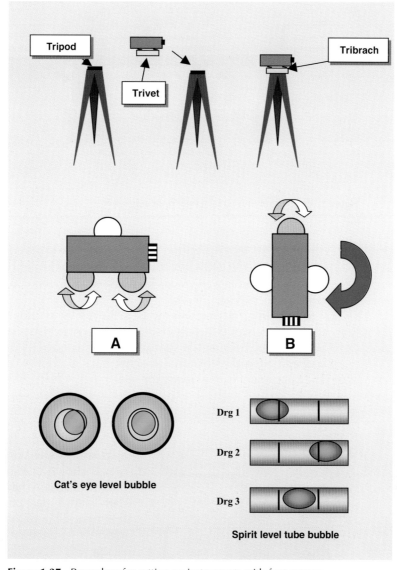

Figure 1.37 Procedure for setting up instruments with foot screws.

Adjusting for parallax

1 Place your hand or a book about 300 mm away from the objective lens.
2 Rotate the eyepiece in either direction whilst looking through the lens.
3 The stadia lines will appear blurred whilst adjusting; however, a point will be achieved when they are very sharp and clear. This is the parallax.
4 Most optical dumpy levels have the same basic operation. Rough sight using the gun sight on the top of the instrument, targeting the surveying

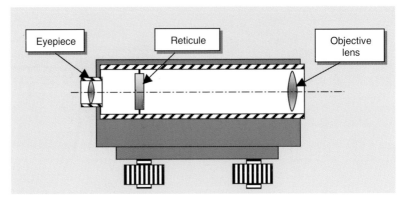

Figure 1.38 Section through a simple optical level.

staff (Figures 1.39 and 1.40). Now view the staff through the telescope. To achieve focus rotate the focus knob in either direction until the image is clear.

5 Fine horizontal adjustment can be achieved by rotating the slow motion screw either way. If there is insufficient adjustment available in one direction, turn the body of the instrument slightly beyond the position of the staff and wind the slow motion screw back to achieve correct sighting.

6 The instrument is now ready for use.

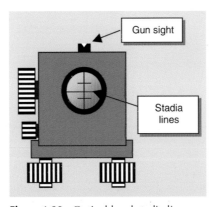

Figure 1.39 Optical level stadia lines.

The other piece of equipment used for levelling is the levelling staff. To help understand staff readings a section is shown in Figure 1.41. (Additional dimensions have been included that do not appear on the actual staff.) It is basically a telescopic measuring stick with numbers every 100 mm. To enable easier reading, the spaces or increments are made up in E shapes and rectangles or blocks. Each arm of the E, each space and each block is 10 mm tall. Therefore the black E is 50 mm tall and the space between the top of the red E and the bottom of the black E is also 50 mm.

Points labelled a, b and c in Figure 1.41 are shown in Figure 1.42 as the surveyor would actually see them through the telescope.

Using the stadia lines have a go at defining the readings – the answers are shown below.

(Answers: 1.485 m, 1.445 m, 1.392 m)

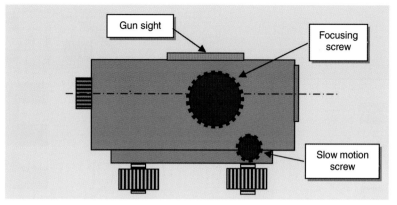

Figure 1.40 Main component parts of a surveyor's level.

Taking readings

There are two methods:

- height of collimation
- rise and fall

The order of the readings will be the same for both methods. The first reading is always the **back sight** (BS). It has nothing to do with the location; it is just the first sighting. The next one can either be an **intermediate sight** (IS) (which is the norm) or the **fore sight** (FS) if carrying out **flying levels** (also known as **line levelling**) (Figure 1.43).

Flying levels (line levelling)

This method is used when transferring a level over a long distance or around obstacles.

1 The first reading could be, say, the invert level at the mains sewer or an OSBM. It would be booked as the **back sight**.
2 The instrument would be rotated and a second reading taken at a suitable point approximately equidistant (at a similar distance) from the instrument. The distances being similar will reduce any errors that may be in the instrument head. This would be booked as a **fore sight**.
3 **Whilst the staff is kept at the same point**, the instrument can be moved to another location and a reading taken from the staff again. This reading will be booked as a **back sight** on the same row as the previous fore sight. It is termed a **change point** (CP). It is useful to write CP in the remarks column to emphasise the change point booking.
4 The process is repeated until the final transfer has been positioned. This is commonly the TBM for the site.

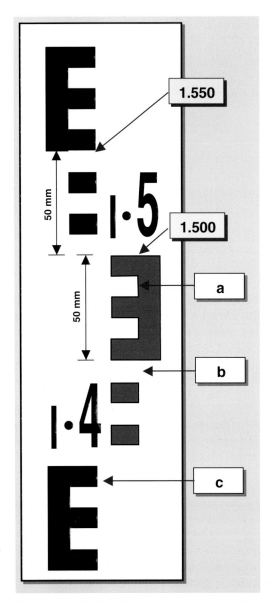

Figure 1.41 Section of a standard surveyor's staff.

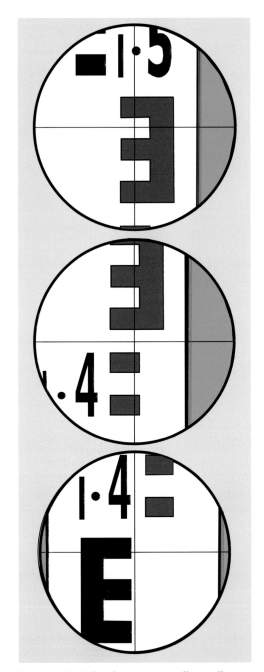

Figure 1.42 What the surveyor will actually see.

1.8 Height of collimation method

Height of collimation (HOC) is sometimes written as height of plane collimation (HPC).

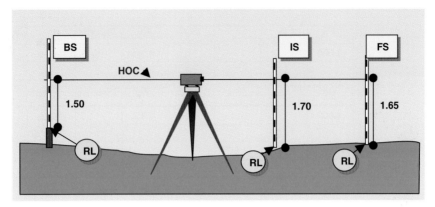

Figure 1.43   A simple levelling survey.

HOC method

1 Set up the instrument in a *safe* position to take readings (be particularly careful when working near roads – wear a high-viz vest and use warning triangles and traffic cones).

2 Sketch the area to be surveyed and include location, name and date, etc., on every booking sheet; identify the points to be surveyed using lower case letters.

3 Use a reference point on which the levelling survey is to be based; this is commonly a bench mark. It can be a TBM or, if the survey is to be referenced to, say, the mains sewer, an OSBM may be used.

4 Enter the level as the reduced level (RL) of 6.350 on the first line of the booking sheet. If the survey is taken against the OS map, then state the RL as OSBM in the remarks box (Figure 1.44 (1)).

5 The assistant will hold the staff on or against the bench mark (Figure 1.30).

6 Aim the optical level towards the staff using the gun sight. Now using the telescope, focus the lens until the image is clear. Fine adjustment to the left or right should be made with the slow motion screw (Figure 1.40).

7 Take the reading to three decimal points using the horizontal stadia line and enter it on the booking sheet in the appropriate column and line (Figure 1.44 (1)). In this example the first reading would be the back sight at 1.500 above ordnance datum (AOD).

  On some instruments additional horizontal stadia lines are marked above and below the centre line: these are used for calculating distances but are now redundant as EDMs are more accurate. Do not confuse your readings, check and compare them with the previous reading on the booking sheet.

8 Move the staff but not the instrument to the next point; then rotate the level, target, focus and take the new reading of 1.700 at point b.

9 Enter the reading on the next row under the IS column and record the position in the notes box (Figure 1.44 (2)).

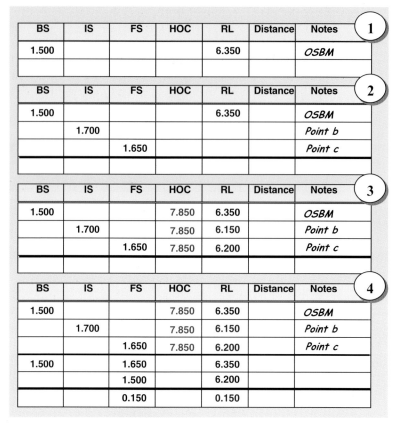

Figure 1.44 Examples of HOC booking sheets.

BS	IS	FS	HOC	RL	Distance	Notes	(1)
1.500				6.350		OSBM	

BS	IS	FS	HOC	RL	Distance	Notes	(2)
1.500				6.350		OSBM	
	1.700					Point b	
		1.650				Point c	

BS	IS	FS	HOC	RL	Distance	Notes	(3)
1.500			7.850	6.350		OSBM	
	1.700		7.850	6.150		Point b	
		1.650	7.850	6.200		Point c	

BS	IS	FS	HOC	RL	Distance	Notes	(4)
1.500			7.850	6.350		OSBM	
	1.700		7.850	6.150		Point b	
		1.650	7.850	6.200		Point c	
1.500		1.650		6.350			
		1.500		6.200			
		0.150		0.150			

10 Repeat the procedure by taking readings and entering them on the booking sheet. Remember to enter in the notes box where the readings have been taken.

11 The final reading is 1.650 at point c, the fore sight. Enter the reading in the FS column on the next row down (Figure 1.44 (2)).

Booking method

1 The first **reduced level** (RL) has been entered as the bench mark. (If a TBM is being used an arbitrary figure such as 10.000 m can be entered.)

2 The first reading was the **back sight** (BS) taken at the TBM or OSBM.

3 The next reading was the **intermediate sight** (IS) entered on the second line in the IS column. There may be one or one hundred and one ISs; therefore enter them on each consecutive line under the IS column.

4 The final reading was the **fore sight** (FS) which has been entered on the next line down from the last IS reading under the FS column.

5 Where possible return to the original point or TBM. The accuracy of the survey (termed **misclosure**) can be measured on the basis of how many

instrument changes have been made and the distance travelled using the formula $\pm 12\sqrt{K}$mm, where K is the length of the circuit in km.

Now complete the booking sheet.

6 Add the BS to the RL on the first row and enter the result under the HOC column. In the example, $1.500 + 6.350 = 7.850$. The instrument has not been moved so the HOC will remain constant and can be entered as shown (Figure 1.44 (3)).

7 Subtract the IS from the HOC on the next row and the result will be the RL.

$$7.850 - 1.700 = \mathbf{6.150}$$

Finally, subtract the FS from the HOC and the result will be the last RL.

$$7.850 - 1.650 = \mathbf{6.200}$$

8 Now carry out the **checks** (Figure 1.44 (4)). Add all of the BS entries and enter the total as shown. In this example there was only one instrument position; therefore there was only one BS. Total the FS entries (again in this example there is only one entry for the same reason). Subtract the smaller number from the larger number and enter the result as shown.

$$1.650 - 1.500 = \mathbf{0.150}$$

Subtract the first and last reduced level the smaller from the greater and enter the result.

$$6.350 - 6.200 = \mathbf{0.150}$$

If both results are the same the survey should be correct. If the results are different the chances are that an arithmetical error has occurred in the RL column. Check those calculations first.

1.9 Rise and fall method

The practical procedure of taking the readings for the R&F method is the same as for the HOC method; however the booking sheet method and calculations differ.

Booking method

1 Enter the levels for the RL, BS, IS and FS in the same way as for the HOC method. However there is no HOC column. Instead there are two columns headed Rise and Fall (Figure 1.45).

2 Subtract the IS from the BS. If the result is positive then enter the result in the **Rise** column. If negative enter it in the **Fall** column. In the example shown subtract the IS 1.700 from the BS 1.500:

$$1.500 - 1.700 = -0.200$$

Therefore enter the result as a **Fall** (Figure 1.45 (2)).

BS	IS	FS	Rise	Fall	RL	Distance	Notes	1
1.500					6.350		*OSBM*	
	1.700						*Point b*	
		1.650					*Point c*	

BS	IS	FS	Rise	Fall	RL	Distance	Notes	2
1.500					6.350		*OSBM*	
	1.700			0.200	6.150		*Point b*	
		1.650	0.050		6.200		*Point c*	

BS	IS	FS	Rise	Fall	RL	Distance	Notes	3
1.500					6.350		*OSBM*	
	1.700			0.200	6.150		*Point b*	
		1.650	0.050		6.200		*Point c*	
1.500		1.650			6.350			
		1.500			6.200			
		0.150		7.850	0.150			

Figure 1.45 Examples of R&F booking sheets.

3 *ALWAYS subtract* the line below from the line above.
4 Repeat the method for the next two numbers. It is always the previous number less the next new number. Therefore

$$1.700 - 1.650 = 0.050$$

This time the figure is positive. Enter it under the **Rise** column.
5 When all of the calculations for the rise and falls have been completed, start with the RL on the first row. Either add or subtract the rise or falls as applicable from the RL and enter the new RL (Figure 1.45 (3)).
   In this example

$$RL - Fall = new RL.$$

$$6.350 - 0.200 = \mathbf{6.150}$$

$$RL + Rise = new RL$$

$$6.150 + 0.050 = \mathbf{6.200}$$

6 When the RLs have been completed the checks should be carried out in the same way as for the HOC method. Now compare the reduced levels from the R&F method with those using the HOC method. They should be exactly the same.

The HOC method is useful for levelling surveys taken from a small number of instrument placings, such as setting out for curb levels. In comparison, the R&F method is more suited to frequent instrument placings, such as on rough terrain or moving around many obstacles.

The disadvantage with the R&F method is that every RL is dependent on the previous RL. If an arithmetical error occurs, every RL from that error will be incorrect. In comparison, the HOC method does not rely on the previous RL. Both methods can be used for any application.

1.10   Change points

Change points are required when it is impossible to take readings due to significant changes in height or an obstruction. Figure 1.46 shows a change point

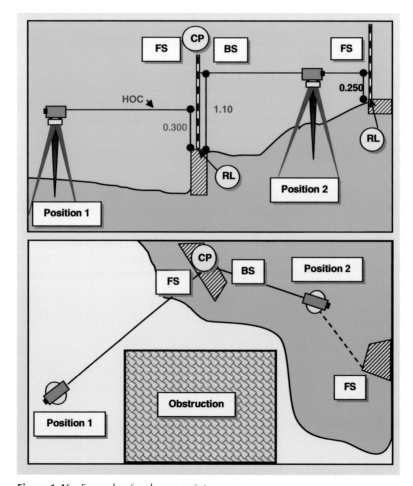

Figure 1.46   Example of a change point.

BS	IS	FS	HOC	RL	Distance	Notes	①
1.650			7.850	6.200		*OSBM*	
1.100		0.300	7.850	7.550		*Point c CP*	
		0.250	8.650	8.400		*Point d*	
2.750		0.550		8.400			
0.550				6.200			
2.200				2.200			

CP = Change Point

BS	IS	FS	Rise	Fall	RL	Distance	Notes	②
1.650					6.200		*OSBM*	
1.100		0.300	1.200		7.550		*Point c CP*	
		0.250	0.850		8.400		*Point d*	
2.750		0.550			8.400			
0.550					6.200			
2.200					2.200			

Figure 1.47   Comparative booking sheets.

resulting from both an obstruction and a change in height. The same survey is shown in plan below, emphasising the obstruction.

CP method

1 Carry out the survey and finish with a fore sight – **position 1**.
2 Keeping the **staff** on the same point **move the level** to position 2.
3 Take a reading on the same staff and book it as a new back sight. Enter the figure on the same line as the previous fore sight (you might find it useful to write the letters 'CP' in the remarks column as a reminder).
4 Continue with the survey in the normal manner.

The same procedure is carried out for either the HOC or the R&F method.

Booking method (Figure 1.47)

1 At the **change point** note that the fore sight and the back sight are both on the same line and share the same reduced level.
2 Calculate the HOC in the normal way:

$$RL + BS = HOC$$
$$6.200 + 1.650 = 7.850$$

3 Subtract the FS from the HOC and enter the RL. Now add the BS to the RL and enter the new HOC on the following row. In the example the HOC

at the change point is 7.850.

$$HOC - FS = new\,RL$$
$$7.850 - 0.300 = \textbf{7.550}$$
$$RL + BS = new\,HOC$$
$$\textbf{7.550} + 1.100 = \textbf{8.650}$$

4   To complete the example the closing FS should be subtracted from the new HOC.

$$HOC - FS = RL$$
$$8.650 - 0.250 = \textbf{8.400}$$

1.11   Inverse levels

Tunnels, bridges and soffits require levels to be taken overhead. They are termed **inverse levels**. The survey would be carried out in the usual way but the staff would be inverted. (Inverse levels should not be confused with **invert levels**). The bookings would be written in the appropriate columns within parentheses () to indicate they are **inverse levels** which are *negative* numbers.

Method

1   Carry out the survey in the normal manner. In the example shown in Figure 1.48 the intermediate sight (IS) is a soffit. Invert the staff so that the base is tight beneath the soffit. (If using a bar code reading level an adjustment must be made to read inverse levels – see the instrument manual).
2   Repeat the procedure for the next IS.
3   The FS is a normal reading.

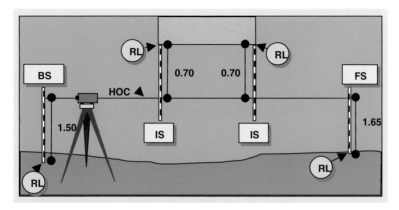

Figure 1.48   Example of an 'inverse' levelling survey.

BS	IS	FS	HOC	RL	Distance	Notes 1
1.500			11.500	10.000		*TBM*
	(0.705)		11.500	12.205		*Point b CP*
	(0.700)		11.500	12.200		*Point c*
		1.650	11.500	9.850		
1.500		1.650		10.000		
		1.500		9.850		
		0.150		0.150		

BS	IS	FS	Rise	Fall	RL	Distance	Notes 2
1.500					10.000		*TBM*
	(0.705)		2.205		12.205		*Point b CP*
	(0.700)			0.005	12.200		*Point c*
		1.650		2.350	9.850		
1.500		1.650			10.000		
		1.500			9.850		
		0.150			0.150		

Figure 1.49 Comparative booking sheets 1 and 2.

Booking method

1 In Figure 1.49, booking sheet 1 shows the HOC method of booking. Note the inverse levels are shown in parentheses indicating they are negative figures. Note that no signs are shown other than parentheses.

In the example the RL is an arbitrary figure of 10.000 = TBM

$$RL + BS = HOC$$
$$10.000 + 1.500 = \mathbf{11.500}$$

The instrument has not been moved so the HOC will remain the same. To calculate the next RL, subtract the IS from the HOC in the normal manner. However, the IS is in parentheses indicating that it is an **inverse** level and therefore a negative number.

$$HOC - IS = RL$$
$$11.500 - (0.705) = \mathbf{12.205}$$

The rule is 'a minus, *minus* a minus': the middle '*minus*' becomes a 'plus' so *add* the inverse level to the height of collimation.

$$11.500 + 0.705 = \mathbf{12.205}$$

2 To calculate the next RL, subtract the IS from the HOC. As with the previous reading the IS is in parentheses, so add the IS to the HOC.

$$HOC - IS = RL$$
$$11.500 - (0.700) = \mathbf{12.200}$$

3   To complete the example, the closing FS should be subtracted from the HOC in the normal manner.

$$HOC - FS = RL$$
$$11.500 - 1.650 = 9.850$$

4   The same bookings have been recorded using the R&F method for comparison – see Figure 1.49, booking sheet 2.

The maths processes are slightly more complicated when inverse levels are considered, though they keep to the same principles. If you are using a calculator, enter the figures as read from the booking sheet; for example, 1.500–0.705 = 2.205. It is a positive number so therefore should be entered as a **rise**.

HOC method

To calculate the reduced level, **subtract** the intermediate sight or fore sight from the HOC. Where the numbers are in brackets **add** them to the HOC.

R&F method

To calculate the rise or fall, **subtract** the second reading from the previous one. In the example the intermediate sight is subtracted from the back sight. If the second figure is in parentheses, remember it is a negative figure. If the result is a positive figure it goes in the **Rise** column and negative figures go in the **Fall** column. When a change point occurs the FS is subtracted from the previous IS; the next two readings will be the BS minus the next IS. The FS and the BS are on the same RL; therefore they cannot be used together.

In Figure 1.49 the RL is an arbitrary figure of 10.000 = TBM. The calculations will be BS minus IS equals Rise or Fall.

$$BS - IS = R \text{ or } F?$$
$$1.500 - (0.705) = \mathbf{2.205} \quad \text{a positive figure therefore a Rise}$$

$$IS - IS = R \text{ or } F?$$
$$(0.705) - (0.700) = \mathbf{-0.005} \quad \text{a negative figure therefore a Fall}$$

If using a calculator ensure the data are entered correctly:

$$-0.705 - (-0.700) = -0.005$$
$$IS - FS = R \text{ or } F?$$
$$(-0.700) - 1.650 = \mathbf{-2.350}$$

To calculate the RLs, start with the original RL and either subtract or add using the rise or fall figures. Each figure is related to the previous RL and not the original RL.

In the example the original RL is 10.000

$$RL \pm R \text{ or } F = RL$$

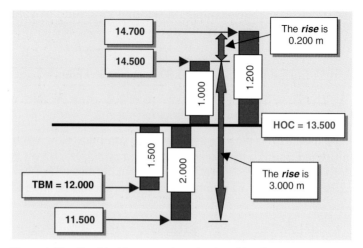

Figure 1.50 Graphical interpretation of reduced levels.

$$10.000 + 2.205 = 12.205$$
$$\text{RL} \pm \text{R or F} = \text{RL}$$
$$12.205 - 0.005 = 12.200$$
$$\text{RL} \pm \text{R or F} = \text{RL}$$
$$12.200 - 2.350 = 9.850$$

If it is unclear what is happening with the rise and fall inverse levels, look at Figure 1.50 and the booking sheet shown in Figure 1.51. Both the HOC and the R&F methods have been calculated and a graphical comparison shown. Note that with the R&F method the RLs are compared with the previous RL. In the example, point b is compared with point c, showing there is 3.00 m difference. Then point c is compared with point d showing a 0.20 m difference. Now look at the previous example and draw the problem.

Worksheet 1

Photocopy the booking sheet in Figure 1.52 and complete it as appropriate. Then use the reduced levels to determine whether a vehicle 2.30 m high can pass below the soffit of the bridge. Spot heights f and i are directly below spot heights g and h, respectively. Include the appropriate checks. The solution is given in Figure 1.53.

1.12 Invert levels

Invert levels must not be confused with **inverse** levels. Invert levels are normally associated with drainage runs, for example the surface of a pipe where the fluids will flow (Figure 1.31). Where there is no access chamber, the crown of the pipe would be used and an additional allowance calculated for the extra depth. For example, the following calculation would be carried out:

BS	IS	FS	HOC	RL	Distance	Notes
1.500			13.500	12.000		*TBM*
	2.000		13.500	11.500		*Point b*
	(1.000)		13.500	14.500		*Point c*
		(1.200)	13.500	14.700		*Point d*
1.500		(1.200)		14.700		
(1.200)				12.000		
2.750				2.750		

BS	IS	FS	Rise	Fall	RL	Distance	Notes
1.500					12.000		*TBM*
	2.000			0.500	11.500		*Point b*
	(1.000)		3.000		14.500		*Point c*
		(1.200)	0.200		14.700		*Point d*
1.500		(1.200)			14.700		
(1.200)		1.500			12.000		
2.750							

Figure 1.51   Comparative booking sheets 3 and 4.

BS	IS	FS	HOC	RL	Distance	Notes
1.210				12.420		*TBM*
	1.324					*a. cor*
	1.355					*b. cor*
	1.370					*c. cor*
1.300		1.390				*Cp d. cor*
	1.380					*e. cor*
	1.385					*f. cor*
	(0.908)					*g. br s*
	(0.910)					*h. br s*
		1.450				*i. cor*

Figure 1.52   Partially completed booking sheet.

Reduced level taken to the crown of the pipe is say	2.750 m
Pipe diameter is	+ 0.100 m
Pipe wall thickness is	− 0.012 m
Invert level is	**2.838 m**

BS	IS	FS	HOC	RL	Distance	Notes
1.210			13.630	12.420		*TBM*
	1.324		13.630	12.306		*a. cor*
	1.355		13.630	12.275		*b. cor*
	1.370		13.630	12.260		*c. cor*
1.300		1.390	13.630	12.240		*Cp d. cor*
	1.380		13.540	12.160		*e. cor*
	1.385		13.540	12.155		*f. cor*
	(0.908)		13.540	14.448		*g. br s*
	(0.910)		13.540	14.450		*h. br s*
		1.450	13.540	12.090		*i. cor*
2.510		2.840		12.420		
		2.510		12.090		
		0.330		0.330		

Figure 1.53   The correct solution to Figure 1.52.

Where a connection to an existing pipe run is to be carried out an access chamber may be formed so the invert level should be established.

1.13   Health and safety

A risk assessment should be carried out before any survey. The following list is not exhaustive:

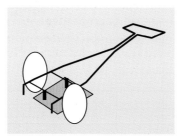

Figure 1.54   Inspection chamber lid trolley jack.

- Weight – Some cast iron access chamber covers used in the road are extremely heavy. They are often sealed with the **blacktop** bitumen road surface, surface rust and small grit, etc. A jemmy bar is useful to free up the cover before using manhole keys.
- Manhole keys can be used to lift and drag the cover; however, this might result in back strain. To overcome the problem specialist lifting tools should be used. The key locates into the cover and the hydraulic lift raises it, enabling the whole operation to be carried out by one person (Figures 1.54 and 1.55).
- If the access chambers are located in the road, warning triangles should be positioned a reasonable distance before the workings; placement will depend on the speed of the traffic and the visibility of the works. (If the works are over the brow of a hill, warning triangles should be placed

Figure 1.55   An assembled chamber lid trolley jack.

before the brow where they can be seen. More than one triangle would be required. Even with the warning triangles on display, you should try to park your vehicle as a last resort barrier between you and the flow of traffic, but not too close, because if there is impact from a speeding vehicle you do not want to be in the line of debris.)

- High-viz jackets or vests must be worn. Where a survey is adjacent to or on a road it is a requirement that all personnel wear them. A hard hat can also help make a surveyor more visible.
- Toe protection footware is also essential, either as rubber boots that can be disinfected when required, or stout leather boots.
- Where the levelling staff is placed in the invert level of a live or used sewer or drain, either cover it with a polythene bag secured with a rubber band or be prepared to disinfect the foot of the staff. Sewers can be home to rats and mice: both creatures can carry disease including Weil's disease. According to the Department for Environment, Food and Rural Affairs (DEFRA) it is very rare in the UK; however, where water is fast flowing an open wound can become infected. Always wear gloves when working in an access chamber. Disinfectant wipes and other hand cleaning products are available. The symptoms of Weil's disease can appear similar to those of influenza: headaches, muscle pain, fever, etc. You should tell your doctor if you have been working in a sewer. Weil's disease can be treated successfully if diagnosed in time; however, it can be fatal in a few days. For more information look on the DEFRA website at http://www.defra.gov.uk.
- Gases in drains can be lethal. If working inside a deep manhole, a gas detector and respirator must be worn. Never enter a manhole unless it has been tested for gases. Workers have died due to illegal substances that have been discharged into the sewers. The subject is beyond the scope of this book.

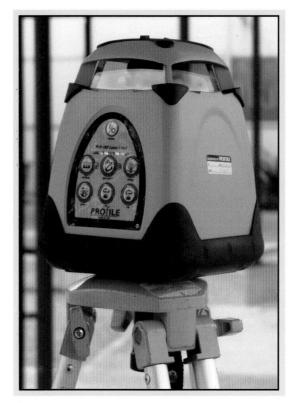

Figure 1.56 An automatic laser level.

1.14   Levelling instruments

Optical and electronic levels

- Rotating laser levels
- Bar code reading levels

Rotating laser levelling

Modern rotating laser levels are almost self levelling (Figure 1.56). The instrument can be attached to a conventional tripod or to a specialist fixing clamp. Some instruments use a series of coloured lights to indicate when the head is level. Adjustment of the tripod legs or clamping device is all that is required as the fine adjustment is automatic within the instrument. When the head has been set up a laser beam rotates at a constant height. Some instruments have a visible red laser spot that can be seen on the surface of the surrounding vertical surfaces. When working in direct sunlight the red spot cannot be easily seen; therefore an electronic receiver is used. Figure 1.57 shows a laser receiver clamped to a levelling staff. This particular instrument has both audio and vi-

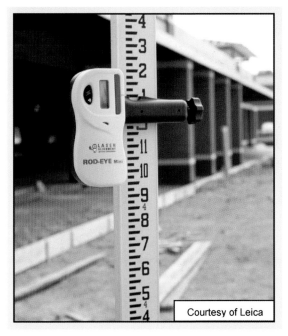

Courtesy of Leica

Figure 1.57 A laser level receiver and staff.

sual settings. The receiver can be fixed to the staff for setting constant heights or marking contours, fixed loosely on the staff using the indent for reading, or using the back of the staff which has finer detail. The advantage of a laser level is that one person can set up the instrument and take staff readings. (A point of caution – whilst it is useful to remotely take readings using a laser level or other sophisticated pieces of equipment, you should be aware of risk of theft.)

Bar code reading levels

Over the past decade, bar code reading levels have become more sophisticated and cheaper. Modern versions will read the bar coded face of the levelling staff, calculate the lineal distance from the instrument and display the results numerically. Some instruments can record the data which can be electronically downloaded and printed out as hard copy results or used directly with specific computer software. The advantages of electronically read levels are the speed with which they can be taken and that the chance of human error is reduced. It is still very important to check the readings to ensure that the staff has been correctly extended, i.e. each consecutive section has been extended in order and fully locked into position. The most sophisticated bar code reading levels have a curvature of the Earth compensator; therefore true readings can be taken over greater distances.

Other levelling devices

These include

- water levels
- Cowley levels
- spirit levels.

These devices are used to transfer or compare levels over short distances. Although water levels are very useful as they are the only piece of equipment that can take levels around corners, they have generally been replaced by rotating laser levels. Cowley levels are also known as 'bricklayers' levels, are simple in design and reasonably robust; as the company is no longer trading the equipment will be of historic interest only. Spirit levels are useful to check short distances or the verticality of frames, walls and the like: they are not suitable for taking levels.

Chapter 2

Topographical surveys

This chapter concentrates on the measurement of what is on the land. It explains the processes and techniques for measuring angles, starting with the objectives, and then setting up the equipment and taking readings. As the techniques become more involved, references are required, such as **whole circle bearings** and **coordinates**, that require more calculations. The illustrated worked examples are broken down into stages, from setting up the equipment, taking the readings and booking them, through to adjustments and drawing up procedures.

When measuring angles, a surveyor may work with degrees, minutes and seconds for greater accuracy. By using fully worked examples, this chapter introduces the reader to the finer accuracy of angular measurement, including the necessary **trigonometry**.

The chapter concludes with the processes of setting out a small building and drainage runs, and setting out stations for traversing. There are worked examples with colour coded calculations showing both open and closed traverses. They are presented in stages and include graphics for easy reference. Throughout the chapter modern equipment, such as a total station, is described and its operation related to the fundamentals of lineal and angular measurement; however, a theodolite and electromagnetic distance measure (EDM) or tapes can be used as an alternative.

2.1 Measuring angles

Angular measurements can be taken either vertically or horizontally. They can be used to calculate the distances between remote objects, the heights of buildings or overhead power lines, without the need to touch them. The main instrument for measuring angles is the theodolite which basically consists of a telescope mounted on a horizontal shaft to enable vertical rotation, and a vertical shaft that can rotate horizontally: these are termed the **vertical** and **horizontal axes** (Figure 2.1). Attached to each axis is a disc divided into degrees, minutes and seconds. Early theodolites were read optically, meaning that the surveyor had to view the readings via another smaller telescope and interpret the finer readings of minutes and seconds. (An analogy would be the old-fashioned slide rule, in contrast to the electronic hand held calculator.) In common with calculators, the electronic circuitry has become increasingly sophisticated, so by the addition of an EDM, which enables distances to be measured at the same time as angles, the electronic theodolite has developed into the total station.

2.2 Setting up a theodolite or total station

If the instrument is being used to measure vertical angles only, it is set up using a similar method to other types of optical level (see Section 1.7, Setting up an optical level). A fixing screw attaches the instrument via a trivet onto the head

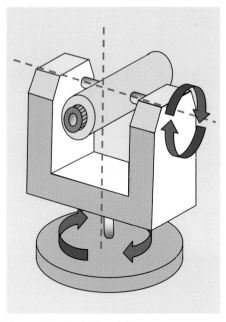

Figure 2.1 Basic theodolite movements.

of the tripod. *Always* hold the instrument with one hand until it is attached to the tripod. Levelling the tripod head by adjusting the length of the legs will provide approximate levelling using the cat's eye level on the tribrach. Final adjustment is made using the foot screws with the instrument body across two screws to level in one direction, and then rotating the instrument body 90° and adjusting the third foot screw only.

When the instrument is being used to measure horizontal angles, a second dimension must be considered. The central axis of the instrument will be at the node or junction of the angle; therefore a reference point (a surveyor's pin, a wooden peg with a nail or marked cross, or a metal rod embedded in concrete) is placed with the instrument directly above. To enable the instrument to be aligned over the reference point (**station**) the tripod must be set up with the fixing screw above the station. A plumb bob and string will help approximate alignment. The instrument is then fixed to the tripod. Some instruments have an optical plummet which is a small telescope that uses a prism to enable the station centre to be seen (Figure 2.2). Other instruments use a red laser beam that projects a spot directly beneath the central axis, enabling easy alignment. By ensuring that the trivet of the instrument is central to the tripod head, a small amount of horizontal adjustment can be made in any direction. When the instrument is directly above the station,

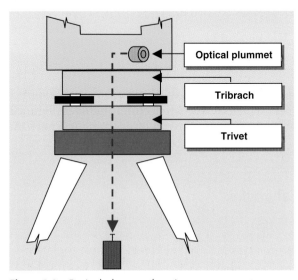

Figure 2.2 Optical plummet function.

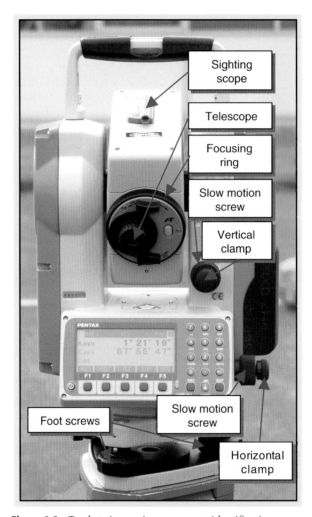

Figure 2.3   Total station main component identification.

check that the body is level. Some instruments use a spirit bubble as previously described, whilst others use electronic bubbles (Figure 2.3): the levelling operation is the same for both types of instrument. Check the instrument for parallax (see Section 1.7, Setting up an optical level).

2.3   Measuring vertical angles

Method

1   When the instrument is set up correctly zero degrees will be directly above the instrument at the zenith (see Figure 2.4).

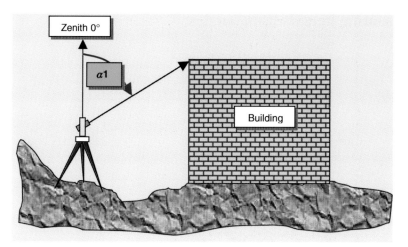

Figure 2.4 Measuring a vertical angle.

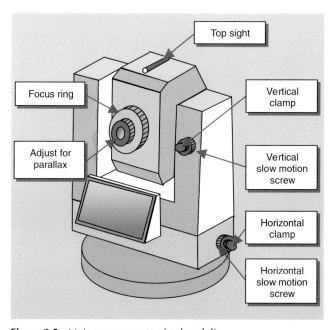

Figure 2.5 Main components of a theodolite.

2 Using the sighting indicator on the top of the instrument rotate the telescope downward until the target is in line.

3 Lock the telescope using the vertical clamp and the body using the horizontal clamp.

4 Focus the telescope using the focusing ring and make any fine adjustments to targeting using the slow motion knob attached to each clamp (Figure 2.5).

5 Read off the vertical angle from the digital display.

2.4 Measuring heights using angles

Set up the total station in an appropriate place and take readings (Figure 2.6).

Method

Carry out the procedure for measuring vertical angles and book the relevant readings starting from the zenith (Figure 2.7).

1 Top of building $= \alpha 1$
2 At horizontal (90° to the zenith) use the EDM to establish the distance from the building
3 Base of building $= \alpha 2$

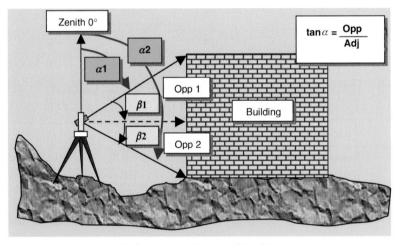

Figure 2.6   Measuring the height of an inaccessible object.

Station	Target		Angle ° ' "	Distance (m)
A	Zenith		000° 00' 00"	
	Top	α1	25° 15' 20"	
	Horizontal		90° 00' 00"	35.250 m
	Base	α2	105° 15' 42"	

Figure 2.7   A suggested booking sheet for angular measurement.

Calculations

1 To calculate angle $\beta2$ subtract $90°$ from angle $\alpha2$
2 To calculate angle $\beta1$ subtract $\alpha1$ from $90°$

Example

To find the overall height of the building shown in Figure 2.6, proceed as follows:

1 Sight the top of the structure, take a reading and enter it in the appropriate box (Figure 2.7).
2 Set the instrument at horizontal (i.e. at $90°$ to the zenith) and record the lineal dimension **distance**.
3 Target the base of the building and record the reading in the appropriate box.

Calculations

1 To calculate angle $\beta2$ subtract $90°$ from angle $\alpha2$

$$105° \, 15' \, 42'' - 90° \, 00' \, 00'' = 15° \, 15' \, 42'' = \beta2$$

Calculate the **opposite** dimension using the formula $\tan\beta = \frac{\text{Opp}}{\text{Adj}}$

$$\text{Opp} = \tan 15° \, 15' \, 42'' \times 35.250$$
$$\text{Opp} = 9.618\,\text{m}$$

2 To calculate angle $\beta1$ subtract $\alpha1$ from $90°$ (Figure 2.8)

$$90° \, 00' \, 00'' - 25° \, 15' \, 20'' = 64° \, 44' \, 40'' = \beta1$$

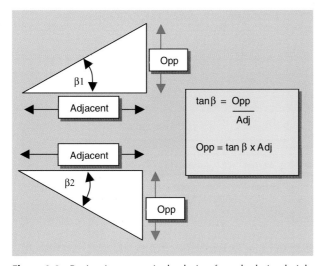

Figure 2.8 Basic trigonometrical solution for calculating heights.

Calculate the **opposite** dimension using the formula $\tan\beta = \frac{\text{Opp}}{\text{Adj}}$

$$\text{Opp} = \tan 64° \, 44' \, 40'' \times 35.250$$
$$\text{Opp} = 74.722 \, \text{m}$$

Add both opposites together and the result will be the height of the building.

$$74.722 + 9.618 \, \text{m} = 84.34 \, \text{m}$$

Answer: The total height of the building is 84.34 m.

The same procedure is used to measure the height of overhead power cables. A reflector prism would be positioned directly beneath the power lines to enable a distance from the instrument to be established.

2.5 Measuring horizontal angles

Set up the instrument directly over the station as previously described. Always take readings in a clockwise direction (Figure 2.9).

Method

1 Set up the instrument.
2 Target the instrument on the origin of the angle and lock the horizontal clamp.

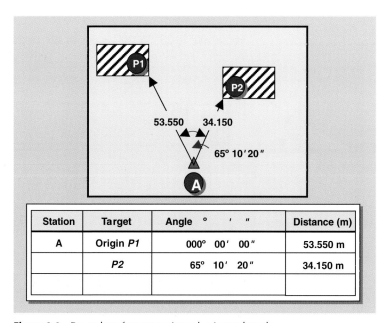

Station	Target	Angle ° ' "	Distance (m)
A	Origin *P1*	000° 00' 00"	53.550 m
	P2	65° 10' 20"	34.150 m

Figure 2.9 Procedure for measuring a horizontal angle.

3   Make any fine adjustments using the slow motion screw and then set the instrument to zero.
4   Unlock the horizontal clamp and rotate the instrument clockwise and target the other angle.
5   Lock the horizontal clamp and use the slow motion screw for fine adjustment.
6   Book the reading.

The horizontal angle at station A between points P1 and P2 is 65°10′ 20″.

To enable the positions of several stations to be referenced, a method of horizontal angles and distances from a given point termed **whole circle bearings** is used.

2.6   Whole circle bearings

To enable a horizontal angle to be referenced a base or origin is required. These can be either:

- magnetic north
- reference object.

Magnetic north is useful when surveying in open countryside. The compass reading can be compared with an Ordnance Survey map; therefore coordinates can also be given. Global positioning systems (GPSs) can be used; this is particularly useful for civil engineering works such as roads, tunnels and rail links.

When taking bearings an important factor to be considered is electromagnetism. When working near high voltage power lines, electric powered rail-lines, and electricity substations, the electromagnetic field they produce may affect the direction of the magnetic compass. Specialist equipment is available that shields the compass, or you can simply walk away from the likely source and monitor the direction of the compass. Although less common, natural magnetic forces or buried services may give false compass readings.

Reference objects (ROs) are useful when working on a single site or in a built up area. Choose a prominent point such as the corner of a building and use that as the zero reading.

Method

1   At station A, target the RO or north and set the instrument at zero. All other points can be taken from that point (Figure 2.10).
2   Rotate the instrument in a **clockwise** direction and target the next station. Figure 2.10 (1) shows station B is 70° from the RO.

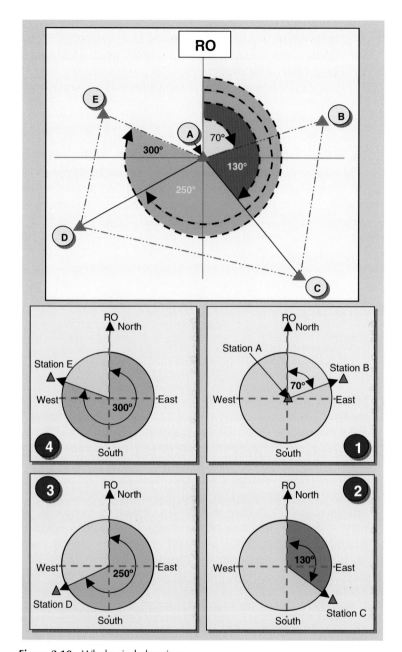

Figure 2.10 Whole circle bearings.

3 Rotate the instrument clockwise and target the next station. The reading is 130° from the RO (Figure 2.10 (2)). Figure 2.10 (3) and (4) show two more stations at 250° and 300° from the RO.

4 We always take **clockwise readings**. Therefore if we are using north as the reference, 90° from the origin must be the east; the west would be at 270°.

Another term for **points** is **bearings**. Continuing with the example, if zero is directly in front of you and directly behind you is 180° from the origin (half a circle) then the origin must also be 360°, the **whole circle bearing**. Try the following examples. Draw a cross with north at the top and label the four main bearings. Then estimate where the following bearings will occur: 116°, 178°, 15°, 285°, 135°.

Drawing office work

Example

A land survey of an irregular plot has been recorded as shown in the booking sheet in Figure 2.11. The surveyor had set up station A over a survey pin

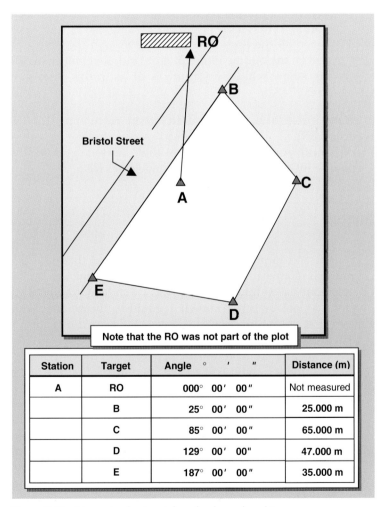

Note that the RO was not part of the plot

Station	Target	Angle ° ' "	Distance (m)
A	RO	000° 00' 00"	Not measured
	B	25° 00' 00"	25.000 m
	C	85° 00' 00"	65.000 m
	D	129° 00' 00"	47.000 m
	E	187° 00' 00"	35.000 m

Figure 2.11 Measuring horizontal angles for a closed traverse.

hammered into the car park and used the adjacent Bristol Street for points B and E. Prepare a scaled drawing to present to the client.

Method for drawing up a survey manually

1 Consider the longest dimensions and select an appropriate scale for the drawing.
2 To scale plot a line from station A to the reference object. The line length will depend on the size of paper used.
3 Using a protractor measure 25° from the line station A–RO and lightly draw a line. Mark off 25m to scale starting at station A. The other end will be station B.
4 Repeat the process by measuring 85° from the line station A–RO and to scale draw a line 65m long. This will be point C.
5 Repeat the process for the remaining points, always using the same line station A–RO as the origin or baseline.
6 When all the lines have been drawn join the points together and that will be the outline of the plot of land. Remember to orientate the drawing (show which way north is or, as in this case, show and label the road).

Whole circle bearings are used when calculating traverse surveys (see Section 2.9).

Azimuth angle is the horizontal angle between grid north and any bearing. Angular measurement should be made **clockwise**.

2.7   Coordinates

Position and location are relative. This means that any given point on the ground is relative to somewhere else that is known or recognised. For example, if someone asked for directions to, say, a railway station, it would be usual to start at a point which is known. It could be the point at which you both are. From that point, direction and distance (linear dimension) would be given, such as 'about a mile in that direction'.

More detail may be required such as 'go to the end of this road, turn right and continue for about half a mile, then turn left at the pub and it should be about 200 metres on the right'.

In surveying even more detail is required: 'about' isn't good enough. If you wanted to provide precise locations, a set of coordinates would be used. They relate to recorded points such as those on maps or in relation to magnetic points. All of the British Isles has been mapped by the Ordnance Survey and recorded on various scaled maps in hard copy and electronically. The coordinates are the same for all maps and originate to the south west of England, so that all locations are above and to the right of the point. If compared to a graph it will mean that all coordinates must be positive in value.

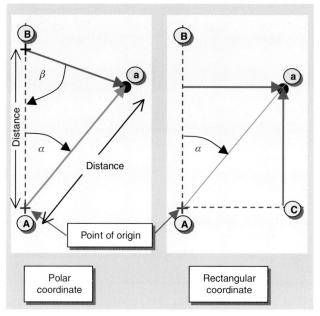

Figure 2.12 Polar and rectangular coordinates.

The two main types of coordinates used in surveying are:

- polar coordinates
- rectangular coordinates.

Polar coordinates

Like a bar magnet, a line has two poles or ends. A third point can be plotted using measured angles to the line termed **polar coordinates** (Figure 2.12). For example the position of point a can be plotted by measuring the internal angle (α) from the baseline A–B and measuring the distance from the point of origin.

Where it is impractical to measure the distance to the point a second internal angle (β) can be taken from the baseline as shown in the figure. The distance between points A and B *must* be measured. The procedure can be used to plot a point many miles away as in cartography (map making) or several metres away as in traversing (see Section 2.9).

The equipment most commonly used for measuring angles is a total station or a theodolite with an EDM to measure the distances.

Rectangular coordinates

Optical and electronic instruments can measure angles to an accuracy of seconds; however, angles cannot be plotted so accurately using a protractor; for

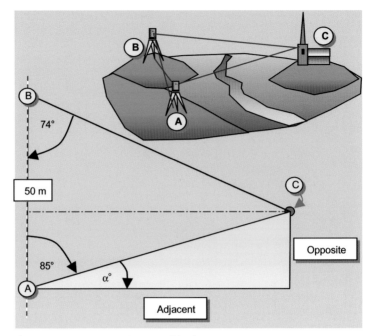

Figure 2.13 Measuring inaccessible distances.

example, the angle shown could only be read to the nearest 30″. However, use of rectangular coordinates enables more accurate points to be plotted: if the side opposite and adjacent to the angle were known, an accuracy to the nearest second could be achieved.

Example

Figure 2.13 shows point C in relation to points A and B using horizontal angles and a measured distance. Calculate the distance of point C from point A in terms of a **rectangular coordinate**.

Method
1 Find the length of side AC. The procedure for calculating the length of line AC using the **sine rule** is shown in Figure 2.14.
2 Convert the polar coordinate to a rectangular coordinate. Rectangular coordinates are straight lines at 90° to the baseline. In the example the **adjacent side** would be the rectangular coordinate (Figure 2.15).

The conversion from **polar coordinates** to **rectangular coordinates** is used with a closed traverse. The adjacent will be the **departure** and the opposite is the **latitude** (see Section 2.9).

Sine rule

$$\frac{A}{\text{sine A}} = \frac{B}{\text{sine B}} = \frac{C}{\text{sine C}}$$

To find angle C subtract angles A & B from 180°

180° = 85° + 74° + angle C°

angle C°= 180° – (85° + 74°)

angle C°= 21°

- **Distance between A & B = 50 m**
- **Side 'A' is opposite angle A**
- **Side 'B' is opposite angle B which is unknown. It will be the hypotenuse of the triangle.**
- **Side 'C' is opposite angle C and measures 50 m.**

$$\frac{A}{\text{sine A}} = \frac{B}{\text{sine } 74^\circ} = \frac{50}{\text{sine } 21^\circ}$$

$$B = \frac{50}{\text{sine } 21^\circ} \times \text{sine } 74^\circ$$

$$B = 134.117 \text{ m}$$

Therefore side AC = 134.117 m

Figure 2.14 Sine rule calculation.

2.8 Setting out

Setting out a small building

Setting out a building is subject to the site requirements. Buildings can be 'infill' between existing buildings, or an estate on, say, a green field site. Either way the objectives are the same (Figure 2.16):

1 to set out the building in the correct position
2 to set out the building at the correct height
3 to erect the building vertically.

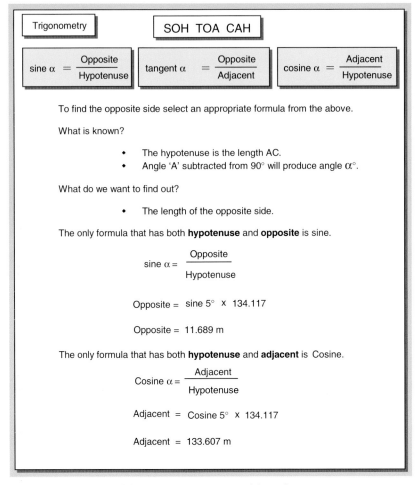

Figure 2.15 Selection of the correct trigonometrical formula.

Equipment

- setting out drawings/plans
- timber pegs and club hammer (plus a metal spike for use in hard ground)
- coloured paints – spray
- a site square and optical level and staff
- tapes (at least two 20 m tapes)
- profile boards
- setting out line

For setting out the positions of foundations, the corners of the **external** walls are used. Foundations will also be required where load bearing internal walls have been designed. They may be a continuation of an external wall as shown in Figure 2.17 (11) or you may have to calculate the centre line of the wall and use the position for the centre line of the foundation. From a practical point

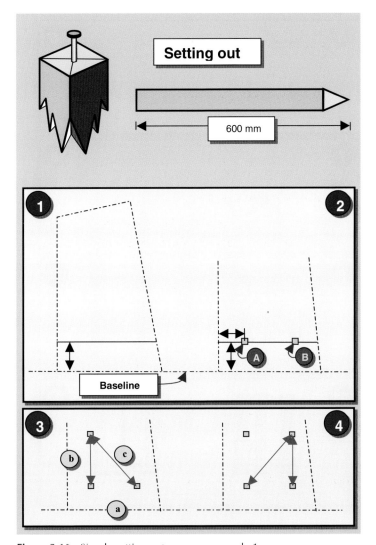

Figure 2.16 Simple setting out process, example 1.

of view, if the foundations are to be mechanically excavated the bucket size will be the same throughout the dig. Where a solid ground floor has been designed the internal load bearing wall may only require a thickening of the concrete oversite. Check the 'section through' on the drawings for the foundation design.

Method using a baseline (Figure 2.16 (1))

1 **Clear the proposed working area of topsoil.** Topsoil can be stored on site for future landscaping purposes. In recent years, planning authorities have been able to restrict the removal of topsoil and spoil from site. Where

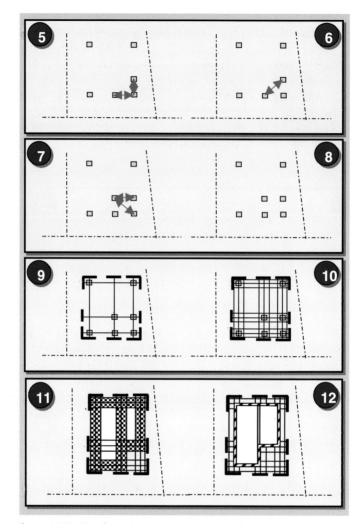

Figure 2.17 Simple setting out process, example 2.

removal is allowed topsoil may be sold and spoil may be used for capping off old landfill sites.

2 **Establish a temporary benchmark (TBM).** This will be referenced against the existing mains sewer as shown on the setting out drawing. As it is unlikely that access to the mains sewer will be available, use the nearest manholes either side of the potential access point and interpolate the access height. For example, if the distance between manhole A and B is 50 metres and the invert levels for A is 7.00 m and B is 6.375 m, then a new manhole with an outfall of, say, 6.500 m could be used (see p. 69, Setting out drainage runs).

3 **Establish a baseline.** When setting out the site, the centre lines of the road would be established first. If the road already exists use the edge of the curb as the baseline (Figure 2.16 (1)).

4 **Set out the position of the external corners of the front wall.** The initial points to be set out are commonly adjacent to the edge of the plot. In this example the dwellings are to be L shaped. Measure in from the boundary to position the first peg A, then peg 'B' (Figure 2.16 (2)). The peg centres will represent the external corners of the walling.

5 **Establish the third corner.** Using two tapes measure the diagonal and the straight side of the proposed building. Where the two points meet, position the third peg. Using this method will ensure the corner is at a right angle. The diagonal can be calculated using Pythagoras' theorem

$$a^2 + b^2 = c^2$$

where a is the distance between the first two pegs and b is the distance between the first peg and the third peg. The diagonal will be c. Find the square root of c^2 and that will be the diagonal measurement (Figure 2.16 (3)).

6 **Establish the fourth corner.** Using the same diagonal measurement and two tapes, position the fourth peg (Figure 2.16 (4)).

7 **Confirm the squareness of the corners.** The technique adopted for positioning the corner pegs should now be confirmed before any further setting out is done. If the four corners are not correct then the remainder will be incorrect. Check that both diagonals are the same dimension.

8 **Establish the 'want'** (the small rectangle is termed a **want**). Using the existing pegs position the corner pegs to form the want. Measure from the corner peg. Now check the diagonal. The diagonal dimension can be calculated using the same method as previously given (Figure 2.17 (5), (6), (7) and (8)).

9 **Then check both diagonals again** (Figure 2.17 (6) & (7)).

Now that all the setting out pegs have been positioned, the profile boards can be erected. Profile boards are horizontal rails 75 mm × 25 mm nailed to softwood pegs. The rails must be longer than the width of the proposed foundation. Corner rails are commonly nailed to a single corner peg (Figure 2.18).

10 **Positioning profile boards.** The boards should be close enough to the proposed building lines to enable them to be used easily but not so close that they could fall into the trench if the sides fall in. About 0.60 m to 1.00 m from the proposed trench would be suitable (Figures 2.18, 2.19 and 2.20 show the relationship between the profile boards, the foundations and the walls). Position a board in line with each row of pegs as shown in Figure 2.17 (9).

11 **Running string lines.** Most surveyors use orange plastic string lines to set out. The first line should run from one profile board over the centre of the pegs in the row and onto the opposite profile board. This will mark the external face of the wall. Repeat the procedure for every row (Figure 2.17 (9)).

12 **Establish inner and outer edge of the foundations.** So far every profile board has the position of the **external** face of the walling. From that point

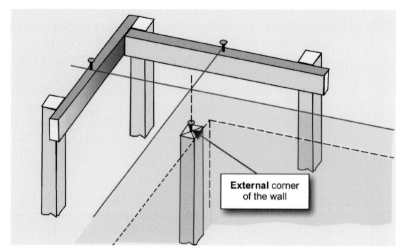

Figure 2.18   Profile boards.

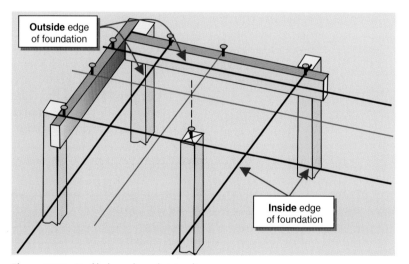

Figure 2.19   Profile boards and string lines.

measure the appropriate projection (P) from the wall face to the outer edge of the foundation. From the outer edge measure the width of the foundation and mark the inner edge on the profile boards (Figure 2.20). Repeat the process for all the profile boards and check all the dimensions again (Figure 2.17 (10)).

13   **Mark out the foundation lines.** This used to be carried out using chalk, lime or even cement powder; however, an aerosol paint marker is more commonly used today. Using the outer two string lines (tramlines) as guides, mark the lines on the surface of the ground. Remember that the 'want' walling does not include the outside corner (Figure 2.17 (11)). When all the lines have been marked on the ground, wind up all the string lines and remove the pegs leaving all the profile boards in position.

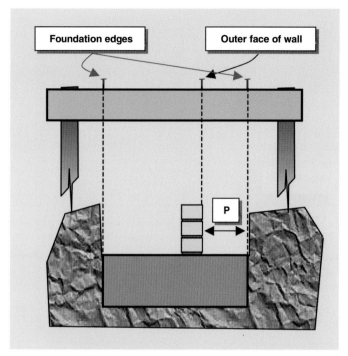

Figure 2.20 Profile board markings.

They will be used again when the bricklayer marks out. The trench will be excavated either manually or mechanically; if the latter ensure that the excavator does not knock the profile boards (Figure 2.17 (12)).

14 **Establish levels.** It is essential that the foundations are level and at the correct height to enable the bricklayer to start coursing the bricks or blocks. If the concrete is not at the correct height a **split course** will have to be laid which is both time consuming and costly. The modern approach is to set up a rotating laser level in the centre of the proposed building and back sight off the site TBM. Clamp the receiver to the levelling staff and check the **formation level** as the excavation proceeds (formation level is the bottom of the trench or excavation). To ensure the top of the foundation concrete is level, either use a series of softwood staves knocked into the centre of the trench with all the tops at the correct level, or a series of small pegs knocked into the trench walls marking the correct level for the concrete. Figure 2.20 shows a traditional strip foundation and Figure 2.21 shows a trench fill foundation with centre staves and side pegs. Only one method would be used in practice. For further information about establishing the correct levels and finishing for foundations see Chapter 4.

General notes

Steel tapes are accurate at 20°C; therefore allowances should be made if the tapes have been left in a closed car in very hot weather or on very cold days

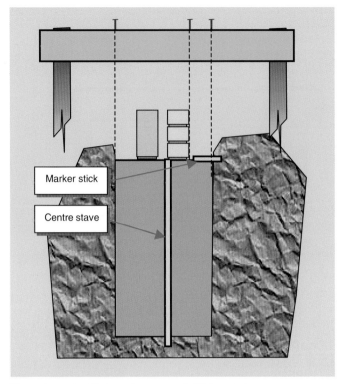

Figure 2.21 Profile board over trench fill foundation.

as they could be inaccurate. An allowance of 1 mm per 10 metres of tape per 10°C difference in temperature should be made. For example, for a 30 metre tape at 0°C this will be

$$-(1 \times 3 \times 2) = -6\,\text{mm}.$$

Therefore an allowance of −6 mm should be made for every tape length measured (i.e. add 6 mm to the total length of a 30 m tape). Do not let the tape sag, or pull so tightly that the tape stretches: a tension of 4·5 kg should be sufficient.

Pegs are usually 50 mm square softwood with a pointed end. They are driven into the ground to mark the position of each station or as working heights. To provide greater accuracy a nail is then driven into the centre of the peg (Figure 2.16). On larger sites setting out pegs are painted with a bright RED 50 mm band, whilst levelling pegs require BLUE paint bands.

Where the project requires more accuracy or the duration is longer, it is more likely to peg using steel rods or angle iron bedded in concrete (Figure 1.5). The pegs must be positioned so that they are not in the way during construction. The TBM should be protected with barriers or tapes to prevent accidental movement.

It is important to check the invert level of the existing drain or the level of the existing sewer: the information may be on the drawing, but if an error has been made it can prove to be very expensive to provide a pumping station to pump uphill. The idea of checking all information given on drawings is *VERY* important. There are housing estates that have been set out and built back to front, new houses where walls have had to be demolished and rebuilt to allow baths to be installed level, a shopping development where a flank wall is over 450 mm out of alignment, and a new multimillion pound public building that required a very costly additional pumping station to pump up into the main sewer, all due to errors of setting out.

Setting out drainage runs

If practical, visually check the sewer pipe where the intended connection will be formed. However, in most cases interpolation between inspection chamber invert levels may be the only option. Using the sewer invert level set up the TBM. The TBM will be used for all setting out purposes as previously described. Unlike setting out for the building drainage runs use **gallows profile boards** set at specific levels for use as **sight rails**. Ideally, two supporting posts provide a more stable sight rail. Figure 2.22 shows a typical sighting rail with the cross rail painted white. Where many trenched services are to be laid, a colour coded tip to every rail ensures that the correct line is used. On the face of the rail the **centre line** of the services is shown, in contrast to the building profile boards. The invert level is noted and also the formation level. There should be at least three profile boards per runnage to provide a simple check against accidental movement (Figure 2.23).

Method

1. Carry out a survey to establish spot heights of the existing terrain, say at 10 m chainage.
2. The designer will require:
   (a) the sewer invert level at the connection point
   (b) the proposed runnage or **chainage** (distance from the sewer connection to the proposed building)
   (c) spot heights at regular intervals along the proposed runnage.
   The information would be gathered using either a height of collimation (HOC) or rise and fall (R&F) booking form as previously described.

Office calculations

1. Computer software can be used to calculate the setting out information. Alternatively, a spreadsheet such as Microsoft Excel can be programmed. A simple programme is shown in Figure 2.24 for a 70 m chainage; it is very easy to extend or reduce chainage as required.
2. The fall of the pipe is subject to the diameter, the pipe material, and the use (see Chapter 12 for further details). In the example shown in Figure 2.24 the fall is chosen to be 1 in 40, meaning that for every 40 m of chainage the pipe will have a gradient fall of 1m. This can be expressed as $1/40 = 0.025$. The chainage (distance) between each spot height is 10 m so the difference

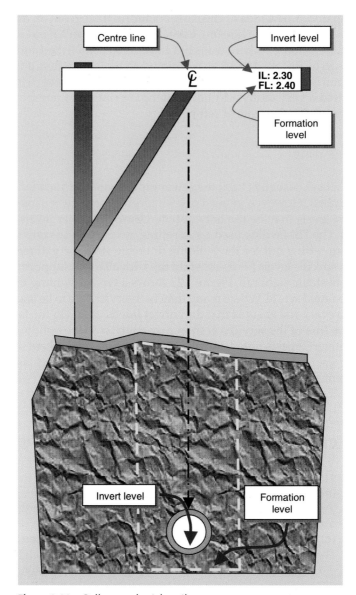

Figure 2.22 Gallows style sight rail.

in height between each spot height will be

$$10 \times 0.025 \, \text{m} = 0.250 \, \text{m}$$

a Enter the RL at the outfall. (The outfall is the lowest end of the pipe run.) Then enter the IL at the outfall.
b Subtract the IL from the RL to give the depth of the trench.
c An allowance for bedding material and the thickness of the pipe is required; therefore deduct 0.112 m to give the formation level.

d At the next spot height 0.250 m should be added to the IL. Then carry out similar calculations as previously described. Repeat at every spot height.
e When all the calculations are complete decide on an appropriate length for the traveller. A traveller is a length of wood with a blade forming a 'tee' shape used to **bone** through. The blade can be sighted through using the sighting rail of the profiles (Figure 2.23). Add to the formation levels at the outfall and the apex (the top manhole) and a midway spot height, and enter as shown.

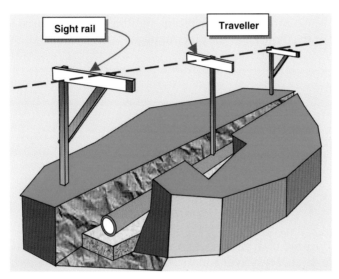

Figure 2.23 'Sighting through'.

Calculation of trench depth

Chainage (m)	0	10	20	30	40	50	60	70
	outfall							peak
G.L. (RL)	106.100	106.210	106.450	106.600	106.820	107.100	107.200	107.450
IL	105.000	105.250	105.500	105.750	106.000	106.250	106.500	106.750
Depth of trench	1.100	0.960	0.950	0.850	0.820	0.850	0.700	0.700
Formation level	104.888	105.250	105.500	105.750	106.000	106.250	106.500	106.750
Sight rail height	106.888							108.750

Fall: 1 in	40
Thickness of pipe [m]:	0.012
Thickness of bed [m]:	0.100
Traveller length:	2.000

Procedure:

1. Enter the RL at outfall
2. Enter IL at outfall
3. Enter the required fall
4. Enter the RL at chainage points
5. Enter traveller length

This spreadsheet is calculating the IL at each chainage point by taking the reciprocal of the fall and multiplying it by the distance between the chainage [e.g. 1 / **40** × 10 = **0.250** m then adding it to the IL.

Figure 2.24 Drainage depth spreadsheet.

Site work

1 Set out the position of the outfall manhole chamber and place the sighting profile board with the top edge level and at the height shown on the setting out sheet.
2 Set out the position of the intermediate sighting rail in line with the proposed drainage run and with the top edge as per the setting out sheet.
3 Finally, set up the apex profile sighting rail in the correct position at the top of the drainage run and the height as shown on the booking sheet.
4 The centre line on each profile sighting rail will be used as the centre line of the trench and the pipe runnage.

2.9 Traversing

There are two types of traverse:

- open traverse
- closed traverse.

Open traverse

The purpose of a traverse is to relate survey data to specific points; for example, to survey a river a measurable framework is required. The framework comprises a series of stations that relate to each other by direction and distance. Normally a reference object (RO) or coordinate is used to record the data against a map or chart.

Method

1 The risk assessment should include personal protective equipment (PPE) appropriate to the area in which the survey will be performed. For example, if working near a deep sided river or deep water, buoyancy equipment should be worn and appropriate life saving equipment, including floats and ropes, etc. should be available, as was the case when surveyors were measuring the Docklands area in London where the water is extremely deep and cold, and the dock sides high and vertical. The risk assessment should be adjusted to suit the conditions encountered.
2 Equipment required:
   - ideally a total station comprising an electronic theodolite and EDM; alternatively, a theodolite and steel tapes can be used
   - a spare set of fully charged batteries for the total station
   - an appropriate tripod on which to mount the total station

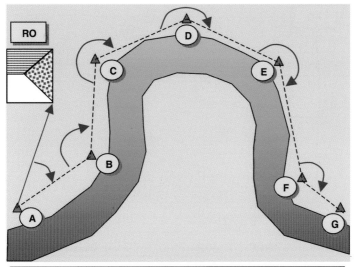

Station	Target	Angle ° ' "	Distance (m)
A	RO - B	037° 00' 00"	Not measured
B	A--C	128° 00' 00"	A--B 35.000 m
C	B--D	265° 00' 00"	B--C 55.000 m
D	C--E	229° 00' 00"	C--D 47.000 m
E	D--F	287° 00' 00"	E--F 35.000 m
F	E--G	132° 00' 00"	F--G 23.000 m

Figure 2.25   Open traverse.

- reflecting prisms mounted on tripods
- booking pad to book readings
- digital camera – useful to photograph the area for reference when you go back to the office
- appropriate clothing, including warm wet weather clothing in case the weather changes (surveying is not a fair weather job): the instruments are designed to be used even in rainy conditions.

3   Set up the first station using either coordinates or reference points.

4   Position other stations as required, ensuring they are visible to the previous and next station (see Figure 2.25).

5   Set up over station A and target the RO. Set the instrument at zero. Rotate the instrument clockwise and target station B. Measure the distance between stations A and B and book the readings.

6   Move the instrument to station B and target station A. This is known as the **back bearing**. From station A rotate the instrument clockwise and target station C. Take and book the reading. Also take and book the lineal dimension from B to C.

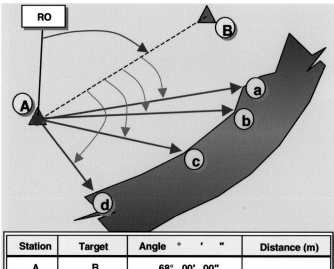

Station	Target	Angle ° ' "	Distance (m)
A	B	68° 00' 00"	
	a	19° 00' 00"	35.000 m
	b	28° 00' 00"	33.000 m
	c	38° 00' 00"	25.000 m
	d	82° 00' 00"	14.000 m

Figure 2.26 Bearings.

As the traverse proceeds, survey readings would be taken from each station using the adjacent station as the reference. Figure 2.26 shows the instrument set over station A and horizontal readings and lineal dimensions to points of note. Each reading is referenced from the line A–B. The booking sheet shows a method of recording the data. Modern total stations have an internal memory that records all the readings; therefore the surveyor does not need a paper record. When carrying out the survey it is useful to produce a sketch plan and photograph the surveyed area. Back in the office, photographs are very useful to pick up points that the survey missed and the sketch plan will help relate the computerised data. Some computer software will enable digital data from the total station to be presented three dimensionally, including coloured rendering.

The website http://www.leica-geosystems.com/corporate/en/ndef/lgs_31465.htm provides further information on the software available for use with a total station.

An open traverse cannot be checked for accuracy or adjustment.

Closed traverse

In a similar way to an open traverse, the purpose of a closed traverse is to relate survey data to specific points termed **stations**. A closed traverse starts

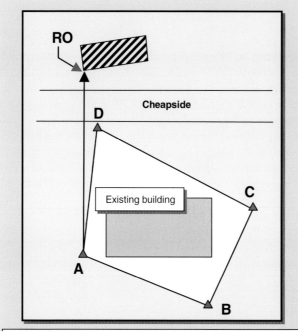

Angle	°	'	"	Distance (m)	
RO Â B	110°	15'	10"	A--RO	Not measured
D Â B	97°	37'	10"	A--B	24.100 m
A B̂ C	96°	23'	10"	B--C	16.000 m
B Ĉ D	84°	34'	29"	C--D	28.010 m
C D̂ A	81°	25'	25"	D--A	16.300 m

Angle	°	'	"	Adjustment	Correct' angles
RO Â B	110°	15'	10"	No adjustment	110° 15' 10"
D Â B	97°	37'	10"	– 4"	97° 37' 06"
A B̂ C	96°	23'	10"	– 4"	96° 23' 06"
B Ĉ D	84°	34'	29"	– 3"	84° 34' 26"
C D̂ A	81°	25'	25"	–3"	81° 25' 22"
	360°	00'	14"		360° 00' 00"

Figure 2.27 Closed traverse.

and finishes at the same station, enabling a check and adjustment for accuracy. Figure 2.27 shows a simple closed traverse. It is normal practice to label the stations anticlockwise so that all the internal angles can be taken clockwise from left to right.

Method

Prepare a risk assessment.

- If working on or near the public highway, signage will be required alerting motorists to the possibility of surveyors crossing the road. The signage should be in the form of warning triangles with a white background and red border with the word 'SURVEYING' in large black lettering. The signs should be placed about 50–100 m from the intended work area according to the speed of the traffic, or before the brow of a hill, or before a bend in the road. The intention is to make the motorists aware before they can see the surveyors and possibly slow down if required.
- All surveyors must wear high-viz jackets or vests as required by the Highways Act.
- Set up a series of stations in appropriate positions that enable vision between them. If there is no option other than to block the footpath, an alternative route should be provided for pedestrians. Permission should be obtained from the local authority for temporary footpath closure.

Procedure

The procedure is the same as that used to take an open traverse.

Example

The following example includes the adjustments and appropriate calculations required to convert booked readings to coordinates. Figure 2.27 shows a simple four station closed traverse. The reference object is the corner of a building. The stations have been labelled anticlockwise.

The design of the booking sheet is a matter of personal preference. The data should be easy to enter and easy to read after returning to the office.

Some total stations are able to store all the data electronically ready for downloading directly into plotting software, so the surveyor does not have to book any readings.

Procedure
1 Set up the instrument directly over station A and a reflecting prism over station B.
2 Set the horizontal circle to zero and lock the horizontal clamp. Target RO.
3 Release the clamp and rotate instrument clockwise and target station B.
4 Book the reading in degrees, minutes and seconds.
5 Book the lineal value (distance) using the EDM. Note that there will be a difference in height between the total station and the point used for the lineal measurement. Calculate the true distance using trigonometry. The sloping distance will be the hypotenuse and the true distance the adjacent. On modern instruments the sloping and true distances are

displayed. The difference between the true distance and the sloping distance will affect the accuracy of the traverse.

6   Move to station B and repeat the procedure, reading from station A clockwise to station C.

7   When all the readings have been taken and booked, carry out a quick check for permissible error. If the error is excessive the survey should be repeated.

8   *Check 1.* Add all the booked angles and compare them using the formula

$$(2n - 4) \times 90°,$$

where n is the number of stations

$$(2 \times 4 - 4) \times 90° = 360°$$

Check 2. The permissible error using an instrument with a 20′ accuracy is

$$\pm \sqrt{n} \, \text{minutes,}$$

where n is the number of stations

$$= 2 \, \text{minutes permissible error.}$$

9   Note the method of writing the name of the angle. It starts with the target to the left, then the station from which the reading has been taken, and then the target to the right.

10   The included angles in the example total 360° 00′ 14″ so the error is well within the acceptable 2 minutes.

11   It is not known where the error occurred. It may have been all at one reading or small amounts over more than one reading; therefore it is necessary to spread the error over all the readings. In the example the readings are all relatively similar. If there had been one or more very large angles, the error would be distributed proportionally.

12   By adjusting the angles to the required total they become **corrected angles** (Figure 2.27).

13   Using the corrected angles, calculate the whole circle bearings (wcb). Start with the wcb AB

	110° 15′ 10″
Add	180° 00′ 00″
Wcb BA	290° 15′ 10″
Add angle ABC	**96° 23′ 06″**
	386° 38′ 16″
Deduct	360° 00′ 00″
Wcb BC	26° 38′ 16″
Add	180° 00′ 00″

	206° 38′ 16″
Add angle BCD	**84° 34′ 26″**
Wcb CD	291° 12′ 42″
Add	180° 00′ 00″
	471° 12′ 42″
Deduct	360° 00′ 00″
	111° 12′ 42″
Add angle CDA	**81° 25′ 22″**
Wcb DA	192° 38′ 04″
Add	180° 00′ 00″
	372° 38′ 04″
Deduct	360° 00′ 00″
	12° 38′ 04″
Add	180° 00′ 00″
	192° 38′ 04″
Add angle DAB	**97° 37′ 06″**
Wcb BA	290° 15′ 10″

The only whole circle bearings required are:

AB	110° 15′ 10″
BC	26° 38′ 16″
CD	291° 12′ 42″
DA	192° 38′ 04″

14 Inspect the wcb values and determine which quadrant they will be in.
   - Positive value departures run west to east →
   - Negative departures run east to west ←
   - Positive latitudes run south to north ↑
   - Negative latitudes run north to south ↓

15 Where appropriate, deduct the whole quadrant angle(s). For example, wcb AB is in the second quadrant; therefore deduct 90° (Figure 2.28).

$$110° 15′ 10″$$
$$90° 00′ 00″$$
$$20° 15′ 10″$$

Wcb BC is less than 90° and therefore unchanged. The wcb CD is in the fourth quadrant so therefore deduct 270°; DA is in the second quadrant so deduct 180°. The results are termed **reduced bearings**.

The reduced bearings for the example are as follows:

AB	20° 15′ 10″
BC	26° 38′ 16″
CD	21° 12′ 42″
DA	12° 38′ 04″

16 Using trigonometry, the lengths of the adjacent and opposite can be calculated. The distances between the stations are all the hypotenuse of the triangles; therefore the following trigonometric formulae would be

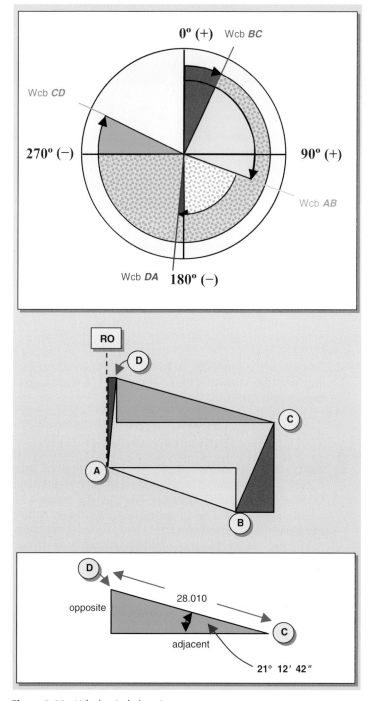

Figure 2.28 Whole circle bearings.

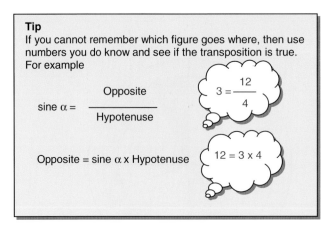

Figure 2.29 Mathematical tip.

used:

$$\text{sine}\,\alpha = \frac{\text{Opposite}}{\text{Hypotenuse}}$$

$$\text{cosine}\,\alpha = \frac{\text{Adjacent}}{\text{Hypotenuse}}$$

17 Transpose the formulae so that the unknown is on the left of the equals sign and the known is to the right of the equals sign (Figure 2.29). The transposed formulae will be:

$$\text{Opposite} = \text{sine}\,\alpha \times \text{Hypotenuse}$$

$$\text{Adjacent} = \text{cosine}\,\alpha \times \text{Hypotenuse}$$

18 Carry out the calculations for each coloured triangle shown in Figure 2.28. Use the reduced angle and the hypotenuse as shown above. Look at the signage indicating whether the result will be a positive or negative number. The green triangle has been fully calculated as an example.

   The line is CD; therefore the hypotenuse is 28.010 m.

$$\text{Opposite} = \text{sine}\,21°12'42'' \times 28.010$$

$$\text{Opposite} = 10.134\,\text{m}$$

$$\text{Adjacent} = \text{cosine}\,21°12'42'' \times 28.010$$

$$\text{Adjacent} = 26.112\,\text{m}$$

Note that the opposite is running from south to north; therefore it is a **positive** figure and the adjacent is running east to west making it a **negative** figure. The 'adjacent line' is termed the **departure** and the 'opposite line' is termed the **latitude**. Departure is always presented first followed by the latitude. Therefore the results will be −26.112 m, 10.134 m.

   Calculate the other results (the answers are shown Figure 2.30)

	Deg	Min	Sec	Line length	Depart +	Depart -	Latitude +	Latitude -	Dep diff	Trav perim	Dep corr	Lat diff	Trav perim	Lat corr	Correct' depart	Correct' latitude	Dep coord 100.100	Lat coord 200.100
AB	20	15	10	24.100	22.610			-8.343	0.107	84.410	0.031	0.188	84.410	0.054	22.579	-8.397	122.679	191.703
BC	26	38	16	16.000	7.174		14.302		0.107	84.410	0.020	0.188	84.410	0.036	7.154	14.266	129.833	205.970
CD	21	12	42	28.010		-26.112	10.134		0.107	84.410	0.036	0.188	84.410	0.062	-26.148	10.072	103.686	216.041
DA	12	38	4	16.300		-3.565		-15.905	0.107	84.410	0.021	0.188	84.410	0.036	-3.586	-15.941	100.100	200.100
				84.410	29.784	-29.677	24.436	-24.248							0.000	0.000		
					-29.677		-24.248											
					0.107		0.188											

Figure 2.30 Closed traverse calculations.

19 When all the departures and latitudes have been calculated, total up all the positive departures and then the negative departures and deduct the smaller total from the larger. In this example the result shows there is an error of +0.107 m going from west to east (departure); therefore it is a positive error which means a deduction of 0.107 m will be required to balance out the error.

20 Repeat the process with the latitudes. The example shows an error of +0.188 m going south to north (latitudes); therefore this is also a positive error. Again a deduction of 0.188 m will be required to balance out the error.

21 To enable the error to be spread proportionally over the closed traverse the following formula should be used:

$$\text{Departure correction} = \frac{\text{Departure difference} \times \text{Line length}}{\text{Traverse perimeter}}$$

The departure difference is calculated by subtracting the smaller departure from the larger departure. In the example it is the negative from the positive, resulting a value of +0.107 m. The traverse perimeter can be calculated by adding the lengths of all the lines. In the example it is 84.410 m (Figure 2.30).

22 Repeat the procedure for the latitude corrections using the following formula:

$$\text{Latitude correction} = \frac{\text{Latitude difference} \times \text{Line length}}{\text{Traverse perimeter}}$$

23 Enter the results in the appropriate negative columns. To complete the corrections subtract the departure corrections from the departures and the latitude corrections from the latitudes. The result should be two columns of numbers showing the corrected departures and the corrected latitudes for each line.

24 The process is used to provide coordinates for setting out based on an origin. For example if station A had a departure coordinate of, say, 100.100 m and a latitude of, say, 200.100 m, it would be written: 100.100, 200.100. Some surveyors use a prefix letter E100.100, N200.100 indicating the easting and the northing. Note that the departure is shown first followed by a comma and then the latitude. In the example, station B is 22.579 m east and −8.343 m north. As previously mentioned, the lines running west to east are positive (departure) so therefore the coordinate for B will be 100.100 + 22.579 = 122.679. The lines running south to north are positive; therefore −8.343 m means it must be going the other way, i.e. from north to south. When using maps and coordinates only use the **eastings** and **northings**: they are both positive. If a negative number has been calculated it indicates the line is going in the opposite direction. There are no westings or southings. (See also Chapter 1, Section 1.2.)

Chapter 3

Building surveying

This is the third chapter on surveying, this time concentrating more on the role of the building surveyor. The section has been divided into two parts, starting with the role of the building surveyor and introducing the range of surveys that they undertake. Part A concludes with practical guidance on how to carry out a conditions survey. This includes the health and safety issues to be considered before the survey as well as practical problems that can be encountered during the survey. The content is not exhaustive and should be considered more as an outline guide.

Part B concentrates on the structural problems in buildings. Many structures have cracks in the walls, ceilings and floors and a structural survey should be commissioned. This can be an emotive issue for a client. A structural survey is very different from a conditions survey where a crack will only be reported. In contrast, the structural survey should ascertain what has caused the crack and whether structural repair is required or simple cosmetic treatment. The subject is very complex so this section covers only the fundamentals, ranging from the equipment required, heath and safety issues through to guidance on what to look for during the survey. The science behind the reasons for the crack has been considered, leading the reader to other sections of the book as required.

Common faults and reasons for the faults have been included, together with legal requirements of the employer/client and the surveyor such as 'duty of care' and the Party Wall Act 1996. The chapter concludes with a measured survey, describing its function and method of approach. As in previous chapters, discussion of suitable personal protective equipment (PPE), risk assessment and suitable equipment are presented. The chapter closes with the method of carrying out a measured survey.

PART A

3.1 Types of building survey

There are four main types of building survey:

1 valuation survey
2 conditions survey
3 structural survey
4 measured survey

3.2 Valuation surveys

Where a client wishes to buy or sell property, insure property or declare property as an asset, a qualified valuation surveyor would be employed. Usually the surveyor will be local to the property with a good working knowledge of the current values of similar property, or the client may have faith in one particular surveyor who will carry out investigations in the area. Valuation surveyors may work for an estate agent or private practice. The valuation would start with visiting the premises, recording the number of rooms, approximate age of the structure, size of the building plot and the overall condition of the fabric and decorative finish. It is not as thorough as a conditions survey, however.

When the data has been collected, the surveyor would compare any similar property in a similar location. It is true that value is often based on 'location'. Two identical houses within a very small area can differ in value by many thousands of pounds. For instance, a house in a quiet road near a railway station would be more sought after than the same house on a noisy main road. Valuations should be based on what someone is prepared to pay, not how much it would cost to build. Insurance valuations may therefore be many thousands of pounds different from a selling valuation.

Valuation surveys do not report structural defects or the condition of the building unless it would have a significant effect on the value. If that is the case, the valuation surveyor should indicate that a structural survey is required. Typically minor cracks in walls would not be noted. However significant cracks, roof irregularities, or noticeable dampness should be reported.

3.3 Conditions surveys

There are several different types of conditions survey:

- house buyer's report
- dilapidations survey
- maintenance survey
- party wall survey

Where a client is considering buying a property, a **house buyer's report** would be commissioned. The surveyor would report on the condition of the fabric and the decorative order of the structure. It is not a structural survey; therefore any significant cracks in the structure should be reported for further inspection. With modern living the role of the conditions surveyor has changed. As recently as the 1980s it was possible for the surveyor to gain access to the floor materials and, where practical, to lift the occasional floor board to inspect the void below. However, modern flooring materials such as sheet chipboard prevent lifting, and many floor coverings such as sheet vinyl, fitted carpets and the recently popular laminate flooring cannot be lifted. Fitted furniture, especially in kitchens and bedrooms, is commonplace and the surveyor will not be able to gain access behind it.

'Do it yourself' (DIY) has produced many problems for the surveyor. Untrained or unqualified people often remove structural components and then cover up the damage. A good example was a friend who removed the struts and collars that support the roof of his mid-terraced property. He then screwed loft sized chipboard sheet to the ceiling joists, pushed fibreglass insulation between the rafters and covered them with hardboard. They were then finished with wood chip paper and emulsion paint. The electrical sockets had been connected as a spur from the mains electrical circuit. Access to his new room was via the loft hatch and a metal ladder. A sheet of ordinary annealed glass had been fixed in between the rafters using secondary sash window sections and a lot of mastic. Apart from the roof window, the new loft room looked very good indeed. However, the work had been carried out illegally – Building Regulations require permission from the local authority before any structural changes can be made. The house was sold without any reference being made by the buyer's surveyor to the illegal work that had been carried out.

Another area where the DIY enthusiast or unregulated 'builder' can cause problems is in the conservatory. Many conservatories that are exempt from the Building Regulations have been added to dwellings. However, any work that has been carried out must not *contravene* the Building Regulations of the main structure. For example, new electrical sockets and wiring must comply with Approved Document P: Electrical Safety which came into force in January 2005, but those in a conservatory are exempt under Regulation 9. This and other approved documents can be viewed freely on the website http://www.planningportal.gov.uk.

Building over a drainage run is also fairly common. The original concept of a conservatory was a lean-to structure with a transparent or translucent roof forming a room where people could sit – a room in the garden. The modern concept using the latest materials is to create another living room without the need for planning permission or Building Regulation approval. There are regulations regarding the size of the conservatory as planning issues, but many people ignore them and take the risk that they will not be found out.

Patios, driveways and new footpaths can enhance a property. Although they do not require Building Regulation approval or Town and Country Planning approval (unless in a conservation area), they must not contravene the Building Regulations of the main structure. The most common problem is where a new patio, driveway or footpath has been laid over the existing layer. The original finished level should be a minimum of 150 mm below the damp proof course (dpc) of the building. By adding a new layer the minimum is commonly breached.

This problem has increased since the recent regulations concerning disabled access and mobility issues came into force. Ramps are being constructed to allow wheelchair users and less mobile people access to existing buildings. However, ramps should still be in compliance with the current Building Regulations Approved Document C, which is to maintain a minimum dpc at 150 mm vertically from the finished ground level. There are methods for complying with both requirements. For example, Figure 3.1 shows an example of a public building that has contravened the Building Regulations since complying with disabled access requirements. A solution to the problem would be to apply an impervious layer such as a rendered plinth, using an additive to make the render waterproof, or apply a sheet metal splashguard using sheet lead cut into an appropriate mortar bed and tucked between the new ramp and the wall. The main reason for requiring a dpc at a minimum of 150 mm above any finished ground level is to reduce the risk of damp penetration from rebounded rain.

Figure 3.1 Dpc problem.

3.4 Dilapidations surveys

Dilapidations surveys are commissioned by landlords and commercial clients. For example, a landlord who owns a flat might require a formal report on the condition of the premises before the next person rents and occupies it. The report would list any damage to the fabric including decoration, and comment on the need for renewal or maintenance. It would form part of the contract, most probably in a rewritten format. When both the landlord and the tenant (person(s) renting) sign the contract they will have agreed as to the condition of the premises on that day and this would therefore form documentary evidence if a dispute arises at a later stage. For example, the dilapidation report may state that the hand basin in the bathroom has a chip near the cold water tap. When the tenant terminates their contract another dilapidation survey would be carried out and compared with the previous survey. The chipped basin would appear in both surveys so the tenant would not be liable for the damage.

On long term lettings, dilapidations surveys would probably be carried out every five years or so. The landlord would arrange for a new survey to be carried out to enable the condition of the fabric and finishes to be monitored. Repainting ceilings, walls and woodwork may be part of the landlord's or the tenant's obligations.

The dilapidations report is commonly referred to as the **schedule of dilapidations** and will be used as a basis for future approved alterations. For example, a long term leaseholder may want fitted wardrobes installed in a bedroom. The landlord may give permission for the work to be carried out; however, when the contract comes to an end, any damage, such as screw holes and change to decoration, must be made good and the property returned to the condition stated in the schedule of dilapidations. If then the tenant fails to do this, the landlord can have the work carried out at the tenant's expense. The final scheduled survey would confirm the work has been carried out as per the terms of the contract.

The dilapidations survey may also be used for maintenance programming – the condition of heating appliances, electrical wiring and sockets, sanitary ware and so on. The survey could highlight a fault of which the occupant was unaware, such as insect infestation in a roof void, so remedial work could be started before the problem became serious. Upgrading or maintenance of appliances under a landlord's legal obligation could be monitored during the survey. Examples of this might be the condition of the hot water system, electrical circuitry and any gas appliances.

Owners of commercial premises need to plan maintenance as part of their budgeting programme. A schedule of dilapidation may highlight the fact that the covering to a flat roof is coming to the end of its useful life and further repairs would not be cost effective in the longer term. The client could then commission a full survey of the roof, including upgrading of insulation and utilising modern roofing material such as polyester (see Chapter 5). The surveyor would prepare drawings and a specification from which contractors could then prepare their quotes. The client would then know how much the work would cost and be able to include it in the company's budget.

3.5   Maintenance surveys

Regular maintenance surveys, although costly, can save a company money. Landlords and commercial businesses normally carry out maintenance surveys on a regular basis. They enable a programme of planned maintenance to be prepared. For example, electrical circuitry may become overloaded when new equipment is installed. Typically, as more office equipment becomes electronic there is a need for additional sockets instead of adaptors and gang socket adaptors. Painting of window frames, door frames, fascias and soffits also requires regular inspection and maintenance (see Section 3.7, Windows).

3.6   An approach to carrying out a conditions survey

Like all surveys, conditions surveys require a methodical approach. If you are working for a firm, a risk assessment should have been carried out and filed with the management. It is unlikely that a risk assessment would be carried out on each specific survey, although the employer has a **duty of care** to all employees, visitors and the public regarding potential risks. Building surveying by its very nature can be hazardous and experience is essential to recognise the potential hazards.

The type of personal protective equipment (PPE) will vary according to the type of property. The following PPE is suggested for both domestic and industrial jobs:

- high-viz vest and/or jacket
- hard hat
- stout shoes (with toe protection)
- rubber gloves and latex gloves
- medicated hand wipes
- dust mask
- eye protection (polycarbonate lenses or cover glasses if you need to wear glasses, or protective glasses)
- ear defenders.

The PPE may look extreme, but the following scenarios illustrate where the equipment would be used.

The client may require a conditions survey for a large store with a stockroom or stockyard where fork lift trucks may be operating. A **high-viz vest** or **jacket** will give the forklift driver more chance to see you and therefore hopefully avoid you. Wearing a **hard hat** may protect your head if you need to climb under beams or enter areas where the ceiling or soffits are low. It also increases your visibility for machine operators and prevents impact to the head from falling smaller items.

When working in a stockyard adhere to any floor markings such as lines. Some yards have a system where pedestrians stay one side of the line and the fork lift trucks and lorries stay the other side. Drivers will not expect a pedestrian and therefore may not see you.

Stout footwear is essential, especially with a minimum standard of toe protection. Whilst climbing to gain access, heavy items may drop on your feet. Although materials should be stacked appropriately, they may not be, and so may be unstable. Nails sticking out of packing materials are another hazard, so metal soled shoes are a good idea.

Rubber gloves are essential if inspecting drains or working where materials are stored. Some chemicals can cause dermatitis or other skin/blood problems. Spillages do not usually have warning labels. Vermin such as rats, mice and pigeons all produce effluent and urine, so **latex gloves** are a minimum level of protection. When you have finished the survey it is wise to at least wipe your hands thoroughly with **medicated hand wipes**, or wash if facilities are available. Viruses and bacteria can be transferred easily to steering wheels, keyboards and so on, and the sandwiches at lunch time taste better without them.

Some areas to be surveyed may be very dusty because of powdered materials such as cement, lime and plaster. Usually any hazardous areas should be identified with the appropriate hazard warning triangles; however, derelict buildings may not be well signed. Accessing areas where pigeons have been fouling, such as roof voids, is particularly hazardous. Pigeons carry over 40 diseases that humans can catch. According to Dr. Trisha Macnair (http://www.bbc.co.uk/health/ask_the_doctor/psittacosis.shtml), psittacosis pneumonia is caused by the bacterium *Chlamydia psittaci* carried by pigeons and many domesticated birds such as budgerigars and parrots; in extreme cases this respiratory disease can be fatal. Therefore a suitable clean new **dust mask** should be worn.

Wearing **eye protection** when looking up or surveying a dusty environment is essential. Some major building sites insist that all visitors and workers must wear eye protection at all times whilst on site – the risk of eye damage is known.

The survey may require you to enter rooms where noisy machinery is operating. Unless entry is prohibited owing to the level of noise, which the signage should make clear, wear **ear defenders**. There are normally yellow warning triangles or blue compulsory signage if there is a recognised hazard; therefore ensure you are fully equipped.

Some further brief points on health and safety are now listed:

- If you are carrying out a survey on your own, **tell someone where you have gone** and about what time you should return. This is especially important when surveying empty property. Having a mobile phone is very useful, but if you have an accident either you may not be able to use it or the signal may be poor, especially when you are working in basements or steel framed structures.
- If you think there may be **asbestos** or another hazardous material and you are not qualified or experienced, or equipped with the necessary PPE, it may be better to ask a specialist to survey that particular area and make a note on the survey.
- Treat all **electrical wiring and fittings** as potential hazards. This is particularly important where an unoccupied premises has been vacated in anger. The author has seen bare live wiring where appliances have been taken away from dwellings that have been repossessed. Electrical wiring and switching gear may be very old and the insulation fragile. During your survey you may disturb wiring that can become dangerous, especially in roof voids. Following on from an earlier point, DIY enthusiasts may not have wired new sockets or switches correctly and they are therefore potentially lethal. See Chapter 12 for further information.
- Use **appropriate equipment** for the job, for example a surveyor's ladder as opposed to standing on the handrail of the stairs. Ensure that where you intend walking is safe: a flat felted roof may look strong enough to walk on, but may have rotted decking that will not support your weight. *Never* under any circumstances walk on corrugated asbestos or cement sheet roofing panels: both materials become weak with age and if you do fall through the sharp edges can do untold damage. Use **duck boards** or **roof ladders** and ensure they are secure before placing your weight on them.
- Carry **spare batteries** for your torch. Carrying out surveys in unoccupied premises, especially during the winter months or in areas with no natural light, can soon discharge torch batteries.

- Use a **plastic easy-wipe board** or take a roll of tape and a marker pen with you. If you need to switch something like gas or electricity off during your inspection, leave a sign, preferably taped to the valve or switch, stating that you are working in the area and need to keep the service dead. Decades ago it was easier with electricity as it was possible to take the fuse out and carry it with you. Now with miniature circuit breakers (MCBs) it is not so easy.

Equipment

The equipment required by the surveyor will vary. A simple domestic conditions survey may require the following items:

- Surveyor's ladder – a small sectional ladder that can be transported in the boot of a car and should enable access to loft spaces and flat roofs on single storey buildings. Other surveyor's ladder types include a telescopic ladder that when fully extended will provide access similar to the sectional ladder. Where access to greater heights, such as above eaves level on a two storey building, is required, it is recommended that use of a tower scaffold or access platform should be considered. Use of a good pair of binoculars can often solve the access problem and is much safer (see Section 3.7)
- Torch – a hand held torch and a headband torch are particularly useful, especially when inspecting roof voids and under stairs
- Digital camera
- A folding rule (minimum 2 m long)
- Steel tape (3 m long)
- Plastic coated steel or glass fibre tape (50 m long)
- Hand held laser distance meter – lasers use only light, whereas sonic distance measures use sound waves that can be deflected giving false readings
- Manhole cover keys
- Gas and electrical cupboard keys
- Moisture meter
- Clipboard and squared survey paper – helps when drawing
- Plumb bob and spirit level – a boat level is adequate for most occasions
- Stiff wire – a wire coat hanger is useful
- Drill and masonry bit – a good quality battery drill 18 V or better with percussion drive. If mains electricity is available, a 110 V transformer and hand held pneumatic drill may be more useful; however, it is bulky and you may not be close to where you have parked
- Endoscope and digital camera – you will need either mains electricity or a small generator to power the optical fibre light source
- Mirror on an articulated shaft – available from car spares and good tool shops
- Claw hammer and bolster chisel – if you prefer, choose an electrician's version with a large rubber protection flange around the stem of the chisel

- Screwdriver – currently available are screwdrivers that contain a small flat blade, a medium flat blade and crosshead blades located in the handle meaning that only one tool is required
- Mains electrical socket testing plug and meter
- Drain testing equipment – air test and smoke testing. Other testing would be part of a specialist survey or structural survey
- Signage – either a plastic easy-wipe board or taped paper and a marker pen to enable you to leave a note where applicable (see above)
- Magnetic compass
- A simple bar magnet.

At this point it should be emphasised that not all the equipment is needed for every type of conditions survey. Particularly in situations where you have to carry everything with you and travel by public transport, it may be useful to hire heavy and bulky equipment, and have it delivered to the survey location and collected afterwards.

3.7   Conditions survey procedure

It is likely that you will be working for the potential buyer of the property and not the current owner. Therefore do not cause damage during your survey, such as by pulling up carpets, prodding suspicious wood until it falls apart, etc., otherwise you may have to pay for the damage to be rectified.

Checklist

As previously mentioned, a methodical approach to the survey is important. A common approach is to prepare a checklist. This will ensure that the surveyor does not forget anything. It is embarrassing, costly and unprofessional to have to return because you have forgotten to survey something. The type of building will dictate the details; however, the following checklist should be useful. After the checklist each issue is developed further, and what to look for and why are discussed. Surveyors should have a questioning and reasoning approach, not only to what they see, but also to what cannot be seen.

The checklist should contain the following points:

1   **The address of the property** – This can be the postal address including the post code. This is particularly important where the road is very long or the name of the road is common to many areas, such as 'London Road'. There are websites, such as http://www.streetmap.co.uk and https://www.royalmail.com, that enable a post code to be entered and then display a detailed map of the area.
2   **The date of the survey** – This is very important as there may have been a previous survey carried out, or there may have been alterations after the day that you have carried out your survey.

Figure 3.2   Original timber framed buildings.

3   **A description of the property** including the approximate age of the
original structure and any significant additions or alterations – this will
provide confirmation for the client. The façade may not be the same
as the structure. Buildings frequently have facelifts as fashions change.
Figure 3.2 shows an example of mid eighteenth century cottages. The
original structure is 'Essex boarded' on timber studwork. It then became
popular to create the illusion of a stone building by smooth rendering the
front façade (known as **stucco**) and scribing lines to give the appearance
of jointed stones.

Figure 3.3 shows an example of a beautiful Victorian villa. Note that
the brickwork on the main structure is Flemish bond, indicating the
walls are solid 1B (1B means the wall is one brick thick; see Chapter 5).
The villa was built at the end of the nineteenth century; a two storey extension has been added in cavity wall design complying with the current
Building Regulations. Also note the change of materials for the windows
from painted softwood vertical sliding sash single glazed to modern
PVCu vertical sliding sashes with double glazed sealed units. The roof

Figure 3.3 A sympathetically designed modern extension.

has been re-covered with man-made slates. The whole extension has been designed in sympathy with the original design using similar look-ing materials with high tech performance.

4 **The weather conditions** – It is very useful to note the temperature and weather conditions, both on the day of the survey and over the recent past few days or so. If a survey is carried out in, say, July after sev-eral weeks of very dry and hot weather, the likelihood of recording any dampness is significantly reduced. If the seller has recently decorated, any staining may have been covered up. Conversely, if the soil is clay and there have been several months of rain, structural cracks may have closed up completely due to ground swell. In winter months the heat-ing may be working and is therefore easily tested for leaks, noises and efficiency, whereas in warmer months the surveyor will have to rely on signs (see Section 3.8).

5 **The name of the surveyor** – to help with any future queries, especially if there are several surveyors in a firm.

6 **The type of survey to be carried out** – it is important for the surveyor to know the extent of the survey and its purpose.

7 **External inspection** – carry out the following checks:

   a Stand away from the structure and view the roof line – look for continuity of materials and colour.

b Look at any protrusions through the roof covering, such as pipes and chimneys.
c Look at the guttering and inspect joints and connections with the down water pipes.
d Look at the fascias and soffits (if applicable) for regularity and decoration.
e Look at the upper floor windows for shape, verticality and finishings.
f Look at the panel of masonry or surface finish between the windows and check for any cracking or discoloration.
g Look at the panels between the upper windows and the lower windows (the **spandrel panels**) for any signs of cracks, discoloration and for verticality.
h Look at the ground floor windows for shape, verticality and finishings.
i Look at any doors, door frames, porches or openings. Check for cracks, discoloration, shape and verticality.

If applicable, carry out the same checks on all elevations. It is also worth reviewing any adjoining property, especially if the property is terraced or semi-detached. If you find cracks in the façade, sagging roofs, or irregular colouring of the fabric, look at several nearby buildings. There may be a fault due to poor workmanship, design or exposure. The building may have trees close by, or signs that trees have been removed. Look for any inspection chamber covers for an indication of any adjacent water runnages, such as drains and sewers. Look at adjacent property for signs of age. Has there been recent major building work that may have caused the cracks? For example, an infill property may be built very close to your survey structure. If there has been demolition close by, have there been any shock waves that have fractured water runnages? Water can erode soil structures and cause subsidence over relatively short periods of time.

8 When entering through any doorway, check the **door** for closure and fit. Is the door **binding** or **sticking**? Binding is where the gap on the side of the hinges or butts is insufficient and the timber of the door squashes onto the timber of the frame or lining, making closure difficult or impossible. Sticking is a term used when the opposite side to the hinges or butts touches the frame or lining. Possible causes of binding or sticking include ground movement displacing the walls and/or floor. In certain circumstances, movement can cause part of the wall to turn slightly from the vertical, creating a twisted frame or lining which causes the door to become **in wind**. Look for other signs of structural movement, such as cracks in the wall above the door or at ceiling level. Another cause could be the workmanship of the original fixer. If the original wall has not been built vertically and the frame or lining follows the wall, the door will be in wind (twisted like a propeller). Check the door stop and rebate for regularity (Figure 3.4). Also check the latch and, if applicable, the lock.

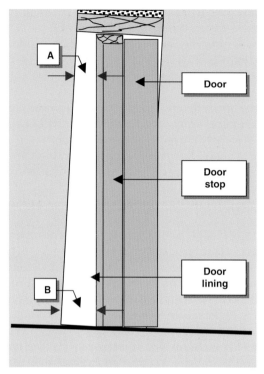

Figure 3.4 Door frame in twist.

9 Look at the **ceiling** for cracks, staining, patching and gaps – new property or recent building work may have been saturated during construction so when drying out it will shrink.

10 Look at the **tops of any openings**, such as windows or doorways, for cracks. If you have noted cracks on the outside façade, particularly look for signs on the inside.

11 Look at all accessible **walls** for cracks, bulges and verticality.

12 Look at **wall coverings** such as ceramic tiles, papers and vinyl for colour, staining, fading, etc. Check for mould growth as an indication of moisture levels.

13 Look at **picture rails** and **dado rails**, if applicable, for continuity. Doorways may have been filled in or new ones formed, and rails commonly show scars. Likewise look at the architraves: are they the same profile as the rest of the building?

14 Look at **skirting boards** for continuity for the same reason as the rails. Check that the gap between the floor line and the lower edge of the skirting is regular.

15 Look at **windows** for any gaps, staining, condensation in between sealed units (if applicable), condition of gaskets or putty, and the operation of every opening window. Check **window locks** if applicable.

16 Look at any **heating equipment**, such as radiators, gas fires and electric storage heaters. Check the spandrel panel behind the radiator, if fitted below a window, for cracks.

17 Look at the **floor finishing** for colour differences, depressions, impressions or repairs, and for level.

18 Look at all **electrical fittings** including lights, sockets and switches. Operate light switches to check that they work – do not take it for granted.

19 Check all **sanitary ware** for finish, staining and cracks.

20 Check all **water operations**: flush the WC, operate all taps on the bath, basins, sinks, showers, bidets, etc. Listen for possible water hammer and watch the water discharge for potential blockages.

21 Access the **roof void**, if applicable, and look at the insulation. If practical, enter the roof void and inspect at eaves level for daylight. It is especially important to observe the health and safety risk assessment. If there are no boards for safe entry or the insulation covers boards, it is recommended that you enter the fact on the survey report as '*inspection not possible due to obstruction*'. If viewing a loft area from the hatchway only, switch off all lighting and look for any points of daylight (obviously this will only work during the day). Use a powerful torch beam to inspect the roof members that are visible. Field glasses/binoculars may be useful for closer inspection, especially if insect infestation is suspected (particularly in older property).

22 Check the **utility intakes** – water stopcock, gas valve, electrical intake.

23 Check the **underside of the stairs** for movement and insect infestation, especially in older property.

24 Check **basements and coal cellars** – the fact the structure has a coal cellar indicates its age. Common problems are water and/or dampness related.

25 Lift all **inspection chamber covers** and look for signs of ground movement or blockages.

This checklist highlights what to check, but is not exhaustive. However, it should be helpful in most conditions survey situations. Depending upon the type of survey, additional issues or points should be considered, so the checklist is now followed by a more in depth description of what to look for and what to do. Like Sherlock Holmes, the surveyor may have to consider many things that are not seen and therefore use a series of deductions based on experience.

Roofs

Check the roof line. The most common roof designs are shown in Figure 3.5. When writing the description you may have a roof that includes more than one design.

When looking at the roof line of a pitched roof inspect the ridge. Are all of the ridge tiles soundly bedded and the mortar intact? As you look at the roof line look at the roof covering. There are many different types:

1 plain tiles
2 single lap

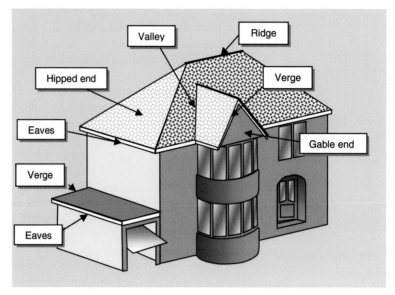

Figure 3.5 Roof terms.

3 interlocking tiles
4 interlocking pantiles
5 slates
6 thatch
7 stone
8 sheet materials
  - corrugated asbestos
  - corrugated metal
  - sheet metal
  - bituminous felt
  - polyester felts
  - asphalt
  - glass

There are many different designs and finishes to the edges of roofs. Roof terms are shown in Figure 3.5 and include the **eaves** where the roof adjoins the wall, the **verge** where the roof abuts the span of the roof, and the **apex** or **ridge** where the top of a roof meets.

Not all roofs have a soffit at the eaves. Although this is a poor design, the house in Figure 3.6 has roof tiles that finish in line with the face of the wall with no overhanging soffit. The original guttering was galvanised iron screwed directly to the face of the wall plate and this has since been replaced with plastic half round gutter on gutter brackets.

Common problems with flat roof verge details are the lack of overhang C or the tightness of the angle of the roofing felt as shown at point A in Figure 3.7. (Also see Chapter 4.) The verge has been formed by nailing a triangular fillet to the edge of the roof decking. The roofing felt apron is

Figure 3.6 Flush roof verges.

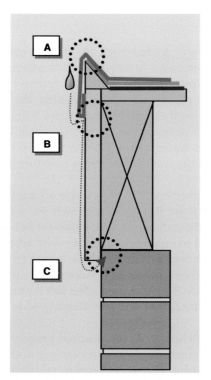

Figure 3.7 Flat roof design faults.

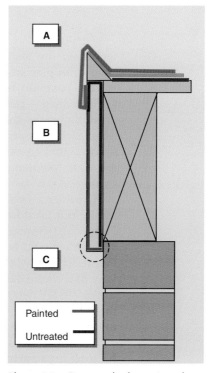

Painted ▬
Untreated ▬

Figure 3.8 Common bad practice when installing a flat roof.

Figure 3.9 Bad practice on flat roof verge.

tacked to the fascia board and folded back on itself. The fascia board B should be primed and undercoated before the felt is attached; however, it is very common for the carpenter to fix prepared (planed all round) timber and then a painter may prime and undercoat after the fascia has been fixed. Figure 3.8 shows the area that is painted, the area that is untreated and the area that is vulnerable. The fascia has been fixed flush with the face of the wall and encourages any water running down the face to continue onto the wall. Maintenance is on the surface, enabling moisture to penetrate the wood.

Figure 3.9 shows an example of poor design and workmanship: look at the photo and jot down a list of the things wrong with the work. In contrast, Figure 3.10 shows a sufficient overhang and provision for better maintenance. The arris should be removed to enable the paint to flow around the corners. Figure 3.11 shows a typical section of timber with the arrises left on. The paint is very thin where the sharp corner remains so is vulnerable to thinning and failure due to UV deterioration. On the right, the timber has the arris removed and the paint continues around the corner without thinning so is less susceptible to UV degradation.

A common defect with pitched roofs on attached properties can be **hogs backing** over party walls. Figure 3.12 shows an example of the roof line over the party wall in terraced housing. The cause is commonly that the roof timbers have started to sag and the section of roof over the party wall rests on it and so remains at the original level.

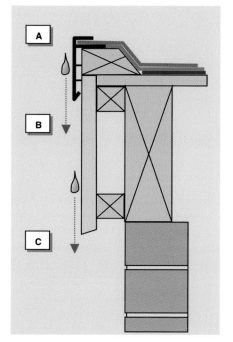

Figure 3.10 Well designed flat roof verge.

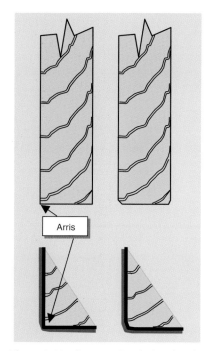

Figure 3.11 Comparative design details.

Buildings on exposed plots are more susceptible to **wind** pulling the raised tiles or slates; in extreme cases strong gusts can remove the whole roof. On your report you should mention the defect and suggest that the roof requires further inspection. Similarly, the problem can occur at gable ends on a roof, with similar consequences where the verge tiles have deflected (Figure 3.13). The extent of the defect should be quantified and the location noted, for example by orientation: *'Roof – (south elevation left flank) verge tiles over gable end displaced, possibly due to settlement/shrinkage of roof timbers. It is recommended that further more detailed investigation is carried out.'*

Protrusions

Inspect chimney or other protrusions using binoculars. Chimney stacks penetrating the roof covering can be the cause of roof leakage. If the roof covering is tiles or slates it is common for either a tiled fillet to seal the gap between the stack and the tile or a metal sheet termed **flashing**. Where the stack is brick, the flashing would be dressed into the mortar joints. The mortar would be raked out (scraped out to a 10 mm depth whilst still wet) and the lead, copper or zinc sheet dressed into the joint. Mortar is then pressed into the joint providing a weather resistant seal. There are normally sheet metal **soakers** beneath the tiles to catch any water that has bypassed the flashing. (There are several good tech-

Figure 3.12 Hog's back over a party wall.

nical text books that cover the detail such as *Barry's Introduction to Construction of Buildings* by Emmitt and Gorse, pages 576, 577.)

An alternative method of dressing a chimney joint is to bed cut tiles in a strong mortar around the chimney as in Figure 3.14. In Figure 3.15 the brickwork shows signs of spalling, the pointing needs repairing, and the plastic flashband has a relatively short life span so should be considered only as temporary.

For soil vent pipes and gas flues, specialist tiles and proprietary fittings, such as a neoprene ring with a stainless steel jubilee band that covers the roof to pipe joint, can provide excellent weathering on pitched roofs. Flat felted or asphalt roofs present more of a problem, and it is common for the roofer to bond felt as an upstand around the pipe. Inspect the seal between the pipe and the roofing material if roofing felt has been bonded to the pipe.

Guttering

Most but not all roofs will have guttering. Although now rare, guttering in Regency times (the early 1800s) was made from zinc sheet. As the

Figure 3.13 Problems with verge tiles.

Tiled fillet

Possible wet rot in soffit and unpainted fascia

Figure 3.14 Tiled chimney fillet plus fascia and soffit problems.

Figure 3.15 Temporary repair to chimney fillet.

metal oxidises it becomes brittle and the gutters are susceptible to fracture if ladders are rested on them. Other metals used for guttering include cast iron, pressed steel, aluminium alloy, lead and copper. Timber guttering can still be found in certain conservation areas, with charred oak or elm being the most common. Modern guttering materials include extruded PVCu.

When gaining access to flat roofs do not rest the ladder on guttering. The extra pressure from the ladder may crack the gutter; however, it is more likely that the top of the ladder will slip. Do not place a ladder on a flat roof unless a **spreader board** is being used beneath the foot of the ladder. Point loading from the ladder stiles can be too great for the roof decking and in extreme cases a rotted deck may give way under the weight.

Guttering is often neglected, and sanded tiles with airborne dust and regular watering enable grass and other plants to grow in the gutter (Figure 3.16). The consequence can be overflowing and cascading of water from the blocked gutter, or the blockage can be washed into the rainwater pipe causing a vertical blockage. The latter commonly breaks old cast iron pipes when the saturated blockage plug freezes.

Look for cracks, especially at joints. They can be the cause of damp patches on walls. Rainwater pipes normally have a **shoe** to deflect the water into the centre of the gulley grating. Figure 3.17 shows the consequence of a missing shoe. The original grey plastic pipe bracket appears to have been fixed at the wrong point, too close to the end of the pipe. The replacement black plastic rainwater pipe has been cut too short and without a shoe. The cement **haunching** has been poorly applied and requires removal and replacement. Note the algal growth and the general condition of the sand cement mortar at dpc level. It is also possible that the rainwater has been entering the ground

Figure 3.16   High level lawns?

Figure 3.17   Drainage and gulley problems.

Figure 3.18 External inspection.

via gaps in the drain leading to subsidence that has caused the concrete to crack.

Inspect the guttering and down water pipes. Are the fixing brackets adequate and the fittings correctly fixed? Figure 3.18 shows a section of rainwater pipe in cast iron with the upper length and offsets completely missing. Check for plant growth or algae on the face of the wall, indicating leakage from guttering over a long period. Whilst looking at the guttering inspect the fascia and soffit boards if applicable.

Fascias and soffits

Figure 3.14 shows signs of wet rot and poorly applied paint. The quality of paint and surface preparation are both important. However, in this case it is likely that the paint has been applied directly over unprepared (and possibly unwashed) paintwork in damp conditions. The consequence is poor adhesion between the paint layers and premature peeling. On the report it would be noted that *'the fascia and soffit boards are in poor decorative order with possible wood rot'*. From the surveyor's point of view it is impossible to state categorically that it is only wet rot and ascertain the extent of the rot from standing on the ground. The surveyor's job is to point out the defect and suggest it be further investigated. Many properties have overhanging eaves formed by the foot of the rafters. The detail provides protection from driving rain and aesthetically softens the lines of the structure by introducing shadows on the façades. Where a board is fixed vertically to the rafter it is termed a **fascia**. Commonly the guttering will be fixed to the board using gutter brackets, or in the case of the old cast iron gutter, direct fixings through the gutter (Figure 3.14). To slow the rainwater down so that it falls into the guttering a **tilting fillet** or **sprockets** will be fixed behind the fascia (see Chapter 7, Figure 7.5).

Since the 1940s, sanded bituminous felt (**sarking felt**) would cover the rafters before the tiling battens were fixed. The principal function was to prevent wind driven snow or rain blowing under the tiles and entering the roof void. Other sarking materials include Kraft paper (a brown paper sandwich bonded with nylon reinforced bitumen) and polythene sheeting. Kraft paper becomes fragile with age and hence easily torn. Sand and dirt blown under the tiles build up and eventually tear the paper as shown in Figure 3.19. The

Figure 3.19   Party wall roof detail.

problems caused by sarking felt lining are mostly due to condensation. Water storage cisterns are typically located in the roof void. Many of the older ones are galvanised mild steel with either no lid or perhaps a board cover. In hot weather the water in the cistern evaporates, creating a high water vapour content. When the roof void cools during the night the water vapour condenses and can cause dampness, especially on roof fixings. (This was a problem with early gang-nailed trusses in the 1970s where the fixings were lightly galvanised mild steel and the gang-nail plates had pressed out teeth and edges.)

Windows

There are many different window designs and materials. The main items to check are as follows:

- **Verticality** – Is the window frame vertical? If not, check the plaster line on the internal wall: has it been filled or is there an unequal gap? If part or parts of the window open, carefully operate and observe if there is any movement of the main framing. If softwood framing there could be wet rot around the fixing or, if old, the fixings may have rusted through. Never try to force an opening window. Apart from the risk, you may not be able to close it again and there is also a chance that structural movement has caused the window to become load bearing. Opening the window may cause the wall above to slip further.
- **Weathering** – Frames can be made in hardwood timber and are commonly finished in a wood stain. Modern finishes include microporous paints and stains. The better quality ones have ultraviolet (UV) stabilisers and filters within the finish and act in a similar way to suntan lotion. If we stay exposed to the Sun's rays our skin will burn: UV rays such as UVA and UVB literally burn the surface. With timber, UV rays from the Sun cause bleaching and long term damage termed **weathering**. If exposed to the weather and water, softwoods are susceptible to wet rot. Figure 3.14 shows examples of possible wet rot and peeling paint on a fascia and soffit.

Glazing – and identifying rot in painted timber

Painted softwood windows, for example, require regular maintenance. Figure 3.20 (A) shows the correct maintenance of a bottom rail of a casement window. The paint has been continued onto the glass from 2–3 mm both sides, forming a good seal. In contrast (B) has been finished at the edge of the putty/glass. The putty could be linseed oil or butyl putty. Either would be suitable for a softwood timber rebate.

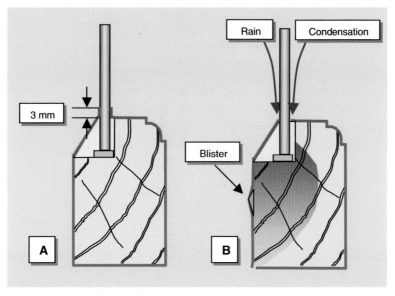

Figure 3.20   Wooden window paint comparisons.

The frame should be knotted and primed before the putty glazing is applied. After the **putty fronting** (the term for that form of glazing) has been completed, it should be left for a few days to form a skin prior to painting. The putty should then be primed and left to dry before undercoating and applying two top coats of finishing paint. It is the finishing paint that continues onto the glass to form the seal. The term used for applying the paint around the glass is **cutting in**. Figure 3.20 (B) shows the paint seal has not been completed. Consequently the putty has dried out and cracked, shrinking away from the glass. The main cause is hot weather heating the glass, as a result of which the oil in the putty evaporates. Because the crack is surrounded by non-absorbent material, capillary action takes place and moisture is pulled into the crack (Figure 3.21). Frost action converts the water to ice, forcing the crack further apart. When the crack is wide enough, rain will enter. If the window had been primed before the putty fronting, water would not have been able to enter the timber, so only the putty would have failed and needed replacing. In this example, the window had not been primed and consequently water entered the wood. Saturation of the timber will result in wet rot.

When the Sun heats the painted surface the water within the window section will try to convert to vapour, pushing the paint away from the wood and forming blisters. The UV rays from the Sun will harden the paint, making it brittle and causing it to crack. The paint loses adhesion and starts to peel, further exposing the wood to the elements and natural decomposition. The maintenance survey should identify the problem and make recommendations

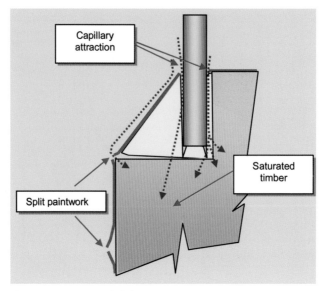

Figure 3.21 Materials failure.

for such remedial work as may be possible to prevent further moisture ingress and allow the timber to dry out fully. Decayed wood can be removed and resin hardeners and fillers used to replace rotten wood.

Another example of dilapidation is shown in Figure 3.22, despite the softwood door frame having been fitted correctly with a vertical dpc to protect the back of the frame from moisture via the brickwork. However, the failure has been caused by the shrinkage of the cill member. A gap has formed, allowing moisture to enter the end grain (the most vulnerable part of timber because the tracheids – similar to the ends of drinking straws – are exposed). As the timber soaks up the moisture, the Sun's action will cause the paint to peel, enabling more water to enter and causing eventual wet rot. Figure 3.23 shows an exploded view of the mortise and tenon joint between the cill and the door jamb.

Cracks in masonry

See Section 3.10.

Spandrel panels

See Section 3.10.

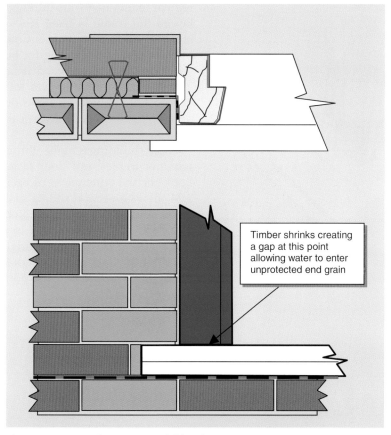

Figure 3.22 Door frame potential dilapidation.

Doors, door frames, etc.

Doors should fit their openings without binding or rubbing. Shut the door and inspect the gap. It should be uniform about 2–3 mm. If the closing edge has been partly planed or the top edge has an unequal gap, they may be signs that the door has **dropped** or the wall has moved. If the latter, see Section 3.10, p. 123. If the former, look at the type of butt (large hinges), the positioning of the butt, the number of butts and the weight of the door: they could all cause the door to drop. An example where the weight could be the cause is a house converted into flats or has a loft conversion making the building three storeys. Some lightweight doors require upgrading for fire requirements, and cement sheet (on older doors it may be Asbestolux) or sheet boarding has been attached without upgrading the butts and screws.

Door frames with timber cills or thresholds are susceptible to rot where the jamb meets the cill or threshold. Poor detailing, workmanship and a lack of

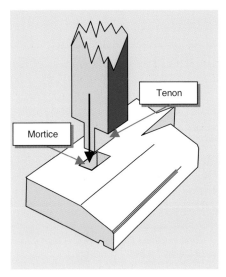

Figure 3.23 Exploded view of a door frame cill joint.

regular maintenance are the biggest causes of rot. Before the frame is fixed it should be knotted and primed as a minimum. Commonly, the cills or thresholds are left long for building in as the work proceeds. Hardwood is normally used as it is less susceptible to wet rot attack. The cill or threshold should be stood or bedded over a dpc (Figure 3.22) with a vertical dpc behind the jambs.

When buildings move through subsidence, window frames and door frames may distort. Look for the squareness of each corner. Figure 3.24 shows a vertical sliding sash that no longer closes properly at the head. Note how the brick arch above the window has dropped to the left. The building shown in Figure 3.25 has moved significantly: look at the spandrel panel between the upper window and the portico head of the porch.

Figure 3.24 Masonry movement in lime mortar.

Figure 3.25 Severe historic masonry movement in lime mortar.

Stained ceilings may result from leaking pipes. Check with a damp monitoring meter such as a Protimeter to see if the ceiling is still wet (Figure 3.26). An overloaded floor may cause the ceiling board joints in the plasterboard to open or straight line crack across the ceiling. A common cause of overloading is the 'home office' where a bedroom has been converted into an office. Overloading the floor with several three- or four-drawer filing cabinets full of paper exceeds the design live load of $1.5 \, \text{kN/m}^2$.

Elevations

The survey should continue with inspection of the windows and wall condition, working down from the top from left to right, one floor at a time. The building in Figure 3.27 is over 600 years old; however, it has only recently suffered from subsidence as the result of groundwater movement. Figure 3.28 shows a support wall that has been overloaded. Figure 3.29 shows differential movement most probably caused by using steel lintels and lightweight blockwork. The two materials will expand at different rates, therefore causing a hairline crack. The blockwork has been finished in a cement render which has also cracked. Figure 3.30 is a close up of the area showing that the masonry paint applied to the smooth render has cracked allowing moisture to become trapped behind the impervious paint layer. Capillary action will pull more moisture through the crack, and the effects of freezing and air frost action will further separate the two surfaces. It is possible for the render to become detached; this is termed **blowing**.

Figure 3.31 shows typical efflorescence in mortar, most likely due to salts in the sand whereas Figure 3.32 shows typical efflorescence in bricks resulting from salts in the clay. Efflorescence is generally cosmetic, although rarely a form known as **crypto-efflorescence** can result in structural damage. Here salt crystals form in the pores of the brick close to the surface. As the surface of the brick dries, crystals form and exert pressure within the pore. When the tensile strength of the pore walls is exceeded, material is **spalled** from the brick. The damage appears similar to frost spalling; however, to occur, crypto-efflorescence requires a warm drying breeze and a regular source of soluble salts in water. Solid brick walls that have no tanking or waterproofing on the surface against the earth can be subjected to continual wetting from one side whilst the outer surface is being dried by warm air.

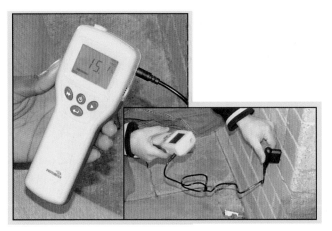

Figure 3.26   Testing moisture content in masonry.

Figure 3.27   Recent severe masonry movement caused by drought conditions.

When all elevations have been surveyed, continue with the rooms, starting at the entrance. Check every door for closure and note if they bind at the hinges (common with a poorly hung door or if there has been structural movement). Check the function of door latches and locks. Look at the door stop with the door in the closed position. Are there large gaps or has the door twisted? Check the door lining or jamb (see Chapter 4 for terminology) with a spirit level if

Figure 3.28 Inadequate design and overloaded masonry support.

required. If the structure is old, it is quite possible that the door frame was not vertical originally or the building has settled over the years. Look for additional signs of movement in the surrounding walls. Look at the head of the door and check that it fits squarely in the frame. Again, very old buildings move over the years; therefore a misshape does not necessarily mean subsidence or failure. Not all buildings were vertical to start with. The Georgian building shown in Figure 3.33 is an estimated 100 mm off vertical; however, the brickwork shows no sign of movement or repair. The late Regency/Victorian buildings shown in Figure 3.34 start about 125 mm apart at the first floor level and the building on the left leans to touch the structure on the right at roof level. In both buildings the face brickwork looks sound with no sign of repairs or cracks.

Returning to the internal part of the buildings, look at the skirting board if present and note the alignment. If there is a large gap between the floor and the skirting board the floor may have subsided. In modern solid floors the most likely cause will be poor compaction of the hardcore sub-base (see Chapter 4 for details). Another possible cause could be subsidence when ground movement has changed the load bearing capacity of the sub terrain (see Sections 3.9 and 3.10).

Figure 3.29 Differential movement.

3.8 Services and heating systems

Wet heating systems can be difficult to survey when they are not operating; therefore look for indications of problems as follows:

- **Water staining** on carpets below radiator valves.
- **Blistering paint** and possible **wood rot** on the skirting board beneath and adjacent to the radiator, indicating continuous water presence. This is an important indicator if the carpet has recently been replaced as an aid to selling the property. If that has been the case, check that the radiator is filled with water and has not been drained down by gently tapping the radiator panel and listening.
- **Verdigris** (green powder or crystals caused by acidic action on copper) could indicate weeping of the valves.
- **Drips** around the underside of radiator valves. Slow seepage can saturate a carpet and the floor beneath. If the floor is standard grade chipboard,

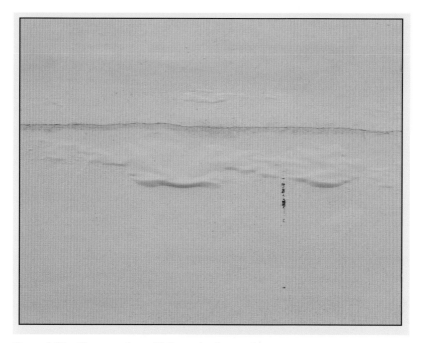

Figure 3.30   Close-up view of failure of paint coating.

Figure 3.31   Efflorescence in mortar.

Figure 3.32 Efflorescence in clay bricks.

it will **blow** which means it will expand and lose all supportive strength. Check the operation of the valves – do they operate?

- Look at the **boiler** if possible. Other than back boiler types which were part of a gas fire, probably in a lounge or dining room, most other gas fired boilers should be accessible for identification. Have a look at the inside of the cover plate or panel for a date of manufacture. If this is not available, take a note of the make and model number for investigation via the manufacturer's website. If the boiler is out of production, an e-mail to the technical department will often establish the possible date of manufacture, although it may be a museum piece if it is that old. Look at the method of venting. Is it a balanced flue, room sealed flue or fan assisted flue? Figure 3.35 shows where the UPVC fascia cladding has been scorched by the flue gases from a balanced flue gas appliance. The vent should be located away from opening windows or carports and anything combustible. The example shown suggests the gas appliance fitter has contravened the Building Regulations.
- The style of the **radiator** may indicate the approximate age of the system. However, an old radiator regularly maintained can be better than a new radiator poorly fitted and neglected; therefore this is not necessarily an age issue.
- **Solid fuel boilers** are less common, although still in use. Their efficiency is difficult to assess; therefore visual inspection of the connection to the chimney, the outlet and controls would be the full extent of the checks. Look for signs of water staining on adjacent materials and signs of soot.

Figure 3.33 Not all buildings are built vertically 1.

- Unless you are a **qualified heating engineer** it would be normal practice to state that the heating is to be inspected by a competent person which, with gas boilers, means someone who is CORGI registered.
- Other heating systems include **ducted warm air systems** where a series of ducts (metal tubes) have been built into the structure and a fan provides air pressure. The heating would be gas or electric, and the fan blows the air through the heat exchanger into the ducts. The system can be used to cool a building by using the fan without the heating. From a surveyor's aspect, testing would entail checking that the outlet grilles are clear and clean with a good flow of air. If possible, you should ask for the heating unit to be operated as, unlike wet systems, the effect is almost instant. As with the previously described gas appliances, it should be checked by a competent person.
- **Electric heating** can be radiant heat from permanently fixed (hard wired) electric fires or storage heaters. Portable heaters are not fixtures so would

Figure 3.34 Not all buildings are built vertically.

Figure 3.35 Non-compliant gas boiler installations.

Figure 3.36 Modern electric storage heater.

not be considered. Look for isolation switches such as fused spurs near or next to the storage heater and not the design. Old storage heaters required their own electrical cabling as the wattage exceeded the ring main limit. New storage heaters have lower wattages and can be wired onto a 30A ring main (Figure 3.36; see also Chapter 12, Section 12.4).

- Inspect **electrical fittings and switching gear**. Anything naturally dark brown with round pins must be further investigated as it indicates very old equipment. There are still examples about where elderly people may have lived in a property all their lives and still have the original electrical system. Brown means 'Bakelite', the plastic that sockets and switches were made of up to the mid to late 1950s. Large sockets with round pins suggest it would be the 15 A radial system which is explained more fully in Chapters 10 and 12. The cables are likely to be rubber coated and now dangerous so the whole system will need replacing. As for the gas boiler, the surveyor should state that a competent person should carry out a full check on the electrical system. This is true of all surveys where electricity is available. The surveyor should inspect the consumer unit for age. If a fuse box then it indicates that it is very old and the system needs replacing. Be aware though that the previous owner(s) may have carried out or had others carry out upgrades. For example, if the building is, say, 1960s and has a new consumer unit containing a bank of MCBs, the wiring may still be the original. In those circumstances a competent registered electrician should carry out further inspection. This would be noted on the conditions survey. With gas and electricity, the utility provider will attach a label with inspection dates. Gas boilers should have service books or documents to state when the appliance has been inspected or installed, the person or company who carried out the work and the date. Some documents may have the CORGI registration number of the installer. This is particularly important for a leased premises because it is a legal requirement of the letting.

- **Water and sanitation**. Water pipes should be cross bonded with the appropriate sized earthing cable and clamps. Metal baths and radiators can all become **live** if, say, the heating system electrical wiring touches a metal pipe. All the pipes are connected via the water they hold and are therefore potential electrical conductors. Test all taps for operation. Visual tests will not establish if they do not operate; therefore you need to operate them. Flush the WC to establish that it operates. Look for isolation valves near every appliance or fitting and check for leaks. Find the stopcock and check that it will operate. Stopcocks can seize up through neglect and be very expensive to replace.

PART B

3.9 Structural surveys

The most thorough of all the building surveys would be carried out by specialist surveyors or structural engineers. Where a building has signs of distress, such as cracks, bulges, or distortions to windows or doors, a structural survey would be commissioned.

Equipment

The most commonly used equipment is as follows:

- spirit level 1.0 m long
- plastic tell-tales
- epoxy resin
- brass round head screws and plastic wall plugs
- battery percussion drill and 6.5 mm diameter masonry bit
- eye protection
- suitable length ladders and stabiliser
- endoscope equipment
- extension lead or electrical generator
- plastic bags and ties
- plastic containers for semi-solid matter and liquids
- jemmy

- 25 mm cold chisel and bolster chisel
- club hammer and claw hammer
- graft (a spade like tool for digging in heavy clay soils)
- safety barriers and signage
- personal protective equipment
- hard hat with chin strap and torch
- rigger or wellington boots with toe and sole protection
- overalls/boiler suit
- high-viz jacket or vest
- eye protection
- rubber or leather gloves.

3.10   The actual structural survey

Structural surveys are mostly carried out in response to a specific issue. This may be a crack in a wall that has been reported during a conditions survey, or a client has heard a loud noise resulting in a cracked wall. The job of the structural surveyor is to:

1   establish why the crack or movement has occurred
2   establish to what extent has the crack will effect the stability of the structure
3   establish whether it is a live or dead crack
4   to provide a report on which the client can decide what action, if any, should be taken.

Cracks are categorised into five groups based on the width of the crack:

1   fine cracks up to 1 mm wide (very slight)
2   cracks up to 5 mm wide (slight)
3   5–15 mm wide (moderate)
4   15–25 mm wide (severe)
5   over 25 mm wide (very severe).

In a similar way to the conditions surveyor, the structural surveyor needs to be a sleuth, finding clues to the cause of the failure. Cracks in walls can be an emotive issue. They may be insignificant where the cause has been due to moisture movement, such as shrinkage of concrete blockwork as it dries out. A common reason when saturated blocks used in an inner leaf of a cavity wall eventually stabilise at about 3% moisture content (MC) including the mortar, and solid wall construction about 5% MC. It can take up to a year in normal conditions for masonry to dry out; however, if the structure has been 'handed over' and the occupant turns up the heating, the wall areas beneath the windows (radiators are commonly positioned beneath windows) will dry out much faster than the remainder of the wall causing localised tensile stresses and possible cracks.

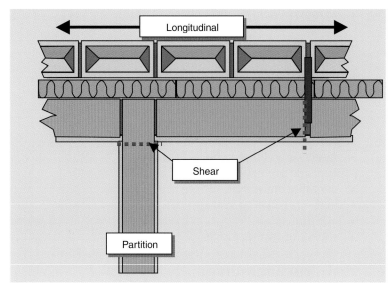

Figure 3.37 Common results of poor detailing and workmanship.

Poor detailing of masonry can often be the cause of cracking. Figure 3.37 shows a typical wall junction on plan. The outer wall is clay brick in a cement sand mortar 1:5 (one volume masonry cement and five volumes soft washed sand). The inner leaf is an aircrete block with a low density of about 400 kg/m³. Rigid steel twisted wall ties have been used at the correct centres (900 mm horizontally and 450 mm offset vertically as shown in the detail). Full cavity insulation has been used. The cause of the cracking could be due to differential movement. This is when materials with different rates of thermal expansion or moisture content are fixed together without allowance for movement. In this case the outer leaf will expand in direct sunlight in hot weather. The inner leaf will be isolated from the thermal gain so therefore not expand. The rigid steel wall ties do not flex and so cause vertical tears in the inner leaf blockwork and plaster finish (Figure 3.38). The correct wall ties would be wire type or flexible steel (see Chapter 5).

Another cause of cracking in walls is a change in stability. If the subsoil changes in its capacity to carry a load, there is a likelihood of movement. There are four main reasons for subsoil movement:

1 change in moisture content – perhaps the most common reason
2 lateral support has been removed – caused mainly by excavation or removal of a structural element
3 mining or subterranean erosion
4 chemical changes within the soil or soil water.

Structural movement can take place over many months or even years; therefore monitoring will be required. The weather and temperature are both important factors and should be noted on the survey report. For example, in an area where the ground is made up mainly of clay and there has been a long period of drought, it may have shrunk causing the foundation to subside. The subsequent

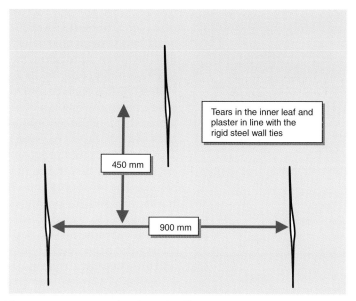

Figure 3.38 Incorrect detailing of wall tie type.

cracks may be, say, 5 mm wide. However, if the survey had been carried out during a long wet period the ground could have swollen, causing the same crack to close and therefore appear as a hairline termed a **very slight** crack.

Monitoring movement can be carried out using a simple thin piece of glass bonded to either side of the crack. If after a period of several months the glass has cracked, it indicates it is a **live** crack. For more detailed monitoring a calibrated tell-tale would be more suitable. The tell-tale is formed of two pieces of clear plastic either screwed or bonded to the structure on either side of the crack (Figure 3.39). One part has a grid of small squares and the other a red cross. Whilst fixing the tell-tale the

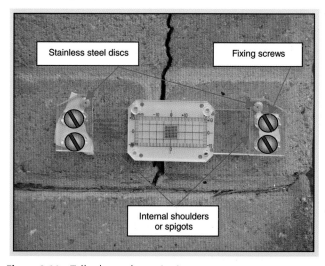

Figure 3.39 Tell-tale crack monitoring.

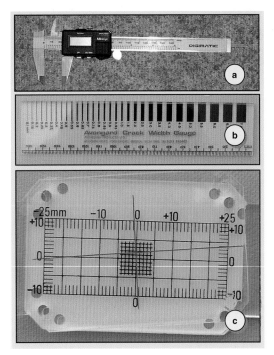

Figure 3.40 Crack monitoring instruments.

two sections are held accurately together at zero with four plastic studs. To ensure it is fixed horizontally use a boat level. The studs should then be removed to allow the tell-tale sections to move independently as required. The calibrated tell-tale enables horizontal, vertical and rotational movement to be gauged. An enlarged detail of the calibration shows both vertical and rotational movement (Figure 3.40c). The period of monitoring would normally be decided by a structural engineer and should be recorded on an appropriate sheet (Figure 3.41). It is important to record the date of each monitoring occasion. Noting the weather conditions on each occasion can also be of help. If the monitoring has been over a relatively short period and the groundwater levels have changed very little, the movement may be negligible. Alternatively, using a digital camera to photograph the calibrated section with the date/time log on the camera can be a quick and useful method of recording.

If the tell-tale has metal monitoring discs attached, a calliper can be used (Figure 3.42). Accuracy to one tenth of a millimetre can be achieved by measuring between the two internal shoulders or spigots of the gauge.

Less obtrusive monitoring can be achieved using three metal discs (Figure 3.43). A digital calliper (Figure 3.40a) can be used to measure the

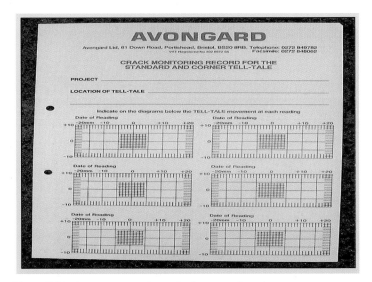

Figure 3.41 Long term crack monitoring record card.

horizontal distance between the discs and the diagonal. They are bonded to the wall using an epoxy resin adhesive which can be easily removed using a sharp cold chisel leaving virtually no damage.

3.11 Change in moisture content

Moisture has a significant effect on clay substrates. For example, the substratum may be saturated due to a broken sewage pipe (Figure 3.44). The localised saturated area when frozen can be enough to push lightly loaded foundations upward, whilst the remainder of the structure prevents the substratum freezing. Figure 3.45 shows the consequence of localised clay heave breaking a strip foundation.

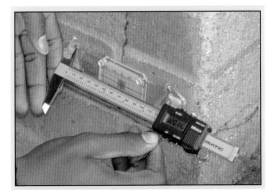

Figure 3.42 Measuring the lineal movement using a digital calliper.

Note the three
stainless steel discs

Figure 3.43 Less obtrusive crack monitoring using three metal discs.

The movement of groundwater can migrate small particles, causing voids to form. When the water levels fall the voids drain allowing the soil to compress and leading in the long term to subsidence. The cause of groundwater movement can be localised land drainage, trees, water abstraction or extreme weather changes such as droughts or flooding. Underground rivers can be the source of freshwater springs. Figure 3.46 shows one of several properties that have significant foundation problems caused by an underground river eroding the fines from the soil (the photograph has been disguised to provide anonymity). Alternatively, land near the coast or tidal rivers can rise and fall with every tide. Generally though, designers will be aware of the problems associated with tidal movement and should design accordingly (see Chapter 4).

Trees are frequently blamed as being the cause of subsidence. A mature tree, especially a deciduous tree (deciduous trees drop their leaves once a year in autumn), requires vast quantities of water. During the autumn and winter when traditionally there are long periods of rain and snow, the groundwater level (known as the **water table**) will rise (Figure 3.47; note the comparison with the measuring stick bottom right of picture) and the tree remains dormant. However, during spring and summer months when the leaves are forming and photosynthesis is taking

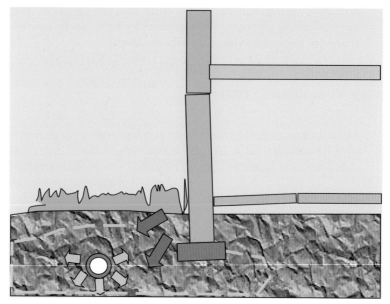

Figure 3.44 Subsidence caused by a broken drain.

Figure 3.45 Heave caused by a broken drain.

place, the tree extracts water from the ground (Figure 3.48). During a drought, although the tree will drop a proportion of its leaves to reduce the amount of water needed and increase its chances of survival, the root system will still extract moisture from the ground. Where structures prevent the Sun's heat from evaporating the groundwater, the ground will remain moist for longer and so attract tree roots seeking moisture (Figure 3.49). The root system will

Figure 3.46 Structural movement caused by underground water.

Figure 3.47 Water table movement in winter.

expand and the soil surrounding the root will shrink as the groundwater decreases, thus causing a high spot concentrating the imposed load to become a point load. Figures 3.50, 3.51 and 3.52 show examples of tree root damage. Alternatively the root can find a fissure (gap) within the concrete or mortar and expand, forcing a crack to form.

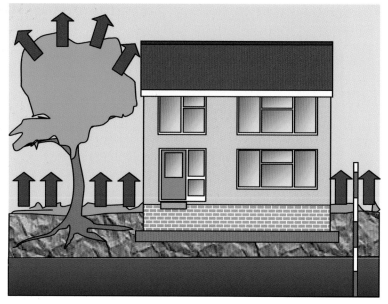

Figure 3.48   Water table movement in spring and summer.

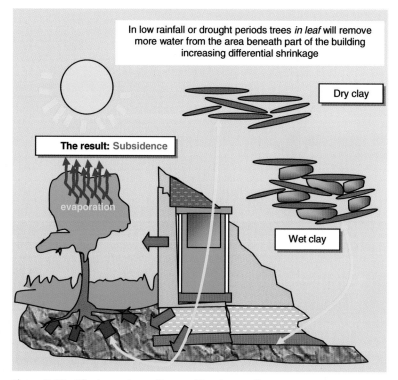

In low rainfall or drought periods trees *in leaf* will remove more water from the area beneath part of the building increasing differential shrinkage

Dry clay

The result: Subsidence

evaporation

Wet clay

Figure 3.49   Effects on clay soil caused by changes in water content.

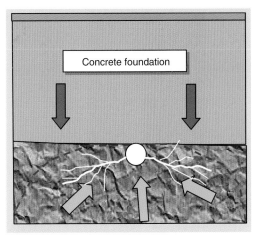

Figure 3.50 Effects of large shrub or tree roots on strip foundations, stage 1.

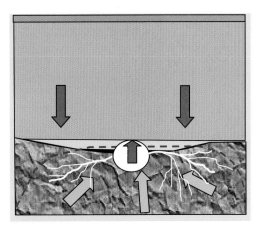

Figure 3.51 Stage 2 as water is removed from the supporting soil.

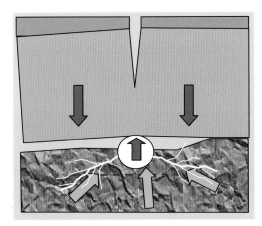

Figure 3.52 Stage 3: lack of support and foundation failure.

In recent years a better understanding of the benefits of trees has been established. Instead of removing trees, tree cultivation is preferred. A tree surgeon can trim the crown of a tree over several years to reduce the water requirements of the tree gradually. The root system can be trimmed in a similar way, resulting in less damage than complete removal.

Fibrous substrata such as peaty soils will expand or contract with water movement. They tend to be close to rivers or the coast. Silt, alluvium and running sands are associated with rivers and or lakes. Alluvium is the deposits of fine particles that have been eroded by water or from **run off** from adjacent land (see also Chapter 4, Section 4.2, Short bore pile and beam).

Try an experiment.

1 Dig a tennis ball size lump of earth and place it in a clean jam jar.
2 Add water and shake until all the soil has become suspended in the water. Allow to stand and settle for at least two hours.
3 Wait until the water has completely cleared. Note how the soil has settled into layers – strata. The lowest strata will be made up of the heaviest particles and stones with layers above containing smaller particles. There will probably be distinct colour bands as the layers deposit. The topmost layer will be silt and/or clay depending upon what part of the country you live in (Figure 3.53). Leave the sample for a week or so until the water has evaporated leaving a crusted layer on the surface.

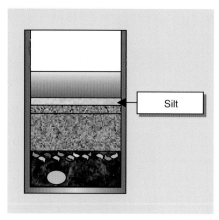

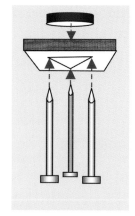

Figure 3.53 Field settling test.

Figure 3.54 Home made test equipment.

4 Has the surface shrunk forming cracks? If so it is likely that it is clay.
5 Knock three 100mm long wire nails into a block of wood and stand it on the dried soils (Figure 3.54).
6 Place a weight on top of the wood increasing the imposed load (Figure 3.55).
7 Check to see if the nail heads have penetrated the topmost layer.
8 Carefully add water to the dried soil sample and watch.
9 Has the imposed load started to sink through the top layers?

The experiment shows the effects of moisture on a soil. The load bearing capacity is reduced when the soil becomes **plastic** (this means that soil has become mouldable). Figure 3.56 shows a similar effect that has taken place on a cliff. Several metres below the surface groundwater levels increased after heavy rain causing the substratum to become plastic. The strata above could no longer be supported and the cliff face subsided. Figure 3.57 shows the effects of the ground movement on a building near the top of the cliff.

3.12 Lateral support has been removed

Gravity is the force that pulls all matter vertically downwards giving it **weight** – the force is also termed **load**. However, there are other forces that act on a structure. For example, if a wall has unequal weights (an **eccentric load**) attached to it, they will cause a rotational force to act on the wall. Figure 3.58 shows chimneys built into the party wall. The techniques made the most of the thermal efficiency of the brickwork, conducting heat from the fire and convecting it into the rooms. It is common for modern home owners to have central heating, so the open fire has become redundant. If the house owner has the chimney breast

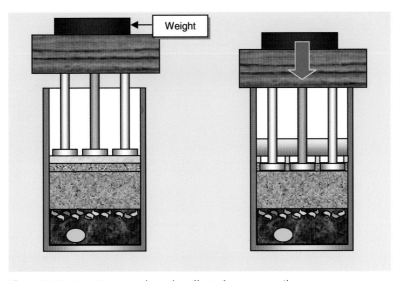

Figure 3.55 Experiment to show the effect of water on soils.

Figure 3.56 Landslip caused by water lubricating subsoils.

and fireplace removed from the room(s) (possibly on ground floor and first floor where there is another fireplace and smaller chimney breast), this still leaves a heavy mass of the shared chimney stack in the roof void and going through the roof. It is usually impractical to remove part of the chimney because the original design spread the load over the whole area, so various methods of supporting it

Figure 3.57 Substantial structural movement caused by the landslip.

Figure 3.58 Chimney built in to the party wall.

Figure 3.59 Bad practice results in insufficient support to remaining chimney.

have been devised. Some local authority building control departments will allow a metal **gallows bracket** to be bolted onto the wall to support the load of the chimney on the side where the chimney breast has been removed, but not all local authorities agree that it is safe. As mentioned previously, the DIY enthusiast can cause problems. They may go ahead and remove the chimney breast without seeking approval and legal permission for the work (Figure 3.59). Figure 3.60 shows a section through the same detail. There are timber beams cut into the party wall and supported on two ceiling joists. A spreader beam transfers the extra loading from the chimney above onto the beams and joists beneath. The joists in turn were supported on a cast in-situ concrete lintel which had cracked under the excessive extra load. The consequences could be catastrophic if the other home owner were to move and the new owners carry out a similar project.

The law changed when the Party Wall Act 1996 came into force. By law both parties who share a common wall or boundary should advise the other party if they want to work on or near the wall or boundary. For detailed information look on the website www.communities.gov.uk and use 'search' for Party Wall Act. Section 5 states the Duties under the Act, emphasising that there is no legal enforcement contained in the Act if either party does not follow the requirements, without having to seek help from a court injunction. The Act, although simple in essence, frequently requires professional assistance to operate. Building surveyors should be fully conversant with the details of the

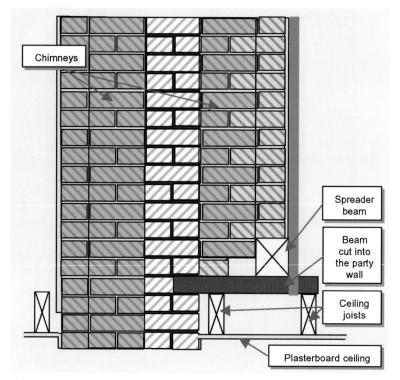

Figure 3.60 Sectional drawing of Figure 3.59.

Party Wall Act 1996. Some surveying firms specialise in party walls because of the complexity of applying the Act.

3.13 Party Wall Act survey

Before any work whatsoever is carried out on a party wall or boundary, it is essential to consult the other party and discuss the proposed works. If necessary (and it can prevent major problems later), a professional surveyor should be appointed. If the project is simple the same surveyor can act as advisor to both parties; however, the other party may want to appoint their own surveyor and costs must then be paid by the party initiating the works. The advantages of appointing a professional surveyor are as follows:

- They will survey the existing structures and boundary and photograph where appropriate.
- They will produce a full report in the form of a **conditions survey**, noting any visible defects such as cracks, decorative order and damp patches of the existing structures for both parties.
- The full extent of the works and protection required will be contained in the report, including the methods of protection and support. This is

especially important where excavations near to or below the level of foundations are required. Protection of existing services should also be noted, such as drainage runs, sewers and gas pipes. Overhead services such as telephone wires or power cables can enter the air space above the property and therefore should be duly noted in the document.

- The time parameters for the project – the expected date for the work to proceed (a minimum period is given in the Act) and the duration of the works.
- It is normal to state that the party going ahead with the works will pay for any damage caused during or by reason of the works as deemed by the completion survey.
- Names and addresses of both parties involved and any interested parties such as the mortgage holder or landlord. The names and firms/practices of the surveyors partied to the contract.
- They will advise and interpret as to whether the work being carried out will have any effect on the property and the extent of protection required.

When the report has been prepared (termed 'the award') and the evidence confirmed as being true, both parties will sign to show acceptance. The document is very important as it may be used in court if the project goes wrong. This is the main reason for appointing a qualified and experienced party wall surveyor. If, for instance, a building is to be demolished, the adjacent property owners would be wise to have their own surveyor to prevent conflict of interest. Technically the surveyor acts impartially with both parties and advises on the basis of the award. However, in a difficult situation it may be similar to having the same lawyer acting on behalf of the prosecution and the defence.

When the works have been completed, another survey (the completion survey) and report would be prepared, noting any damage caused by the works. Any damage caused during the works should be notified to the acting surveyors and ideally photographs taken at the time of the damage or soon after. A reasonable contractor would take action to fully protect the adjacent property throughout the full works; however, in the real world not all contractors are professional and damage occurs.

The Party Wall Act 1996 is in addition to the common laws of damage and trespass with damage, rights to support, and so on.

3.14   Measured survey

As the name suggests, the objective is to physically measure the existing structure. The client may want to extend their existing building, refit or modify it. The building surveyor would carry out a series of surveys dependent on the client's requirements. For example, an existing shop or hotel may require an internal refit. The surveyor would measure the dimensions of the rooms, noting the fabric and positions of doors, windows, radiators, electrical sockets and so on. When the survey is complete the booked information would be drawn up to a suitable scale. Shopfitting contractors usually prepare full size drawings of the new fittings onto **rods**. Nowadays, computer aided drawings can be produced to scale for setting out purposes. Measurements are taken on the horizontal and vertical axes. The horizontal plane is normally an arbitrary height about 1.20 m from the finished floor level (based on its being a comfortable height for the survey or to work at).

3.15 Approach to carrying out a measured survey

A method statement and risk assessment should be prepared before the survey is carried out. If you are to survey a working construction site, the method statement and risk assessment should be supplied to the main contractor to enable them to be incorporated in the overall assessment plan.

The potential risks should be evaluated. See the necessary PPE listed on p. 89 for the conditions survey. The list is not exhaustive and additional specialist inclusions may be required, such as for specific hazards of asbestos, fumes, dust, etc., or working at heights, over water, adjacent to fast moving traffic and so forth.

The surveyor's equipment will be similar to that used on a conditions survey plus:

- rotating laser and tripod
- water level.

As with all surveys, the date, location and the name of the surveyor must be given. If you are working for a company, it is likely that a job number will be allocated so that the costs of the survey can be gathered for invoicing to the client later.

The purpose of the survey will dictate the extent of information and accuracy of measurement. For example, a measured survey for ceiling tiles or carpet tiles will require relatively approximate dimensions as the materials are usually cut in on site. In contrast, a stone mason who is to clad the foyer walls of a hotel will require accurate dimensions of all relevant surfaces.

The actual measured survey

1 Start by quickly looking at the whole area to be surveyed.
2 Then decide where to start.
3 Draw a location plan of the area to be surveyed.
4 Draw proportional plans for each room and label each doorway with a code such as D 01. When drawing a room plan, it is likely that there may be more than one A4 sheet per room. Code each sheet to enable them to be arranged like a jigsaw when back in the office. Coding the doors will help with identification later. Use the same code on the location plan. By

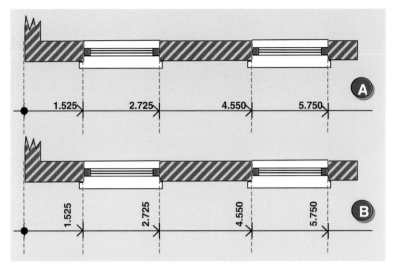

Figure 3.61   Running dimensions.

analogy with a jigsaw, the location plan would be the picture on the box. For the drawing use the following:

- squared A4 surveyor's paper to enable parallel lines to be easily drawn (plain paper could be used instead)
- draw and write in pencil as it is easier to erase when or if required
- if the survey is an internal one, start in a corner of the room
- if you are right handed, start in the left hand corner and work clockwise around the room
- the system of lineal measurement is taken and recorded using the **running dimension** method.

Running dimensions

Start from the left and record a small ball as shown in Figure 3.61. The first measurement will be defined by an arrowhead and a straight line. The measurement should go on top of the line as close as practical to the arrowhead. Where possible, it is useful to place the running dimensions outside the room for clarity. The next dimension will be taken from the ball to the second point. Remember to write above the line and near the arrowhead. If there is insufficient space, write the dimension at right angles to the line or use another arrowed line to show where the dimension should be. This method of booking is more accurate than measuring each item individually, and errors can be reduced. For example Figure 3.61 A shows a straight runnage using the running dimensions method, whilst B shows the same runnage with the dimensions shown at right angles. Note that they all start from the line and ball and go outwards. Figure 3.62 C shows individual dimensions and D the same runnage with an error in red. If the surveyor draws up the survey based on running di-

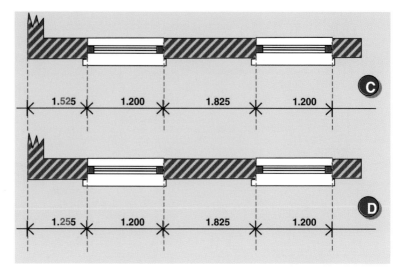

Figure 3.62 Comparison of dimensions showing potential mistakes.

mensions starting at the ball, even if the first dimension has been written down incorrectly it will only affect the first dimension. The overall dimension or other dimensions will not be affected and are therefore correct. In comparison, if the individual dimensions method were used, the overall dimension would be 0.27 m too short and the positions of all the windows and walls would be incorrect.

Diagonal dimensions

To ensure accuracy of the positions of walls, columns and openings, diagonal dimensions should be taken. Figure 3.63 shows two points of origin to enable a series of triangles to be formed. Note the dimensions are *always* on top of the line near to the arrowhead. This is particularly important when several lines are close together. If the rule is followed it is easy to know which dimension relates to which line. If there is not enough space to write the dimension in, encircle the dimension and draw a curved arrowed line showing which line it refers to; again, indicate the dimension *above* the line. Using a curved line prevents it being mistaken for a dimension arrow.

Measuring heights

When comparing heights, a nominal point can be used termed a **datum**. Commonly, when surveying buildings, the datum will be 1200 mm above finished floor level. Datums are particularly useful when surveying rooms and corridors. The example in Figure 3.64 shows that A and B have the same dimension, as would a parallelogram. The third measurement recorded would suggest the end C is 50 mm different, but that is all.

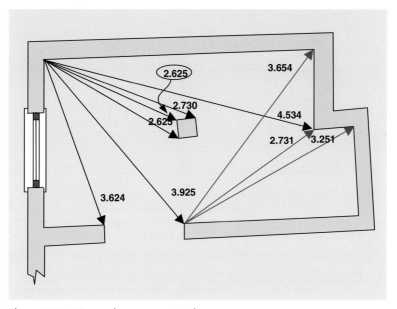

Figure 3.63 Diagonal measurement of a room.

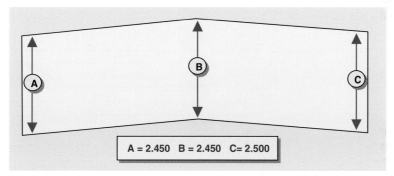

Figure 3.64 Is the floor or ceiling level, or neither?

Does it show if:

1 the corridor has a level floor and the ceiling becomes higher at the right-hand end as in Figure 3.65?
2 the corridor has a level ceiling and the floor becomes lower at the right-hand end as in Figure 3.66?

Set up a datum at point A and measure and book the height from the datum to the finished ceiling, followed by the height of the datum from the finished floor. Transfer the datum using the rotating laser or a water level and repeat the measurements at points B and C. These points are known as **spot heights**, meaning that each pair of measurements refers to one point only. If the horizontal surfaces appear relatively constant, then only one or two spot heights are required per room. However, corridors or older buildings may require several

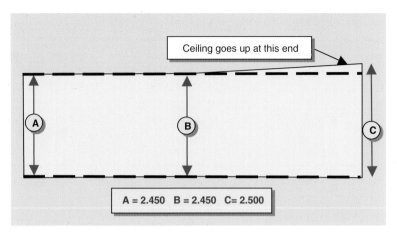

Figure 3.65 Option 1: the ceiling is not level.

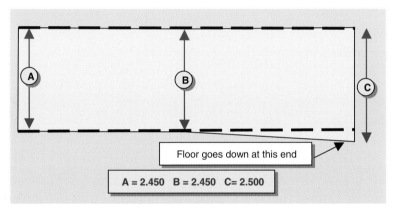

Figure 3.66 Option 2: the floor is not level.

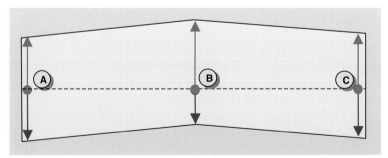

	Point A	Point B	Point C
Datum to: finished ceiling level	1.250	1.300	1.210
Datum to: finished floor level	1.200	1.150	1.220
	2.450	2.450	2.500

Figure 3.67 Using a datum will determine the correct levels.

spot heights per wall length. Figure 3.67 shows the same measurements based on a datum, therefore giving the correct solution to the problem.

Method

1 Mark a datum (a useful tip in working buildings is to place masking tape on the wall so that the datum is not permanent).
2 Using a water level or laser level, transfer the datum to other points to be surveyed.
3 Measure from the datum mark to the floor.
4 Measure from the datum mark to the ceiling.
5 These readings are referred to as **spot levels** or **spot heights**.

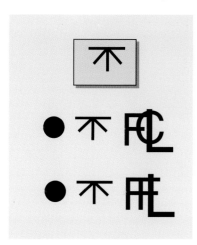

Figure 3.68 Different datum and spot height markings.

When noting spot heights, it is important to state which ones they are on the survey drawing. The surveyor draws a spot with the datum sign to indicate from where the height has been taken, followed by the initials FCL to indicate finished ceiling level and FFL to indicate finished floor level (Figure 3.68). It is important to state whether the levels are finished or not. For instance, shopfitters may survey a building shell before the floor screed and finishes have been placed. If that is the case, then FL would be used to indicate floor level. If the structural soffit has been measured then either SL (soffit level) or CL (ceiling level) would be noted. At a later stage the fixers will place their datum and mark out the finished floor and ceiling levels.

Details

Where specific points of detail are required, such as the position of a power point, switch or service cupboard, a method known as **dims** can be used. There are several ways of booking dims, slash ends or arrowheads being the most commonly used (Figure 3.69).

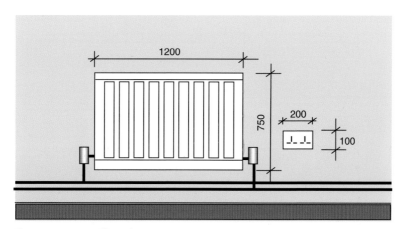

Figure 3.69 Detailing 'dims'.

Chapter 4

Foundations

4.1   Introduction

This and the next three chapters introduce the fundamentals of construction. They have been divided using an elemental approach, starting in this chapter with the foundations. Perhaps the most vulnerable part of a building and certainly the most costly to repair, foundations are not infrequently incorrectly designed or constructed. In the construction industry there is no regulation or requirement for a builder to have any qualifications whatsoever. The designer must comply with regulations, and reputable companies have been trying to ensure they have qualified and experienced staff to undertake the work. Why, then, are there so many claims on insurance for foundation failure? This chapter starts by defining the functions of foundations and some of the issues relating to the effects of water, the most common challenge to the builder. The various designs of foundation are discussed in relation to soil types and types of loading. The introduction to the use of spreadsheets will enable the reader to carry out simple calculations quickly and efficiently. The more complex calculations would normally be carried out by a qualified structural engineer using professional computer software.

Foundation function

The purpose of a foundation is to transfer the imposed load of a building or structure onto a suitable substratum.

Considerations

Loadings

There are three main types of load:

1   **Dead loads** are the static or constant weight of the structure made up from the walls, floors, roof, etc.
2   **Live loads** (also known as **imposed loads**) such as furniture, goods and people, are movable. In addition there are variable loads caused by the weather, such as snow and rain.
3   **Wind loads** can be positive or negative. They are not **dead** or **imposed** and therefore require specific consideration.

Loads can be calculated and expressed as a **force** (weight) that is to be transferred onto the substrata.

Substrata bearing capacity

The **bearing capacity** of the substrata is the amount of weight the ground (substrata) can support.

For a structure to be stable, the force (load) bearing down must be opposed by an equal force pushing upwards. When both upward and downward forces

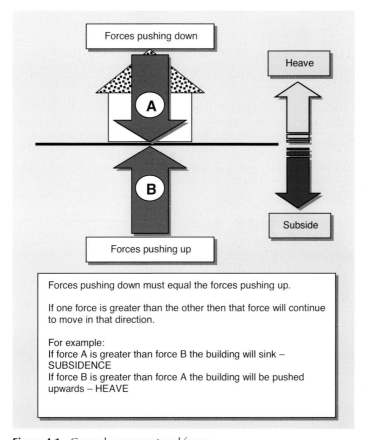

Figure 4.1 Ground movement and forces.

are of the same magnitude (size) they are said to be in **equilibrium**. When the downward force is greater than the upward force, the structure will sink or **subside**. Conversely, a larger upward force than downward force can push the structure upward, causing it to **heave** (Figure 4.1).

What causes heave?

The main cause of movement of buildings is the weather; those on clay soils are most affected. If clay dries out during a period of drought it will shrink, causing uneven support beneath a structure. Figure 4.2 shows how ground around a building has dried out from the effects of the Sun and access to air, whereas the ground beneath the building remains moist; the resulting uneven support for the building may cause the foundations to fail. In the past three decades there have been several severe droughts causing hundred of millions of pounds worth of damage each year. With climate change and a predicted increase in frequency of periods of low rainfall the problem will continue. Building Regulations Approved Document A: 2004, para: 2E4 requires foundations in clay to be a minimum depth of 0.75 m below ground level because it is

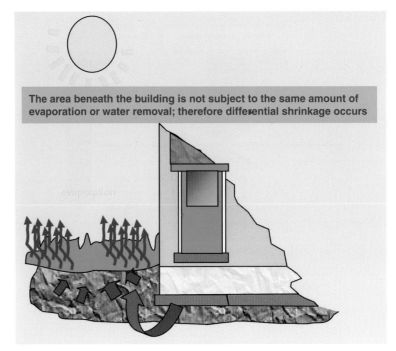

The area beneath the building is not subject to the same amount of evaporation or water removal; therefore differential shrinkage occurs

evaporation

Figure 4.2   Subsidence caused by water evaporation.

considered that the effects of the Sun cannot penetrate down to that level. However, it is the building control inspectors who will decide the required depth on site; this is commonly at least 1.00 m. If tree roots or soft spots are found, the required depth of foundations can increase significantly. In other non-cohesive soils, such as sands and gravels, the minimum depth of foundations can be 0.45 m.

What else can affect the bearing capacity of the ground? Water can have significant effect on the ground; some clay soils can increase in height (**swell**) by up to 75 mm between dry and saturated. As the moisture levels rise, the plasticity of clay increases allowing landslip (see Chapter 3, Figure 3.56 which shows large areas of cliff that have subsided as a result of heavy rain saturating the lower stratum, making it more plastic and allowing upper layers of more dense clay to slide down the cliff).

Another cause of water loss is the presence of trees, in particular broad leaved trees such as apple, oak, beech and poplar, which remove vast quantities of water from the ground in the spring and summer months during their growing period. Water and nutrients are extracted from the ground by the roots and transported through the plant to the leaves from which large volumes of water evaporate; this is known as the **transpiration stream**. Trees absorb carbon dioxide which combines with water in the leaves to produce sugars for plant growth, releasing oxygen as a waste product; this process is called **photosynthesis**. In drought conditions trees continue to evaporate water brought in via their root system. During continual drought the trees will drop many of their

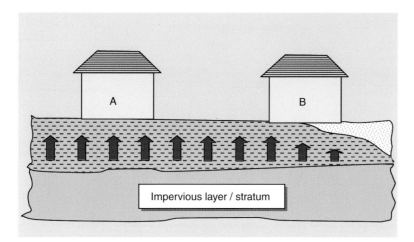

Figure 4.3 Differential ground movement.

leaves to conserve water in the trunk; however, the ground around the tree will have been seriously depleted of water (becoming **desiccated**).

The microscopic particles that make up clay are known as **platelets**. They are like sheets of glass 0.002 mm in diameter and impervious to water. (If they were placed end to end there would be 128 000 across the diameter of a 2p coin.) When the platelets are completely dry they occupy a minimum volume and the clay becomes very hard, shrinks and cracks, as shown in Chapter 3, Figure 3.49. When wetted the clay swells causing **heave** and softens (becomes **plastic**). When clay is saturated, its compressive strength is reduced considerably (by up to 50%) because the platelets become lubricated and easily move over each other. For example, after heavy rain site vehicles may not be able to operate because they would sink into the softened clay. Clay substratum can lose up to 50% of its load bearing capacity and buildings may subside (**subsidence**). In Figure 4.3, structure A may be pushed **evenly** upward and therefore is **unlikely** to be damaged. However, if the ground strata change as beneath structure B, the upward force will be uneven and may result in reduced load bearing capacity causing slip and heave, so structural damage is likely. Frost can also cause heave in saturated clay, although generally only about the top 0.6 m will be subject to movement. Horizontal pressure can cause sub-ground level walling to be moved if inadequate backfill or cavity fill has been used. Figure 4.4 shows the potential problem and Figure 4.5 the consequence.

Other causes of ground movement can include broken drains and changes to water courses (see also Chapter 3). Water abstraction for irrigation, manufacture or human use can have considerable effects on ground stability. Where the land is adjacent to tidal rivers or sea, the ground may swell and shrink with the tidal cycles. This is particularly common in marsh of fibrous peaty substrates.

Old mines which have not been charted, and therefore remain undiscovered in modern times, can be a major cause of subsidence; for example, mines for coal, iron ore, copper, tin, zinc and galena, stone extraction for building in and

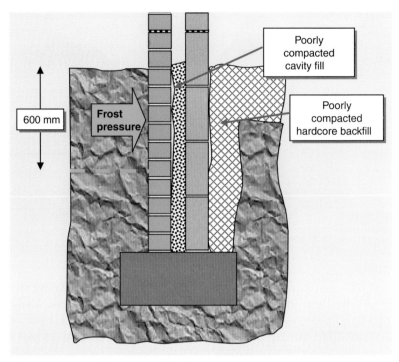

Figure 4.4   Lateral forces caused by freezing water.

around Bath and parts of Derbyshire, or flints in parts of Norfolk (see also Section 4.12).

4.2   Foundation types

Not all buildings need foundations. For example, in areas where the bedrock is close to the surface it is possible to level it off and build directly onto it. Why try cutting a trench into rocks like granite and dense limestone, then to fill it up with concrete mainly composed of granite or limestone? Soils, however, are not as strong as rock so therefore foundations will be required.

Spread footings

These were commonly used on structures prior to about the 1920s. The principle was to broaden the base of the wall using brickwork as inverted corbelling. Beneath the first course, sand, ash or coal dust would be used to provide a smooth surface to the formation level (Figure 4.6).

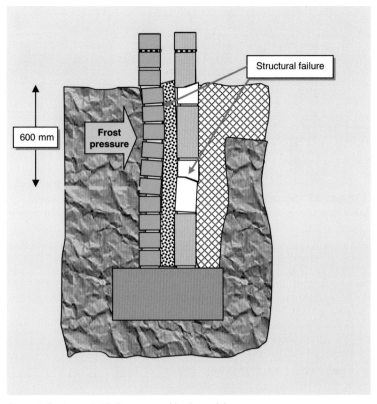

Figure 4.5 Potential failure caused by lateral forces.

Figure 4.6 Spread footings.

Strip foundations

Traditional strip

Foundations are a continuous concrete section supporting load bearing walls. If they are in plain concrete they must be at least 150 mm thick to prevent **shear**

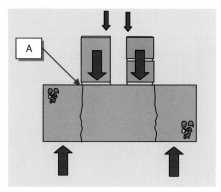

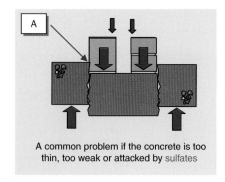

A common problem if the concrete is too thin, too weak or attacked by sulfates

Figure 4.7 Shear failure in a strip foundation.

Figure 4.8 Shear failure in a strip foundation due to various causes.

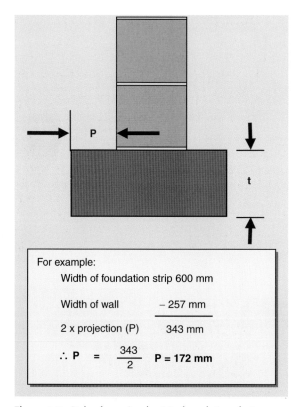

For example:

Width of foundation strip 600 mm

Width of wall — 257 mm
_____

2 x projection (P)   343 mm

$\therefore$ **P** $= \dfrac{343}{2}$ **P = 172 mm**

Figure 4.9 Rules for a simple strip foundation design.

failure at point A in Figures 4.7 and 4.8. To calculate the thickness (t) of the strip the overall width must be known (see foundation calculations in Sections 4.5 and 4.6). If, say, the width has been calculated as 600 mm, subtract the width of the wall and divide by two. The result will be the projection (P) and must be the same or less than t for unreinforced strip foundations (Figure 4.9).

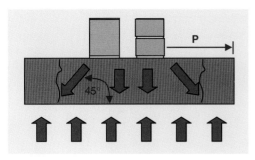

Figure 4.10 Problems if the rules are not followed.

Figure 4.10 shows the downward loading spreading at about 45°; therefore if the thickness of the strip is less than the projection, the concrete may shear. If t is equal to P the force will be spread over the whole underside of the foundation strip (Figure 4.11).

Another cause of shear in strip foundations is sulfate attack. Aggressive soils may contain sulfates, either in crystal form which look like granulated sugar or in a soluble state in the groundwater (Figure 4.12). The sulfates dissolve the cement binder and the foundations become weak and fail. One common remedy is underpinning.

Deep strip

Because drought conditions now occur more frequently, the depth of the strip has increased over the past three decades, especially in clay soils where a required minimum depth of 1.00 m below ground level is now common. Within the loading area there should be no **made ground**, nor wide variation in the type or structure of the substrata, as this will produce irregular loading. The digging of clay for local brickworks in Victorian times would leave large excavations which were subsequently filled with anything to be disposed of: factory waste, rubbish, and used **night soil** (night soil was dried and ground clay used by Victorians to place over their effluent before flushing water toilets became fashionable; at night men would empty the full buckets of clay and effluent and leave another bucket full of dry ground clay for the next day). Likewise, the excavations left after digging out building sand or clay for night soil, would be filled to give made ground. Other made ground could include in-filled Victorian cess pits, in-filled ponds and rivers, or ground where structures have previously been, especially those with basements or coal holes.

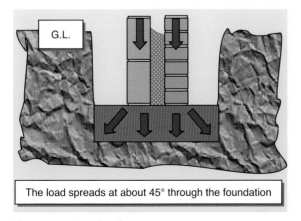

The load spreads at about 45° through the foundation

Figure 4.11 Loading forces.

Figure 4.12 Sulfate crystals in clay.

Deep strip foundations were introduced as building land became more scarce and access to deeper load bearing strata was required. The trenches must be wide enough for the ground worker and the bricklayer to gain access. This means the amount of spoil to be removed is greater than the need for the foundation (Figure 4.13). In loose (friable or granular) soils, such as gravel and/or sand, peat and clay, the walls of the trench need to be supported by boarding. The struts have to be worked around which makes slow going and can lead to dangerous practice (Figure 4.14). Deep trenches in cohesive soils should also be treated with respect. They can change in cohesion by drying out and cracking, or becoming plastic and sliding in. Working in an open trench without support, even at waist height, can be fatal. The weight of soil can be about 1 tonne/m^3, more than enough to crush internal organs. Ground workers often throw buckets of water into open trenches in hot weather to prevent them drying out; covering the trenches with boarding and polythene can also help. Signage and barriers should be placed warning of the excavation.

A typical detail showing the relationship between a solid floor and a deep strip foundation is shown in Figure 4.15.

Trench fill

In the early 1970s the house building boom meant that labour was in high demand and consequently expensive. Machine dug trenches that could

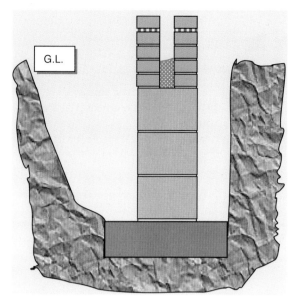

Figure 4.13 Deep strip foundation.

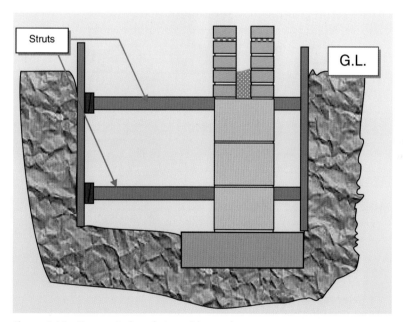

Figure 4.14 Deep strip foundation with trench support.

be filled with mass concrete became popular. It reduced the amount of spoil that had to be dug as no bricklayers needed access in the trench (Figure 4.16). In most cases trench support is not required; the amount of backfill required and spoil to be removed is reduced. The savings offset the extra cost of the concrete. If the loading requires, say, a 450 mm

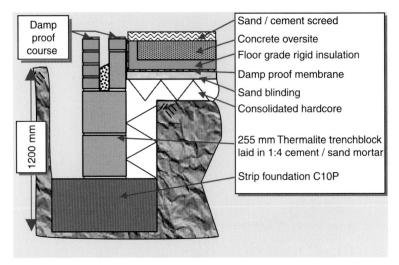

Figure 4.15   Solid ground floor with deep strip foundation.

Damp proof course

Sand / cement screed
Concrete oversite
Floor grade rigid insulation
Damp proof membrane
Sand blinding
Consolidated hardcore

255 mm Thermalite trenchblock laid in 1:4 cement / sand mortar

Strip foundation C10P

1200 mm

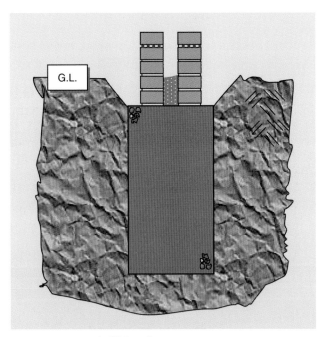

G.L.

Figure 4.16   Trench fill foundation.

wide trench 1.5 m deep, a **backactor** could excavate to those dimensions. In contrast, a traditional strip would require the extra working space (see Figure 4.13).

A typical detail showing the relationship between a solid floor and a trench fill foundation is shown in Figure 4.17.

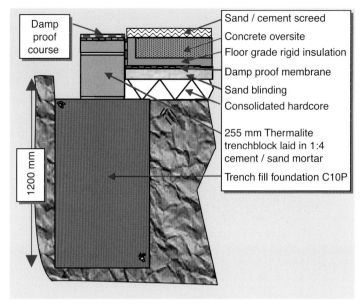

Figure 4.17 Solid ground floor with trench fill foundation.

Wide strip

Where the soil is soft or of a low load bearing capacity, wide strip foundations can be used to spread the load over a larger surface area so that the loading per m² is reduced. (This is similar to standing on the bed: as shown in Figure 4.18, your weight will easily compress the mattress. If, however, you place a strong board (say 600 mm²) on the bed first, although there is additional weight (you plus the board) the load bearing area has increased significantly, so decreasing the loading per m²). It can be seen in Figure 4.19 that it will be very expensive in terms of concrete to achieve the required ratio of P to t, so steel reinforcement is used. However, the density of concrete must be increased to prevent water penetration.

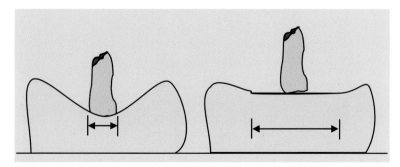

Figure 4.18 Load over area comparisons.

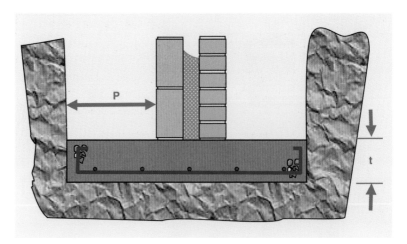

Figure 4.19 Wide strip foundation.

Simple raft

Where the subsoil is very weak the load needs to be spread over a greater area. This is achieved by casting a slab of concrete over the whole ground area and thickening the slab where walls are to be placed (Figures 4.20 and 4.21).

If ground pressures are likely to be excessive at different seasons, reinforcement may be required; this is known as **fabric** when in sheet mesh form.

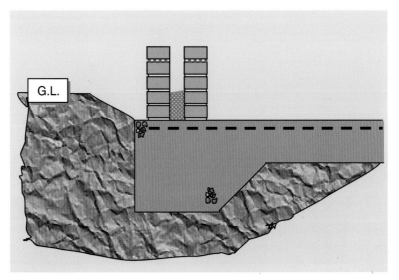

Figure 4.20 Simple raft foundation.

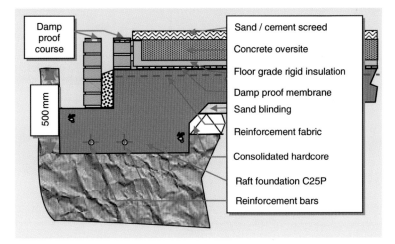

Damp proof course

500 mm

Sand / cement screed

Concrete oversite

Floor grade rigid insulation

Damp proof membrane

Sand blinding

Reinforcement fabric

Consolidated hardcore

Raft foundation C25P

Reinforcement bars

Figure 4.21 Reinforced raft foundation.

Pad

Pad foundations are used when isolated loads need to be supported, such as columns or framed structures. The load would be concentrated in relatively small areas of the pad with large expanses being either non-load bearing or having lightweight loadings, meaning that small slabs to spread the load from the skeletal structure can be cast. Figure 4.22 shows a simple pad below a steel stanchion. Steel reinforcement may be required for higher loadings.

The type of skeletal frame will determine the pad foundation design. For example, a cast in-situ concrete column will require a **kicker** and **continuity bars** to be cast into the pad (Figure 4.23). Steel frame, timber frame or precast concrete framed structures will require holding down bolts to be cast into the top of the pad or sockets to be formed (see Chapter 8).

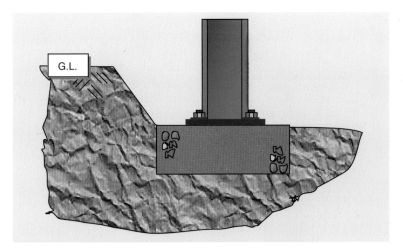

G.L.

Figure 4.22 Simple pad foundation – steel frame.

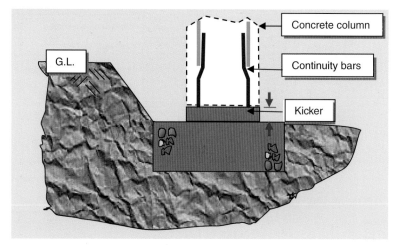

Figure 4.23 Simple pad foundation – RC frame.

Short bore pile and beam

Where the ground conditions will not support strip foundations it may be possible to create a series of concrete columns (**piles**) that can reach down to stronger strata. Short bore piles are typically 2.00–3.00 m long. They can be reinforced with steel if there are any lateral forces present, or just concrete if all forces are vertical. At the top (**cap**) of each pile a connection must be made with a horizontal reinforced concrete section termed the **beam** (Figure 4.24). A typical detail showing the relationship between a block and beam floor and a short bore pile and beam foundation is shown in Figure 4.25; Figures 4.26 and 4.27 show the relationship between the piles and beam and the structure.

The advantages of pile and beam foundations are as follows:

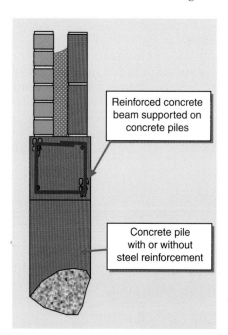

Figure 4.24 Short bored pile and beam.

- in clay soils, if correctly designed, they are not affected by clay heave or shrinkage
- groundwater can percolate between piles easily
- *limited* tree roots have little effect as piles continue further below the surface than any tree root system
- techniques available to overcome groundwater whilst piling, so dewatering not required
- reduced quantity of spoil and minimal ground disturbance
- less concrete required than with the same depth of strip foundation
- precast concrete beams can be used, eliminating casting time on site.

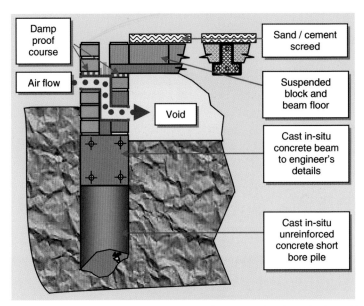

Figure 4.25 Suspended block and beam floor on a short bored pile and beam foundation.

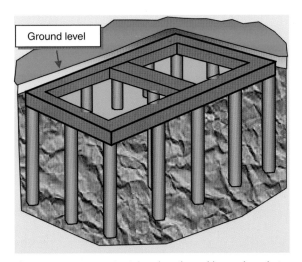

Figure 4.26 Isometric view of a pile and beam foundation.

The main disadvantage of pile and beam foundations is cost. The technique requires specialist plant and labour; however, the extra expense can be offset against the rising cost of spoil removal and time savings. Figure 4.28 shows a piling rig. A long metal screw termed an **auger** is turned into the ground like a corkscrew. The soil is removed when the auger is brought to the surface. A wire brush located next to the auger cleans off the spoil stuck to the blade (Figure 4.29). Figure 4.30 shows the pile reinforcement cage before it is lowered into the pile shaft, and Figure 4.31 shows a pile cast with the reinforcement protruding ready for the pile cap and beam to be attached.

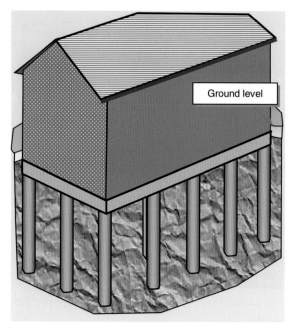

Figure 4.27 Isometric view of a pile and beam foundation supporting a building.

Figure 4.28 Crawling piling rig and auger.

4.3 Foundations on an incline

When building on a hill it may not be practical to level off the ground so stepped foundations would be used. Other than pad foundations, all the previously described designs can be detailed with steps. Figure 4.32 shows a strip or

Figure 4.29 Piling auger and cleaning brush.

Figure 4.30 Pile cages.

Figure 4.31 Pile cage ready to receive pile cap.

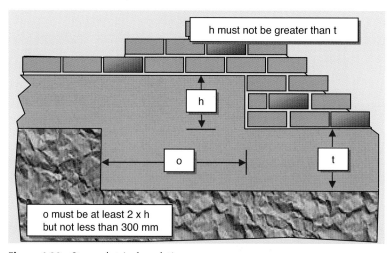

h must not be greater than t

h

o

t

o must be at least 2 x h
but not less than 300 mm

Figure 4.32 Stepped strip foundation.

raft foundation step. The Building Regulations requirement is that the overlap must be at least two times greater than the thickness of the foundation and not less than 300 mm. The thickness of the foundation should be designed to course in with the masonry walling. For example, if cavity walling is to be used, the step could be equal to three brick courses or one block course. The overlap will be $2 \times 225\,\text{mm} = 450\,\text{mm}$. Where pile and beam foundations are to be used, the step should be directly over a pile; however, the exact design will be decided by the structural engineer for the specific application (Figure 4.33).

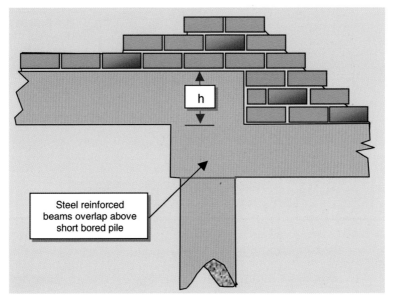

Figure 4.33 Stepped pile and beam foundation.

4.4 How to calculate loadings

Look at the plans for the proposed structure and consider the location of the heaviest section. Figure 4.34 shows a section through a two storey dwelling of traditional construction with a gable roof; Figure 4.35 shows the directions of the imposed loading forces in red arrows and the resistant force in yellow arrows.

Start at the roof and work down:

- Look at which way the roof is spanning. Does it rest on just two walls or is there another load bearing partition wall taking some of the load? In the example the first floor partition is *non*-load bearing; therefore the roof spans the external walls.
- Then consider which way the floors are spanning. If the drawings show floor joists they are in the direction of the span (Figure 4.36). However, some structures have precast concrete, cast in-situ concrete, or even steel decking. In those cases the symbol will be shown on the plans.
- Is the ground floor a suspended floor or a solid floor? A suspended floor will transfer imposed loadings onto the supporting walls, whereas a solid floor will be supported on the ground without transferring any load to the walls.
- Is the first floor supported by a load bearing partition wall? In the example a load bearing partition 100 mm thick is supporting the suspended first floor. Some floors are designed to span between cross walls only (see Chapter 8, Section 8.2).

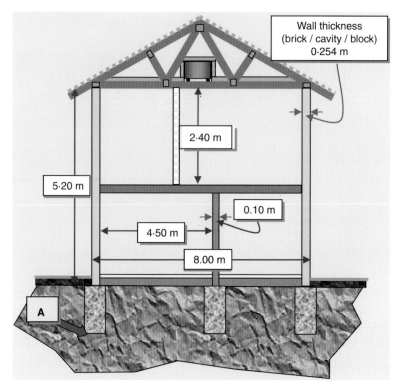

Figure 4.34 Cross section of a simple domestic structure.

- Look at the type of walls. Are they brick and block, or timber frame? It is usual for the inner leaf of a cavity wall to be the load bearing leaf, but this is not true of all walls.
- Are there any internal partition walls being supported by the floors? Partitions walls form rooms; therefore, although they may not be load bearing walls, they will be part of the dead loading. In the example there is a non-load bearing first floor partition wall.
- Are there any water storage cisterns or tanks in the loft area or upper floors? Water weighs 1 kg per litre; therefore a cold water tank can impose a considerable load.
- What is the considered type of foundation?
- What is the live loading? Live loadings on floors are predicted for specific usage such as domestic at $1.5 \, \text{kN/m}^2$ as stated in BS 6399: 1996. Table 4.1, reproduced from Approved Document A of the Building Regulations, provides a guide to the imposed loading for domestic applications with a slightly higher figure of $2.00 \, \text{kN/m}^2$.

Now that the initial observations have been carried out, the location of the maximum loading on the subsoil can be calculated. Ignore any windows or doors, and consider only one lineal metre of the structure as shown in Figure 4.37.

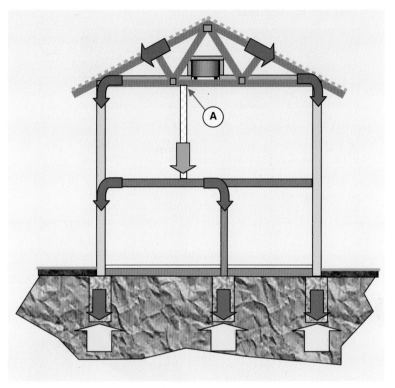

Figure 4.35 Cross section showing direction of forces.

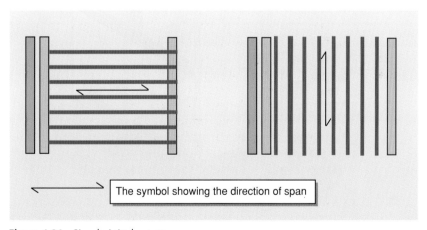

Figure 4.36 Simple joist layouts.

It is important to be familiar with the materials being used; however, the calculations will not be precise. Bricks, for instance, can vary in weight, as can different makes of concrete block; therefore recognised values are used. The calculations are generally carried out by a structural engineer on behalf of the designer.

Table 4.1 Imposed loads[a].

Element	Loading
Roof	Distributed loads:
	$1.00\,kN/m^2$ for spans not exceeding 12 m
	$1.5\,kN/m^2$ for spans not exceeding 6 m
Floors	Distributed load: $2.00\,kN/m^2$
Ceilings	Distributed load: $0.25\,kN/m^2$
	Together with concentrated load 0.9 kN

[a] Crown copyright: Building Regulations Approved Document A: 2004, Table 4.

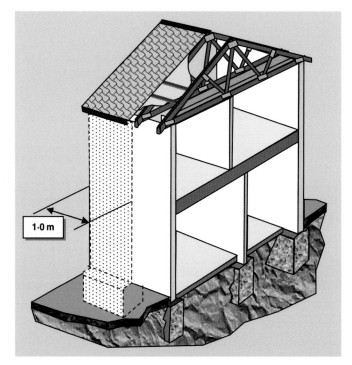

Figure 4.37 Isometric section through a simple domestic structure.

A list of common weights for building materials can be found in BS 648:1964 Schedule of Weights of Building Materials, manufacturers' technical catalogues (many of which can be downloaded from the manufacturers' websites) and Table 1 of BS 6399-1:1996 which gives minimum imposed floor loads.

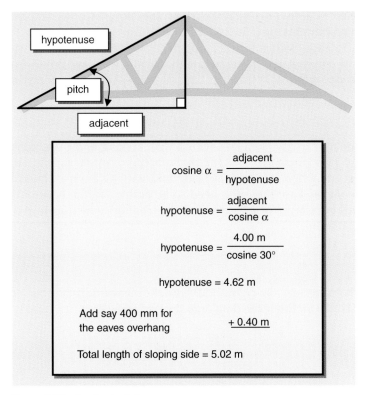

Figure 4.38 Roof calculations.

Example

It can be seen from Figure 4.34 that the span of the dwelling is 8.00 m. Half the load will bear on each wall; therefore only half the roof need be considered. To calculate the weight of the roof tiles, the length of the rafter must be found. Use of basic trigonometry to calculate the pitch or angle shows that it is 30° (Figure 4.38).

- What information is known? *The angle and the adjacent length.*
- What information is required? *The length of the **hypotenuse**.*
- Which formula should be used? $\text{cosine}\,\alpha = \frac{\text{Adjacent}}{\text{Hypotenuse}}$

Transpose the formula so that the unknown is on one side of the equals sign and the known is on the other:

$$\text{cosine}\,30° = \frac{4.00\,\text{m}}{\text{Hypotenuse}}$$

$$\text{Hypotenuse} = \frac{4.00\,\text{m}}{\text{cosine}\,30°}$$

$$\text{Hypotenuse} = 4.62\,\text{m}$$

There is an overhang at the eaves; therefore add, say, 0.40 m to the length of the rafter making it $5.02\,\text{m} \times 1\,\text{m} = 5.02\,\text{m}^2$.

You may want to simplify the calculation process by preparing a spreadsheet and entering the materials methodically, working down from the roof (see Section 4.5).

Roof loading

The roof trusses will be covered with sarking and then battens ready to take the tiles or slates. The tile manufacturers often supply an **as laid** weight per m^2 where the tiles, sarking and battens have all been included; if this is not the case they will need to be added. To the bottom edge of the truss will be a layer of plasterboard (the ceiling) which in turn supports the loft insulation.

Wall loading

The external wall height should now be measured. In the example in Figure 4.34 it is 5.20 m. Enter the data onto the spreadsheet.

Floor loading

Look for the **longest** span of floor bearing onto the wall. Figure 4.34 shows 4.50 m; therefore half the floor load will bear on the external wall and the other half on the load bearing partition wall. Enter the data onto the spreadsheet. In the example a non-load bearing partition wall on the first floor is bearing down onto the suspended floor; therefore measure the height of the partition and divide by two. (It is assumed that the load will be taken equally onto each end of the floor bearing walls – if the wall is significantly nearer one end of the floor joist, or of substantial weight such as brickwork, then more complex calculations would be required.) Enter the data onto the spreadsheet.

If the ground floor is suspended, a similar calculation will be required. In the example a solid floor has been shown so no weight will be transferred to the walls.

In addition to the floor and partition which are termed **dead loads** as they do not move, there are **imposed loads** or **live loads**. Furniture and people are moveable; therefore a given loading from BS 6399 Part 1 has been used. Depending on the use of the building, the minimum loadings start at 2.00 kN/m^2 for domestic dwellings. The floor area multiplied by the loading should be entered onto the spreadsheet.

Natural loadings

Wind

The only loading that has not been considered so far is wind loading. Wind is not a live load. It can exert positive pressure on one part of a structure and at the same time negative pressure (suction) on another part. Calculating wind loadings can be a very complex task which should be left to structural engineers. On larger projects, scaled models of the proposed structure are made, including the surrounding buildings and terrain. The model is then subjected to wind tunnel testing. Computer modelling is also used; however, as with the Millennium 'wobbly' bridge across the River Thames, computer modelling is only as good as the software allows.

The wind in recent decades has exceeded the maximum speeds assumed for design purposes of once in fifty years. On several occasions over the past two decades extreme structural damage has taken place in the sheltered south and south east of England, notably the hurricane of October 1987 and many tornados and high winds since. With global warming the incidence of strong winds is predicted to increase.

For general applications, the wind contour map and topography chart in Building Regulations Approved Document A can be used for guidance. Wind speeds are expressed in metres per second at ten metres above ground level, representing a three second gust. On average, this figure is likely to be exceeded once every 50 years. For example, in a sheltered part of the Home Counties around London a wind speed of 21 m/s is given. More detail can be found in BS 6399 Loadings for Buildings Part 2, and *The assessment of wind loads Part 1: Background and method* published by the Building Research Establishment.

Snow

Snow loadings can vary. Snow may only occur very occasionally, but it must be considered. The assumed **minimum** snow loading can be found in BS 6399 Loadings for buildings: Part 3 as $0.6\,kN/m^2$. Basic snow loadings have been included in the example.

The subject of snow loading is very complex. Consideration must be given to the angle and/or pitch of the roof, the location (whether the structure is exposed or sheltered), wind blown loadings, and so forth (for further reference see the *Handbook of Imposed Roof Loads: a commentary on BS 6399 'Loadings for buildings': Part 3.* Report 247, Building Research Establishment.)

4.5 Using spreadsheets to calculate loadings

To speed up calculations a computer spreadsheet can be used (Figure 4.39). (A basic knowledge of spreadsheets has been assumed.)

Open a spreadsheet and save as 'loadings calcs'. Then fill in the cells as listed below:

- Column A – cell A1 type: Roof materials
- Column B – cell B1 type: Area (m^2)

Roof materials	Area (m²)	Load (kN/m²)	Total load
concrete tiles on battens	5.02	0.51	2.56
sarking felt	5.02	0.05	0.25
fibreglass insulation	4.00	0.01	0.04
trusses @ 600 mm centres	4.00	0.24	0.96
plasterboard and skim	4.00	0.15	0.60
snow loading	5.02	0.60	3.01
Floor materials			0.00
chipboard 19 mm	2.25	0.17	0.38
floor joists	2.25	0.20	0.45
plasterboard and skim	2.25	0.15	0.34
Partitions (first floor)			0.00
timber stud	1.20	0.50	0.60
External wall			0.00
clay brick (inc. mortar)	5.20	1.70	8.84
Thermalite shield (inc. mortar)	5.20	0.74	3.85
live load	2.25	2.00	4.50
storage cistern	1.00	0.12	0.12
foundation	1.00	1.04	1.04
total load per lineal metre			**27.54** kN/m

foundation type: trench fill 0.45 m wide x 1.00 m deep concrete @ 2300 kg/m³				
length (m)	depth (m)	width (m)	density (kg/m³)	load (kN/m)
1.00	1.00	0.45	2300	1.04

[only half the load each end]
2.40 m / 2 = 1.20 m

Figure 4.39 Loading calculations spreadsheet.

- Column C – cell C1 type: Load (kN/m²)
- Column D – cell D1 type: Total load
- Column F – cell F4 type: Length (m)
- Column G – cell G4 type: Depth (m)
- Column H – cell H4 type: Width (m)
- Column I – cell I4 type: Density (kg/m³)
- Column J – cell J4 type: Load (kN/m²)

The data that is entered in column C will be from trade literature or textbooks.

Now enter the following:

1 Cell D2 should have the following formula entered:

$$= (B2 * C2)$$

Then press 'enter'. The data that is in cell B2 has now been multiplied by the data in cell C2.

2 Select cell D2 again. Using the left hand mouse button select the **handle** (a small grey square in the bottom right hand corner of the cell (Figure 4.40) and drag down to cell D23. Press 'enter' and the same formula will have been copied down into each line of cells using the relevant letter. (Always check that the formula is correct and working by entering simple numbers in cells in columns B and C).

3 Enter the following formula into cell J5:

$$= (F5 * G5 * H5 * I5)/1000$$

Dividing the answer by 1000 will show kilonewtons per square metre.

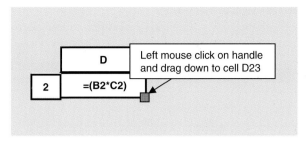

Figure 4.40 Spreadsheet handle.

4 Enter the following formula into cell C23:

$$= J5$$

5 Enter the following formula into cell D24:

$$= sum(D2 : D23)$$

and press 'enter'. The result should be as shown in the example.

Now, by using the example spreadsheet as a template, you can enter your own data in the relevant cells to carry out other calculations. As you become more experienced in spreadsheet design you may use other formulae and protect cells from alteration, specifically the formulae cells. It is a good idea when designing the spreadsheet to create a file and save your work every ten minutes, say. Create a master copy and protect it, or form sheets within the spreadsheet, or create separate spreadsheets. If a major error occurs, you can always revert back to the original protected spreadsheet.

Calculate the mass of a suitable foundation. In the example, 0.45 m wide and 1.00 m deep trench fill has been used. The total loading at the base of the foundation is 28 kN over an area of 0.45 m².

The table in Approved Document A shows the imposed loading per lineal metre, *not* per square metre.

So far we have a total loading for the structure in kN (force) per metre run. The initial ground survey would have identified the soil and subsoil/substratum of the site; CP 6399 (BS 6399) can be consulted for the **bearing capacity** or Table 4.2, reproduced from Building Regulations Approved Document A: 2004, can be used.

Foundation loading

At the design stage the substratum type and bearing strength should be known (see Section 4.7). Based on the ground type a foundation design can be considered. In the example it is known that the soil is 'firm clay' so a trench fill foundation would be suitable. Assuming a minimum depth of 1.00 m (common

Table 4.2 Minimum width of strip footings.

Type of ground (including engineered fill)	Condition of ground	Field test applicable	Total load of load bearing walling not more than (kN/linear metre)					
			20	30	40	50	60	70
			Minimum width of strip foundation (mm)					
I Rock	Not inferior to sandstone, limestone or firm chalk	Requires at least a pneumatic or other mechanically operated pick for excavation						
II Gravel or Sand	Medium dense	Requires a pick for excavation. Wooden peg 50 mm square in cross section hard to drive beyond 150 mm	250	300	400	500	600	650
III Clay Sandy clay	Stiff Stiff	Can be indented slightly by thumb	250	300	400	500	600	650
IV Clay Sandy clay	Firm Firm	Thumb makes impression easily	300	350	450	600	750	850
V Sand Silty sand Clayey sand	Loose Loose Loose	Can be excavated with a spade. Wooden peg 50 mm square in cross section can be easily driven	400	600	Note: Foundations on solid types V and VI do not fall within the provisions of this section if the total load exceeds 30 kN/m			
VI Silt Clay Sandy clay Clay or silt	Soft Soft Soft Soft	Finger pushed in up to 10 mm	450	650				
VII Silt Clay Sandy clay Clay or silt	Very soft Very soft Very soft Very soft	Finger easily pushed in up to 25 mm	Refer to specialist advice					

The table is applicable only within the strict terms of the criteria described within it.
Crown copyright: Building Regulations approved Document A: 2004, Table 10.

in clay soils – see Section 4.12) and a width of 450 mm (0.45 m) enter the data in the 'foundation type' section of the spreadsheet. If you have written in the correct formulae and links, the total load per metre run (lineal metre) should be shown (28 kN/linear metre). Using Table 4.2, the lowest loading shown is 20 kN/linear metre. Go to the next loading up, 30 kN/linear metre. Look for the soil type in the left column, row IV Clay – Firm, and under the column headed 30 a recommended minimum width of 350 mm is given. Enter the data into the spreadsheet to compare the new calculated loading; however, it will not make a significant difference. The results suggest that the foundation width could be 350 mm, giving a considerable saving on the amount of spoil to be removed and fresh concrete required. If the structure is approximately 8.00 m × 6.00 m the saving would be about 3 m³ of concrete/spoil.

4.6 Foundation design theory

Whatever pressures are acting downward they must be equalled by the pressures acting upwards (Figure 4.1). All materials have a **mass**. Mass is a term for the number of molecules that make up the material or substance. The molecules are attracted towards the centre of the Earth by a force known as **gravity**. Generally, all molecules are pulled downwards with an acceleration of 9·81 metres per second per second ($m\,s^{-2}$). For example, a lump of stone with a mass of 1 kg released from the top of scaffolding will fall towards the ground with an acceleration of $9.81\,m\,s^{-2}$; after 3 seconds it will be travelling at a speed of $3 \times 9.81 = 29.43\,m\,s^{-1}$.

Sir Isaac Newton in 1666–67, after watching an apple in an orchard falling towards the ground, developed a theory suggesting that there must be an attraction which pulls the apple downwards. In 1960 the UK adopted Système Internationale (SI) units of measurement and named the unit of force in Newton's honour; in Example 1 we find that a force of one newton is approximately equal to the force of gravity acting on a four ounce apple. However, 1 newton (N) is actually defined as the force that, when acting upon a mass of 1 kilogram, produces in it an acceleration of $1\,m\,s^{-2}$. Gravity produces an acceleration of $9.81\,m\,s^{-2}$ so the force on a 1 kg mass is 9.81 N; however, in construction calculations we use a loading of 10 N/kg.

Example 1 (see Figure 4.41)

An apple has a mass of 100 g (about 4 ounces in imperial units). The mass multiplied by the force of gravity will equal the weight.

$$\text{Weight} = \text{Mass} \times \text{Force of gravity}$$

Express the mass as part of a kilogram

$$100\,g = 0.1\,kg$$

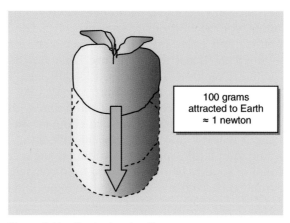

Figure 4.41 Gravitational pull.

For simplicity, assume that gravity produces an acceleration of $10\,\mathrm{m\,s^{-2}}$. Then,

$$\text{Weight} = 0.1\,\mathrm{kg} \times 10\,\mathrm{N/kg} = 1\,\mathrm{N}$$

Example 2

What would be the imposed load (downward force) from $5.02\,\mathrm{m^2}$ of concrete tiles on battens with a mass of $51\,\mathrm{kg/m^2}$?

The calculation is as follows:

$$\text{Area of material expressed in m}^2 \times \text{Mass expressed in kg\,m}^{-2} \times \text{Gravity} = \text{Force}$$

For simplicity, assume that gravity produces an acceleration of $10\,\mathrm{m\,s^{-2}}$. Then,

$$5.02\,\mathrm{m^2} \times 51\,\mathrm{kg\,m^{-2}} \times 10\,\mathrm{N/kg} = 2560.2\,\mathrm{N}/1000 = 2.56\,\mathrm{kN}$$

Answer: the loading is $2.56\,\mathrm{kN}$ (note that no unit of area is shown).

4.7 Ground surveys

There are several methods depending on the size of the project and the topography of the land, both the plot and the local conditions. The historic use of land is an important issue in relation to buried hazardous waste, landfill, mining, quarrying, etc. (see Section 4.8).

The usual ground level surveys are:

- trial pit
- borehole

However, aerial photography using digital cameras and thermal imaging cameras can provide good sub-terrain information by means of hot spots and variation in the colour of foliage. The subject is beyond the scope of this book.

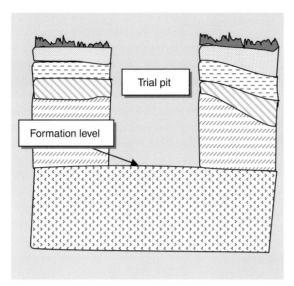

Figure 4.42 Section through a trial pit.

Trial pit tests

Where access is simple, a trial pit is dug either by hand or excavator up to 1–1.5 m deep to expose the substratum. This method enables an experienced surveyor to measure and record the type(s) of strata in-situ (undisturbed – where it is). If the water table is higher than the trial pit formation level, it should be left for a day to stabilise (the pit should be adequately protected if left open overnight both to prevent accidental falls (the contractor has a duty of care under the Health and Safety at Work Act 1974) and to prevent rain filling the excavation and giving a false level). The trial pit should be clear of the intended foundation lines as it will weaken the ground locally. However, caution! As the drawing in Figure 4.42 suggests, the stratum could change in thickness or type within metres of the trial pit. Examples of ground surveys going wrong include a historic rubbish pit in East London not found by the survey that required the design of foundation for two storey sheltered accommodation to change from 1.5 m deep trench fill to at least 2.5 m during trench excavation. In Essex, running sand adjacent to a river was completely missed on the survey which added an enormous bill to the client when the work proceeded.

Borehole tests

On larger projects or where access is restricted, borehole testing can be used. Depending on the size of the proposed structure and the expected soil type, a small borehole can be used, or alternatively where deep pile foundations are required, a piling rig borehole will be needed.

Figure 4.43 Drop hammer soil sampling.

Figure 4.43 shows an impact borehole logger. A heavy metal weight is regularly dropped onto a shaft pushing the bore shell into the stratum. When the shell is filled it is extracted, recorded and set aside for study. The procedure is repeated with an empty shell to record the next level of stratum. Figure 4.44 shows the geologist inspecting the soil sample and bagging it for further chemical analysis in the laboratory. This particular test was carried out next to a river in Colchester to advise a developer of the soil conditions. Note the PPE being worn: high-viz jackets, high-viz waistcoats, hard hats, protective footwear and gloves. Only the technician operating the machine required additional ear defenders and eye protection.

Land near historic rivers can be particularly troublesome due to deposits from the water. Typical problems include **running sand** next to the River Chelmer in Chelmsford and **alluvium** in localised land next to the River Rom in Romford. Both soil types had been missed during the ground survey, but added enormously to the time and cost when the projects proceeded. Running sand, when held in place by its surrounds, can be suitable to build on. The problem is the method of excavation required and how thick the layer is. If an open trench method is used the sand will continually flow into the trench with possible cave in/subsidence of higher levels of strata. The solution used on the specific site comprised a steel sleeve impacted through the running sand into the clay beneath. The contents of the sleeve were augured out using a standard piling auger and a concrete pile was cast inside the sleeve (Figure 4.45).

4.8   Desk top surveys

Before surveying an area it is useful to gather as much data as is known in the office. Maps produced by the British Geological Survey can be particularly useful: these indicate the geological structure of the whole of Great Britain but not the soil types (the **overburden**); they can be found in the National House Building Council (NHBC), the Building Research Establishment (BRE), British Standards and Eurocodes. If the project is small, there are companies which, for a fee, will provide data about the historic and current use of the land. The data has been based on surveys in the area, any major excavations, and local authority records. It should be borne in mind that desk top surveying can be useful as a guide for preparatory work; however, it should not be used as definitive or accurate (Figure 4.46).

4.9 Foundation materials

Concrete mixes

Figure 4.44 Geologist inspecting an undisturbed soil sample.

Concrete has been used since early Roman times. The Pantheon in Rome built in the year 118 AD has a 43 m diameter dome capped in early concrete. After the fall of the Roman Empire, the art of cement making was lost for centuries until, in 1756, John Smeaton used a type of cement to bed the rock used for the Eddystone lighthouse. Shortly afterwards, in 1824, Joseph Aspdin patented cement used to make concrete. The new concrete material looked similar to the popular lime-stone from Portland, so he named his cement 'Portland cement'.

Cements that are used in the production of concrete and mortar are known as **binders.** Their function is to prevent the particles (both fine and coarse) from moving, so it is essential that the materials are well mixed and that the cement paste completely envelops all surfaces.

Concrete comprises:

1 Binder (Portland cement)
2 Aggregates (coarse and fine)
3 Water (it should be potable – drinkable)

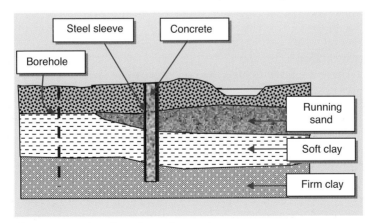

Figure 4.45 Piling through running sand.

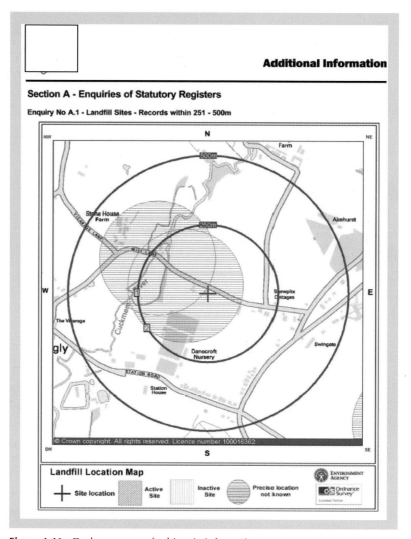

Figure 4.46   Desk top survey for historic information.

(see Figure 4.47). The proportions vary for different types of concrete, but in general terms are based on the number of voids in the aggregates: sand, 33.3%; gravel, 45.0%.

There are three main ways concrete can be specified:

1 nominal mix
2 prescribed mix
3 designed mix.

Nominal mixes

Nominal mixes are produced on a ratio basis such as:

- 1:3:6
- 1:2:4

Figure 4.47 Main components of dense concrete.

The first number refers to the volume of cement, the second number is the volumes of fine aggregate (sand), and the last number the volumes of coarse aggregate. If the coarse aggregate has, say, 50% voids and there are six volumes, then the three volumes of sand will fill them. However, the one volume of cement will fill 30% sand voids.

Prescribed mixes

The required strength is stated in newtons per square millimetre (N/mm^2):

- C7P – mass foundations equal to a 1:3:6 mix
- C10P – as C7P but can be used in wet conditions
- C15P – general mass use, minimum strength for structural unreinforced work
- C20P – general floors and reinforced work equal to a 1:2:4 mix
- C25P – used where the exclusion of water is necessary
- C30P – for moderate abrasion resistance of a concrete floor (equal to $1:1\frac{1}{2}:3$).

The C stands for compressive (the required final strength at 28 days), the number relates to the strength in N/mm^2, and P means the mix is prescribed. A series of mix designs were developed by the Department of the Environment (DoE) and the Cement and Concrete Association that enable the specifier to state the mix design in great detail using a DoE sheet.

Designed mixes

These can be specified. The contractor or manufacturer is responsible for achieving the required strength. This method is now common for foundation concrete. However, the specifier should state which type of cement and any other properties the concrete should have, or any additives required. Alternatively, GEN1 (designated concrete), ST2 (standardised prescribed concrete), or S3d (recommended consistence class) can be used. The mixes shown above are defined in BS EN 206-1 as suitable for housing and other applications.

Where reinforcement is required, the mix strength and density increases to RC30 or FND2 as designated concrete. Concrete floors range from house floors (oversite slab) without reinforcement as GEN1, to a garage floor as GEN3 (Table 4.3). For harder wearing surfaces such as industrial floors, RC50 would be specified. The recommended consistency for the whole range of floors is S2.

Table 4.3 Common design mixes of dense concrete.

Designated concrete	Required strength class	Compressive strength (N/mm²)	Default slump class
GEN1	C8/10	10	S3
GEN2	C12/15	15	S3
GEN3	C16/20	20	S3
FND (all designations)	C16/20	20	S3
RC30	C25/30	30	S3
RC50	C40/50	50	S3
Standardised prescribed			
ST1	C6/8	8	
ST2	C8/10	10	
ST3	C12/15	15	
ST4	C16/20	20	
ST5	C20/25	25	

ST mixes should not be used where sulfates or other aggressive chemicals are present. In such instances FND concrete should be specified.

Where concrete is likely to be exposed to freeze–thaw cycles and possible de-icing salts, such as on drives, roads or footpaths, an air entrained concrete should be used. The concrete should be 'designated concrete' design.

The BS EN 206-1 code provides guidance regarding the minimum cement content required for specific diameter aggregates, and water/cement ratios. The common coarse aggregate diameter is 20 mm, and the water/cement ratios 0.5 and 0.6 require 320 kg/m³ and 280 kg/m³, respectively. Figure 4.48 shows a typical concrete used for foundations.

Aggregates

Coarse aggregates as shown in Figure 4.49 are from quarried glacial deposits formed during the last ice age. Note the smooth surfaces of mainly flints which are common in the south east of England. As the ice thawed, the aggregates were transported by rivers into ancient shallow seas and were rounded off in the process. The small particles that broke off became the sands and with chemical breakdown separated into quartz and silicates that became sand and

Figure 4.48 Saturated dense concrete.

clays. Large deposits of rounded stones with layers of sand and clay will be found in many areas of the south east of England. They tend to be rounded flints, quartz and some sandstone. Many of the exhausted quarries have been used for landfill sites to bury waste (see Chapter 10).

In the UK, over 100 000 000 tonnes of aggregate are used every year. Each year, access to local aggregates is reduced. Many large beds of aggregates cannot be quarried as there are now roads and buildings covering the land. Environmentalists do not want farmland, woodlands and areas of natural beauty changed for ever; therefore planning laws restrict further new quarries from being opened in the south. However, the government is committed to building at least 120 000 new homes by 2016 in their Thames Gateway project which will need an estimated 7 200 000 tonnes of aggregates.

What other resources are available? Basically rocks and waste. Rocks in the form of mountains are being systematically broken up and transported by sea to the south of England and parts of Europe. In the north west of Scotland, Foster Yeoman Glesanda have been quarrying 6 000 000 tonnes of rock per year from a mountain, crushing it and shipping it to the Isle of Grain in Kent (Figure 4.50).

Figure 4.49 Coarse natural dense aggregates.

Figure 4.50 Aggregate wharf and the one of the Foster Yeoman self discharging vessels. Photograph by Samantha Cooke.

From there they transport it by train or road to much of the south east of England. About 5 000 000 tonnes of crushed limestone is brought out of the Mendip quarries in the south west of England by train or road. Crushed aggregates are less workable but have the advantage of greater compressive strength.

Sea dredged aggregates are frequently used on the south coast of England. They have to be washed in fresh water to remove the salt.

Types and shapes of aggregates

There are six main shapes: most are shown in Figure 4.49. Numbers 1, 2 and 3 will give better compaction and workability than 4, 5 and 6. If the rocks have been split they will tend to be angular and therefore difficult to work (of **low workability**). Figure 4.51 shows crushed limestone from the Derbyshire area: note the rough surfaces and angular shapes.

Aggregates used in foundation concrete include coarse aggregates ranging from 5 mm diameter (Ø) up to normally 40 mm Ø. For general concrete foundations (with or without reinforcement) 20 mm Ø and sharp sand would be used. Sand ranges from almost dust to 5 mm Ø. Sizes above 5 mm are defined as gravels and coarse. Sands are classified into zones. Each zone has a specific ratio of particle sizes. The test is carried out using a nest of sieves, a sieve vibrator and balances. The ratios will affect the porosity of the mix, workability and the water content, as well as the cement proportion and/or content.

Figure 4.51 Crushed limestone dense aggregate.

General requirements

Aggregates should be clean and free from sulfates, salts and organic matter (also some reactive silica which causes 'alkali aggregate reaction' – commonly calcium alkali silica which absorbs water and therefore makes the concrete more vulnerable to frost attack or rusting of reinforcement). Flaking of aggregates, clay and silt should be carefully monitored and not exceed the requirements of BS 882.

Aggregates should meet the following requirements:

- be free of any coatings
- be chemically inert and stable
- must not corrode or decompose
- be free from expansion
- will not cause staining.

Aggregates are normally washed before delivery, and this is essential where sea dredged aggregates are being used. Salt is very corrosive and if contained in the concrete it will (when water is present) eventually cause it to fail.

All-in ballast is a mixture of fine and coarse naturally occurring aggregates. These require balancing before they are used for the production of concrete. Ballast should *not* be used for quality concretes.

Density

Density is an important property of both aggregates and finished concrete. Density (**relative density**), formerly known as specific gravity, is the mass of a given volume of aggregate compared with the mass of an equal volume of water at 20°C. This is important as it enables mass to be compared with volume – batch mixing uses the technique where so many kilograms of aggregate and cement produce so many cubic metres of concrete.

Relative density (RD) is divided into three classes to take into account that the aggregate could be porous, so there will be three possibilities:

1 **Oven dry** – no water in it or on it

Oven dry RD

= mass of dry aggregate/volume of aggregate including pores

2 **Saturated but dry** – dry aggregate with all the pores filled to saturation point (also known as **SSD**–**S**aturated aggregate with the **S**urface **D**ry)

RD = (mass of dry aggregate + water in pores)/
volume of aggregate including pores

3 **Apparent relative density** – where the aggregate is considered as a mass and volume not considering the pores

RD = mass of dry aggregate/solid volume of aggregate

The density range of most natural aggregates used for foundation concrete is 2000–2400 kg/m^3; they will have a crushing strength of between 14 MN/m^2 and 100 MN/m^2.

Portland cement

Making concrete

The chemical reaction starts after the water has been added and a paste is formed. The paste goes through two stages:

1 **Stiffening** or **setting** (initial and final). The correct amount of water is required to hydrate cement, any excess will only create air voids when it evaporates and therefore weaken the bonding. When the chemical reaction starts, a latticework of crystals forms. At an early stage (about 2–3 hours after the water has been added) the crystals grow. If the fresh concrete is knocked, frozen, or more water is added, the crystals will rupture and lose their strength.
2 **Hardening** when the useful strength is developed – this is hydration.

The water/cement ratio is important. Listed below are some things to be considered:

- cement requires the correct amount of water to **produce** and **complete** the chemical reaction
- too much water will weaken the final product
- too little water may result in the chemical reaction stopping before it is complete
- concrete needs to be workable – required workability depends on its use
- ambient temperatures may affect the chemical process.

When the water is added the subsequent chemical reaction is **exothermic** and produces heat. In cold weather this heat can help the curing process which ceases at about 0°C (2°C is the cut off figure), but in summer the excess heat can lead to thermal stress and/or differential stresses which can cause thermal

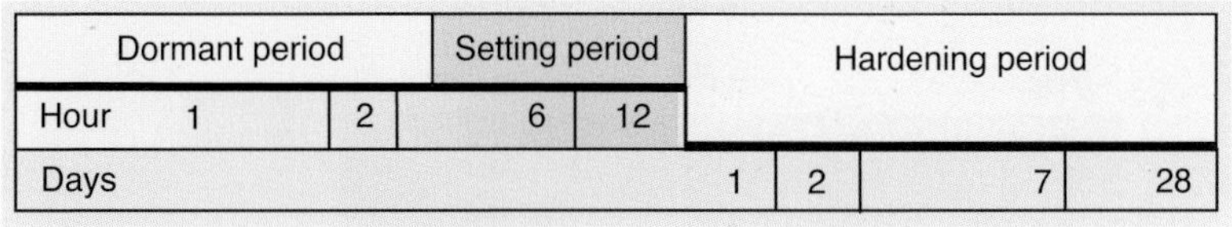

Dormant period		Setting period		Hardening period		
Hour 1	2	6	12			
Days				1 2	7	28

Figure 4.52 Concrete setting periods.

cracking in mass pours. Specialised cements or additives can overcome some of the problems of heat generated by the exothermic reaction.

Hydration period

Figure 4.52 shows the typical hydration period for Ordinary Portland cement.

Portland cement types

Ordinary Portland cement

Ordinary Portland cement (OPC) is the cement most commonly used for concrete, mortars and products such as concrete blocks. There are different versions of OPC – bagged or bulk – giving different strength classes after 28 days' curing: OPC bagged at $42.5 \, N/mm^2$ and CEM I bulk up to $52.5 \, N/mm^2$.

OPC is vulnerable to sulfate attack and therefore if it is known that the soil pH is above 5.5 and therefore acid, OPC must not be used.

The elements calcium and silicon are found in chalk and limestone, clay or shale. Clays often contain large quantities of aluminium and iron compounds which give the characteristic grey colour to OPC. The proportions of these compounds in the cement can be modified to produce specific cement types, such as **sulfate resistant cement**, or **white cement** used for decorative concrete. Clay with a low content of aluminium or iron compounds is used for white cement. The four main compounds obtained from the clay or limestone used as raw material are shown in Table 4.4.

Table 4.4 Chemical composition of Portland cements.

Compound	Short formula	Rate of hardening	CEM I [42.5N] (%)	SRPC (%)	White Portland cement (%)
Tricalcium silicate	C_3S	Rapid	56	64	65
Dicalcium silicate	C_2S	Slow	16	10	22
Tricalcium aluminate	C_3A	Rapid	8	2	5
Tetracalcium aluminoferrite	C_4AF	Extremely slow	9	14	1

Table 4.5 Setting strength comparison between OPC and RHPC.

Time after mixing (days)	Setting strength (N/mm^2)	
	OPC	RHPC
3	13.0	18.0
28	29.0	33.0

The **strength of the cement** is measured at the early age of 2 days and then at 28 days, and should not be confused with the strength of the concrete or mortar tests (Table 4.5).

Rapid hardening Portland cements

Similar to the clays from which it was derived, cement expands when wetted. The quicker the water can activate the cement, the faster the process can work. To speed up the reaction time, OPC is ground down to produce smaller particles and therefore, for the same mass of cement, a larger surface area for the water to reach, giving rapid hardening Portland cement (RHPC). OPC has a surface area of 300 m^2/kg, whereas RHPC has a surface area of approximately 400 m^2/kg – this is known as its **specific surface**.

RHPCs are now defined as CEM I 52.5 (where 52.5 indicates the compressive strength in newtons in given time limits). A slight variation to the formulation is also made. Because CEM I 52.5 also produces heat more quickly than OPC, it is useful in cold weather. The initial and final setting times of RHPC are the same as for OPC. However, the compressive strength after 3 days is dramatically increased.

The disadvantages of RHPC are that, because it produces much more heat, pouring large volumes in the summer months can be problematic. RHPC also has low resistance to shrinkage cracking and sulfate attack.

Portland supersulfated and high alumina cements

These are both used in the construction industry, but *must not* be used for structural concrete.

Sulfate resisting Portland cement

Sulfate resisting Portland cement (SRPC) has a reduced tricalcium aluminate content (C$_3$A less than 3.5%). It is suitable for foundations and certain concrete applications where groundwater has a concentration of sulphur trioxide up to 0.1% or subsoils contain up to 0.5% SO$_3$. This cement is commonly specified for foundations in clay or gravel areas, such as Essex and London, where pockets of sulfates are present. Soil tests or local knowledge, such as from a building control officer, will ascertain whether it is needed.

Low heat Portland cement (BS 1370)

This cement is low in C_3S and high in C_2S and C_3A; it is finely ground but by changing the content of certain constituents slows the hardening down and thus the release of heat. It is used when pouring vast volumes of concrete. Whereas OPC generates about 400 kJ/kg in 28 days, LHPC produces 290 kJ/kg in the same time period.

Masonry cement

Masonry cement is OPC with added chalk or silica and a plasticiser. It is used where a workable mix is required without the addition of lime which can stain certain brick types.

Blastfurnace slag cement

Granulated blastfurnace slag (GGBS) is a waste product from quenching selected molten blastfurnace slag from iron smelting. It can be blended with CEM-1 to produce a sulfate resistant concrete. Portland-slag cement CEM II/A-S contains between 6 and 35% slag. Benefits include good resistance to dilute acids and sulfates, it is cheaper than SRPC and ideal for sea walls, etc.

Commercial concrete manufacture (Figure 4.53)

The bulk materials are stored in bays ready for mixing in a **batching plant**. The bin shown in Figure 4.54 has coarse aggregates and that in Figure 4.55 has the fine aggregates. A loading shovel deposits the aggregates into a hopper which feeds a conveyor belt taking them to the bulk silos (Figure 4.56). Figure 4.57 shows the main parts of the process.

A computerised weighing machine weighs and dumps the aggregates onto a screw drive transferring them to the mixing drum. The batcher can monitor the entire process on the computer screen (Figures 4.58 and 4.59). The left hand dial shows the total amount of cement and the right hand dial the aggregates going to the mixer. In this case the concrete will be resistant to ground sulfates.

Water is also measured by weight: 1 litre of water weighs 1kg, so the whole process can be monitored by weight. This information is important to the lorry driver to prevent overloading.

The workability of the concrete is influenced by the aggregate shape, size, ratio of fine to coarse, and the amount of water it contains. Trial mixes are produced and all the factors recorded. The workability of a concrete is the ease with which it can be placed. For example, a trench fill foundation will require

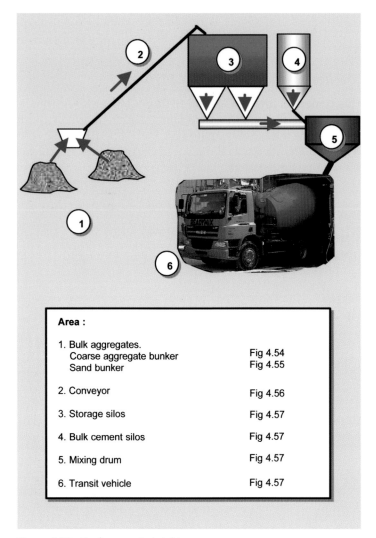

Area :

1. Bulk aggregates.
   Coarse aggregate bunker   Fig 4.54
   Sand bunker   Fig 4.55

2. Conveyor   Fig 4.56

3. Storage silos   Fig 4.57

4. Bulk cement silos   Fig 4.57

5. Mixing drum   Fig 4.57

6. Transit vehicle   Fig 4.57

Figure 4.53   Fresh concrete batching process.

a workable (relatively fluid) mix that flows along the trench with some help from the ground workers. However, it should not be so wet that it self levels. A method of quantifying workability is to state the amount a mix **slumps** (see p. 195, Slump test).

The mixed concrete is transferred to the transit vehicle ready to be delivered to site (Figure 4.57). Although a relatively small batching plant is shown, the process is similar in larger plants. Special mixes can be produced on request with additives to retard or accelerate setting, reduce water content, or using special aggregates. Batching plants tend to be located close to built up areas to reduce the transit time of the mixed concrete. Some remote contracts have the water added to the mix whilst the transit vehicle is en route.

Figure 4.54 Coarse aggregate bin.

Figure 4.55 Fine aggregate bin.

Concrete for foundations

Site procedure

Small quantities of concrete for, say, a house extension may be made manually on site. Generally the concrete will be made to a nominal mix method. The

Figure 4.56 Conveyor and hopper heads.

Figure 4.57 Batching mixer.

strength of the foundation must exceed $7\,\text{N/mm}^2$ so a 1:3:6 (20 mm Ø) mix should be adequate. Table C.1 of BS EN206-1 gives a target cement content for common foundation mass concrete applications.

- 1:3:6 mix should have $190\,\text{kg/m}^3$ of cement
- 1:2:4 mix would have $275\,\text{kg/m}^3$.

Figure 4.58 Materials data monitor.

Figure 4.59 Materials monitor.

Site procedure

All-in ballast is commonly used at a ratio of 1:5 where the supplier has blended the sand with coarse aggregates before delivery. The process on site should be as follows:

1 Provide a flat clean area off the public highway to take delivery. Ideally boards should be used to prevent the aggregates from becoming contaminated with soil.

2 Cover the aggregates if they are to be left overnight as rain can wash the sand from the ballast.
3 Provide a firm level surface for the concrete mixer. The three main types of mixer are powered by:
   - electric motor (it must run at 110 V so either from a transformer or generator)
   - petrol engine
   - diesel engine.
4 A supply of clean potable water.
5 Container to deposit packing and cleaning waste (commonly a skip).

Mixing method
1 Switch on mixer and let it run empty.
2 Using a shovel place five regular shovels full of aggregate into the drum.
3 Place one regular shovel full of cement. (If windy, place the mixer down wind to stop the cement powder blowing in the operator's face).
4 As the mixer blades fold the aggregates and cement together add the water gradually. When the mix falls away from the drum by gravity do not add any more water. (It is useful to use a bucket for the water so that the quantity added can be gauged each time.)
5 Turn out the mix into a wheelbarrow ready for it to be taken to the foundation trench.

Textbooks suggest a gauging box be used to regulate the quantities of aggregate and cement. Whilst obviously it will be technically accurate, it is impractical and rarely done on site.

Testing fresh concrete and aggregates

There are two types of test:

1 On-site
   - field settling test
   - slump test
2 Off-site
   - RD test
   - destructive compressive test using sampled test cubes
   - compaction factor test
   - Vebee test.

Field settling test

On site, fine aggregates (sand) should always be tested for cleanliness. The official test is shown below, but an old jam jar (as long as it is clean and of clear glass) can be used. The object is to determine the percentage of silt held with the sand. Looking at the sand will not help as the silt particles will be less than 150 μm wide. If the sand has been deposited for several days, select a sample

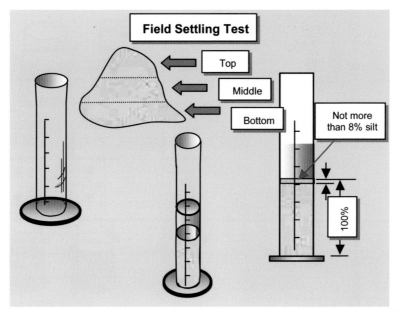

Figure 4.60 Field settling test.

from the top, middle and bottom of the pile and mix in a bucket (if the sand has just been delivered use any sample).

1 Place mixed sample in the glass container and pour clean water to cover by about 25 mm.
2 Cover the top and shake the sample ensuring all of the sand is mobile.
3 Leave on a flat surface for about an hour to settle out. (Half a teaspoonful of salt will speed up the settlement of the smallest particles otherwise wait until the water is clear.)
4 Measure the amount of silt which will appear as a fine yellow/orange layer on top of the sand and express as a percentage of the sand and silt.
5 If the silt percentage is about 8–10% of the total solids then repeat the test. If the percentage exceeds 10% contact the supplier as the sand is too polluted to be used.

The laboratory test is virtually the same; however, it has to be carried out in accordance with the relevant British Standards (see Figure 4.60).

Slump test

To test the workability of fresh concrete on site before it is used, a slump test should be performed. If the fresh concrete has been made off site in a batching plant, it may have been delayed in traffic. It is important to record the time of arrival and check the delivery ticket for the time it was originally mixed. (Remember that the cement will start to form crystals between two and three hours after mixing, depending on the weather, and that a late delivery will

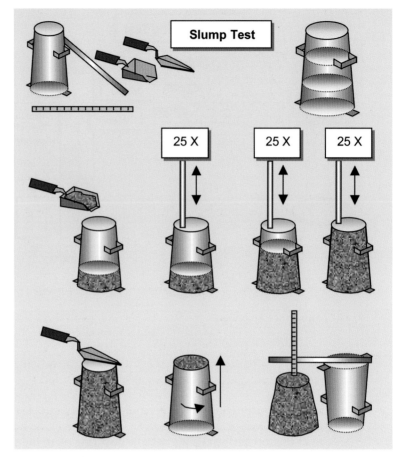

Figure 4.61 Slump test.

reduce the placing time on site.) The slump test will indicate how workable the mix is and whether it has been rewatered en-route. (Reputable ready mixed concrete operators would not carry out this bad practice; however, experience has shown that it does take place.)

Method (Figure 4.61)
1   Deliver a small sample from the transit vehicle into a wheelbarrow or dumper truck – transfer it to a bucket if required.
2   Prepare a flat clean surface and stand the slump cone on it.
3   Stand on the two foot tabs (or ask an assistant to hold the handles to keep the cone firmly in place).
4   Using a scoop or brick laying trowel, deposit the fresh concrete into the open ended cone to about the 75 mm level.
5   Using the tamping rod round end downwards, rod 25 times to remove any trapped air voids.
6   Repeat the procedure at the 150 mm level and tamp 25 times.
7   Repeat the procedure at the 225 mm level and tamp 25 times.

Table 4.6 Concrete workability.

Specified slump class	Slump test from initial discharge	
	Minimum (mm)	Maximum (mm)
S1	0	70
S2	30	120
S3	80	180

8   Finally fill the cone (300 mm) and tamp 25 times.
9   Flush off the top with the trowel, and carefully remove the cone by slightly twisting it vertically to help it release.
10   Invert the cone and stand it next to the sample; place the tamping rod across the wide end of the cone and over the sample.
11   Measure the difference (if any) between the underside of the tamping rod and the top of the sample: this is known as the **slump**.

The greater the slump, the more workable the mix. Slump testing fresh concrete from an initial discharge has an upper and lower limit. Where, for instance, a stiff mix is required a specified slump of up to 40 mm can range from 10 mm to 80 mm at time of discharge. Again a specified target slump of 50 mm to 90 mm could range from 10 mm to 140 mm. Although concrete is a complicated material its manufacture is dependent on many factors; therefore the batcher's experienced eye still is considered the best gauge. Eurocode EN 12350 for testing fresh concrete identifies slump ranges for a specified slump class. Some commonly used slump ranges are shown in Table 4.6.

Some useful British Standards and Eurocodes

EN 1992 (Eurocode 2)	Design of concrete structures
EN 206-1	Concrete
EN 12350	Testing fresh concrete
EN 12390	Testing hardened concrete
EN 13591	Assessment of concrete strength in structures
EN 834-3	Admixtures for concrete
EN 12620	Aggregates for concrete
BS 8500-2:2002	Covers aggregates

4.10   Health and safety issues relating to concrete and cement

Portland cement is alkaline. When activated with water it can burn skin tissue. Cement dust is an irritant and can be breathed in when the operator places the cement into a working mixer. (The risk can be reduced by locating the mixing area in a sheltered place or with the mixer downwind of the materials and operator. This should be part of the risk assessment required by the Health and Safety at Work Act 1974. A first aid box and eye washing bottle should be easily to hand, either in the site hut or, on small jobs, in the van. Appropriate personal protective equipment should be worn. The Act states that it is the duty of everyone to be responsible for their own health and safety and those around them. This means that not only does the employer have a duty of care for all operatives and anyone on or near the work, but the operatives also have a duty to themselves. When working with cement dust in bags, the working area should be sectioned off or far away from the public, including neighbours, to prevent the wind blowing the dust over them or their property. The use of solid hoarding will prevent dust leaving the site (see Chapter 9, Section 9.3).

Personal protective equipment

In recent decades it has become commonplace for concrete workers to wear as little as possible in the summer months, shorts and trainers being popular. However, accidents and poor health have been rife in the building industry, a situation that the Health and Safety Executive is trying to rectify. Larger sites employ safety officers and consider health and safety as a major issue. Training is required and qualified management ensures that health and safety is the number one concern of everyone. In contrast, smaller builders carrying out small works such as building extensions and single houses often employ subcontractors. By the nature of the job it would be impractical to carry out **induction briefs** and **tool box talks**. Many subcontractors may have either limited training other than for their trade, or their training was undertaken many years ago and is now out of date. The Health and Safety at Work Act does not differentiate between one person's health and safety and another's.

All personnel working with cement or fresh concrete should wear the following:

1　Leg cover (trousers or boiler suits).
2　Arm and body cover (shirt or boiler suit).
3　Eye protection (polycarbonate spectacles).
4　Stout site footwear (steel toe protection, water resistant, sole protection). There are many designs and styles ranging from rigger boots to trainer-looking site wear).
5　Gloves (several different types of glove are available). For those placing the concrete, a gauntlet style rubber glove will prevent fresh cement splashing the wrists or falling into the gloves. Cheap washing up gloves will provide some protection. Cotton lined rubber gloves are more comfortable to wear and reduce the wet hand problem caused by sweat.
6　A hard hat should be worn if there is any overhead working.
7　Sun tan barriers. If working outside in direct sunlight for long periods it is now recommended that sun tan barriers are used. The levels of ultraviolet light getting through to England are increasing as part of climate change; therefore it is considered good policy to put a barrier over the skin to reduce the possibilities of skin cancer in later life.

4.11 Problems with foundations

Faulty foundations cost the construction industry and insurance industry in excess of £300 million per year in claims. The greatest failure is caused by tree roots and chemical attack. Tree roots were mentioned in Section 4.1; however because of their importance, further comment is required.

Tree root problems

The following questions need to be asked:

1 What species of tree are present, how old are they, what is the soil condition and what are the weather conditions?
2 Who owns the tree(s)?
3 Were the trees there before the foundations?
4 Were there trees on the site before it was cleared?
5 What is the type of ground/substratum – made up, green field or brown field?
6 What is the topography?

The six issues above are not the only ones that affect foundations. For example, consider the following scenario. A large house and garden have been sold to a developer who intends to demolish the house and build three new detached dwellings. The garden contains several mature apple trees and cherry trees in the area to be occupied by the new buildings. The soil is clay to a depth of 2 m over a bed of ballast of unknown thickness. The builder has a specialist demolition company remove the old house and shallow foundations. (It was common for old houses to be built on foundations only 450 mm deep, even in clay soils.) The houses to each side of the plot both have mature oak trees on their property within 2 m of the boundary. The site is on a hill near to the bottom of a small valley in an area that has been developed since Victorian times.

How will the issues raised above affect the design and the construction of the new development?

1 The mature apple trees have been removing vast quantities of groundwater over a very long period. If the builder removes the trees where will all the water that would have been evaporated by the trees now go? In a clay soil it most likely that it will cause ground heave.
2 When the ground survey was carried out, what was the time of year and what weather conditions prevailed? If, say, the survey had been in March after a relatively dry winter the groundwater table may have been abnormally low. As the site is near the bottom of a hill in a valley there will be a tendency for the water to lie near the site. Was there once a river or brook running in the valley? Has it been filled in or diverted? (Old maps may help, even elderly neighbours are a good source of knowledge. They may remember fishing in the brook, river or pond when they were young.)
3 What about the old foundations? When were they grubbed out (removed) and how deep did the excavator dig? If it was less than the

depth of the new foundations there should not be a problem; however, if not, the soil will now be weaker in that localised area.

4   The oak trees on the neighbours' land will not recognise the site boundary. If the builder cuts significant tree roots that have gone across the legal site boundary, the tree may become unstable and topple during any future high winds. Cutting the roots may harm the tree and cause damage; therefore a specialist should be consulted before any pruning of tree roots takes place. As the trees are oak, they are probably the subjects of tree preservation orders which makes it illegal to prune, remove or otherwise damage them, whoever's land the roots are on. The local authority should be approached for guidance on the matter. It is not a case of trimming the tree roots or branches back and offering them back to the owner.

5   The trees are mature which means they must be at least 50–60 years old, and possibly well over 100, so the builder cannot say they damaged the new foundations. The designer should have taken the tree roots into consideration.

6   We know that the apple and cherry trees were still in place before the demolition took place, so it is unlikely that other roots will be found in the same area.

What options are open to the designer for the new foundations? Clay soil means it is likely to swell and shrink when the seasons change. (Not all clays swell at the same rate; indeed some clays are impervious to water and do not move.) Tree roots are going to seek water; as the climate is changing it is predicted that winters will become less wet or cold and droughts will become more regular.

What foundation will

- allow for near surface expansion of soil?
- not act as a barrier to any ground water movement?

What else does the designer need to know?

- how thick the layer of ballast is below the clay
- whether there are any sulfates in the ground or groundwater.

Chemical attack

What are sulfates and why are they in pockets or localised areas? Sulfates are concentrations of acid rain. Acid rain has existed for as long as water has been on Earth. As volcanoes erupt and faults within the Earth's surface spew vast volumes of sulfur dioxide (SO_2) into the atmosphere the gas combines with moisture in the air to form sulfuric acid – **acid rain**. On clay areas the rain was unable to percolate into the ground and formed ponds. As the water evaporated, the solution of acid became more concentrated. The cycle regularly repeated itself over many years developing into localised areas of concentrated acid rain. Where areas were exposed and dried out completely, crystals formed which look like small bright granules of sugar. During long periods of drought the crystals became covered and eventually trapped beneath the ground. Unless the groundwater can dissolve the hardened crystals they remain unchanged for thousands of years. The process is very slow but is still taking place in areas with high pollution levels such as Eastern Europe.

What will the sulfates do? They have little effect on most plants and materials; however, Portland cement will react with sulfates. A reaction occurs between tricalcium aluminate in the cement and the soluble sulfates in the clay, or flue condensate carried by continuous saturation in water. The reaction produces ettringite (hydrated calcium sulfoaluminate) which occupies a larger volume and is weaker than the cement from which it came.

Sulfates are present in London clay, lower lias, Oxford clay, Kimmeridge clay and Keuper marl. Other sulfates include calcium sulfate (gypsum), magnesium sulfate, and sodium sulfate. Some clay bricks contain enough sulfates to be troublesome. Sulfate attack first hardens the brickwork mortar and then turns it into a white powder. The surface may be hardened by atmospheric carbonation and appear to be normal. Brickwork subjected to frost can increase the problem.

If there are known sulfates in an area, SRPC should be used for masonry under dpc level (**substructure)** and for the concrete foundation. As an alternative, GGBS may be blended with the SRPC to reduce costs and reduce the heat output of the exothermic reaction in hot weather. (see Section 4.9).

Answer to design issues

When the designer knows how thick the layer of ballast is (borehole testing is required) it is likely that a structural engineer will detail a short bored pile and beam foundation with SRPC and possibly steel reinforcement to prevent horizontal shear as the site is on a hill.

The Building Research Establishment and the National House Building Council have both published information about building near trees.

4.12   Soils and substrata

Substrata are classified into five main groups:

1   rock
2   non-cohesive soil – gravel and sand
3   cohesive soil – such as clay
4   peat and shifting sand
5   made ground or fill.

Rock

There are three main rock types:

- igneous
- sedimentary
- metamorphic.

Igneous rocks form as a skin on the molten mantle. They are generally dense and crystalline in structure. Granite and basalt are perhaps the most common igneous rocks in the UK and form the bedrock. Beneath areas where the granite is near to the surface, such as the West Country and parts of Scotland, radioactive deposits of uranium are present. As the uranium decays radioactive radon gas filters through the stratum and eventually into the air. This natural process has only recently caused concern. The gas can percolate through concrete and fill voids beneath floors in buildings. As modern building techniques have reduced draughts and suspended floors have become popular, the gas has tended to become more concentrated within structures and is thought to have increased certain types of cancer. Consequently the Building Regulations now specifically mentions radon gas as a problem that requires installation of specialist barriers and vents.

Where the igneous rocks weather, they fragment into free rocks of various sizes from boulders down to sands. Chemicals within a stratum can be washed through cracks in the rocks at high temperature; the solution is highly corrosive. Over thousands to millions of years, granite can be broken down into clay and quartz sands. Other minerals, some in solution that pass through the cracks, become crystals as the fluid cools, creating mineral veins such as gold, copper, tin, etc. Volcanoes, although now extinct in the UK, are a source of molten magma which rushes to the Earth's surface creating pyroclastic volcanic rock such as agglomerate. The rock looks like natural concrete where coarse grains of previously cooled rocks have been trapped within a molten volcanic deposit that has cooled to become a mixture of dense rock.

Gravel and sand

Where seawater or fresh water has transported boulders and smaller particles, they tend to drop to the bottom (**drop out**) as the energy of the moving water reduces. The process is noticeable on sandy beaches where the waves from the sea deposit sand as the energy of the waves is spent. Another method of transporting the particles from one area to another is by river. Heavy rain may wash smaller particles into the fast current of the river which transports them. As the current slows the larger particles drop out (**deposit**) followed by the smaller particles. River beds, lagoons or shallow seas may dry up during drought periods lasting years, even millions of years, as the climate changes. Calcite within the water may form a type of cement (similar to the furring up of water pipes or kettles). The result is new stone or rock (**sedimentary rock**, meaning it has been formed on the surface from deposits and fragments of igneous rocks), including many types of sandstone.

Shifting sand

Another method of transporting smaller grains is by the wind. As the grains are picked up by the wind they can be deposited in hills or dunes. On a large scale, the area is known as a desert. Basically there are few life forms to decay and form topsoil so the bare rocks can be scoured by sand in the wind. Areas of Britain show that millions of years ago the land mass was on the equator in extreme heat with little to no surface water. The sands formed vast beds of wind rounded grains that over time have been covered by other rocks or topsoil.

Plate tectonics

The land mass is always moving. The skin of solid rock is frequently cracked where the pressure of the molten core of the Earth pushes up through splits (**faults**). As the mass of solid rocks move they collide, either pushing upwards to form mountains or downwards and back into the molten magma. The global process is termed **plate tectonics**; although the plates move very slowly, earth-quakes and freak waves such as tsunami are evidence that this is happening.

Effects of climate

The Earth is currently in the closing stages of an ice age. The global climate is cyclical and over the 4.6 billion years since it was formed there have been several ice ages. The last ice age that covered the northern hemisphere receded about 10 000 years ago, so the oldest topsoil in the UK must be younger than that. Ice rivers (**glaciers**) formed in the valleys of mountain ranges; some were over a mile thick and slowly moved across the UK shifting boulders many miles. It is within the past 40 years that the last remains of a glacier in North Wales have finally melted. The glaciers ground the bedrocks as they passed over them, forming fine deposits known as **glacial flour**. As the glaciers moved their enormous mass pushed the layers of sedimentary rock into hills and valleys, moved the courses of ancient rivers and deposited new layers of sediment when the ice thawed. For example, the River Thames originally flowed into the North Sea near Clacton-on-Sea. The glaciers pushed the river southward and folded the landscape, depositing sands, gravels and clays. Shallow seas covered most of the south east of England millions of years ago when part of the UK was much closer to the equator and marine life was plentiful. As the billions of sea creatures died over a period of thousands of years their skeletal remains fell to the sea bed. Millions of years later the calcium has bonded with the carbon dioxide naturally found in water to form calcium carbonate (**chalk**, a sedimentary rock): there is a thick layer of chalk beneath much of East Anglia running down into northern France. Over the chalk, layers of peat, coal and the evaporite gypsum have formed, together with isolated layers of sandstone. Iron deposits are also concentrated in the sedimentary rocks. In areas of volcanic activity, sedimentary rocks have been folded under other layers and the subsequent heat and pressure changed the sediment into **metamorphic rock**, examples of which include slate, marble and gneiss. They are generally found in the West Country, parts of Wales, northern England and Scotland.

Now we have the basis for the types of substrata in the UK. For more detail regarding the types of rocks, age groups and locations, the British Geological Survey Ten Mile Map is a useful reference when carrying out a desk top survey.

Water covers a large proportion of the Earth. The seas and oceans are reservoirs for the hydrological cycle. As previously mentioned, the tectonic plates are constantly moving albeit slowly. The effects on the seas and oceans are to circulate minerals, soluble sulfates and salts. Where the climate is hot, the water evaporates, concentrating the solutes, for example in the Dead Sea. Where the sea bed is pushed out of the sea or dries up, the bed becomes encrusted with salt. If the sand is densely packed and the salt crystals fill any voids, an impermeable crust forms as has happened in the North Sea. Beneath the sedimentary layers the remains of fossilised plants and animals which lived in hot boggy conditions millions of years ago, have been trapped. Their remains probably were covered by sandstorms forming a lid which later became part of the shallow sea bed. If they had been on the surface the tissue would have been decayed by bacteria, the fluid would have evaporated and eventually the ultraviolet rays from the Sun would have turned the bones and fibres into dust. Enormous pressure and heat has converted their tissue into oil and gas. Since the early 1970s Britain has drilled into the cavities under the sea and removed the oil. Eventually the massive volume of gas was also seen as a natural resource and for several decades North Sea gas and oil have been recovered.

Under the layers of sedimentary rocks, coal has formed from ancient forests and boglands. Coal, natural gas and oil are known as **fossil fuels**. They all contain carbon as they were all derived from living plants or animals.

4.13   Safe loads in subsoils

Rocks such as limestone, shale and hard solid chalk will take heavy loads; they will also prevent water permeating through so the topsoil is likely to require drainage.

Soft or weathered chalk or thin layers of substratum may have inadequate safe bearing capacity. Chalk and limestone are susceptible to chemical and water erosion. Soft areas or pockets of chalk can be dissolved, creating cavities known as **swallow holes** that eventually collapse and cause the ground to subside. Acidic water can dissolve limestone creating cavities as large as a cathedral known as **caves**.

Dry compact gravel or gravel and sand subsoils are adequate for foundations as they will drain well. However, if the water table is high (i.e. the gravel is submerged), the permissible bearing capacity is halved. Sand, when damp, compacted and uniform, holds together reasonably well and is a satisfactory foundation, especially if it is held in place by steel sheet piling driven in to a firmer substratum below.

Clay is subject to movement for the top 900–1200 mm due to water/moisture changes.

Peat and loose waterlogged sand are very poor subsoils, and specialist advice on the type and depth of foundation will be required.

Building Regulations Approved Document A: 2004 lists seven types of soil plus subsoil conditions and practical field tests to provide guidance regarding soil type (see Tables 4.2 and 4.7). The loading that the substratum can withstand is known as the **bearing capacity**. The presumed bearing capacity of subsoils under static loading is given in Table 4.7.

To enable a greater load to be supported, the width of the foundation or footing can be increased, thus increasing the surface area for support.

Table 4.7 Bearing capacity of subsoils.

Subsoil	Type	Bearing capacity (kN/m²)
Rocks	Strong igneous and gneissic rocks in sound condition	10 000
	Strong limestones and sandstones	4000
	Schists and slates	3000
	Strong shales, mudstones and siltstones	2000
Non-cohesive soils	Dense gravel, dense sand and gravel	>600
	Medium dense gravel, medium dense sand and gravel	<200 to 600
	Loose gravel, loose sand and gravel	<200
	Compact sand	>300
	Medium dense sand	100 to 200
	Loose sand	<100
Cohesive soils	Very stiff boulder clays, hard clays	300 to 600
	Stiff clays	150 to 300
	Firm clays	75 to 150
	Soft clays and silts	<75
	Peat and made ground	To be determined by tests

Crown copyright: Building Regulations Approved Document A: 2004.

Table 4.8 Particle sizes of different soil types.

Soil type	Particle size (mm)
Gravel	2.00 or greater
Sand	2.00–0.06
Silt	0.06–0.002
Clay	Less than 0.002

Note that Tables 4.2 and 4.7 should be used as a **guide only**. It is *essential* to carry out physical soil tests on site (see Section 4.7). Where ground is found to be very soft clay, peat, silt or made up, specialist advice will be required. To identify the smaller soil sizes sieves should be used and reference made to Table 4.8. Basically, there should not be any made ground or wide variation in the type or structure of the substratum within the loading area.

Chapter 5

Walls and openings

This chapter on building elements covers the design of walls and openings, including the detailing for sound and thermal insulation. The scientific rationale for the designs is given, including the movement of sound and heat; the issue of cold bridging and the process for calculating U values will enable the correct type of lintel to be selected. Practical detail is included which can be referenced to building and structural surveying. The chapter concludes with the basics of fire insulation.

5.1 Wall types

There are two main groups of walls:

- load bearing
- non-load bearing walls.

Load bearing walls support the roof and floors, and may provide stability for other walls. They can be of:

- solid construction
- cavity construction
- composite construction.

Traditional construction

It is difficult to define 'traditional construction' of domestic housing in the United Kingdom. Essentially the architecture has been **vernacular**, which means that it is styled to the location. For example, historically a 'traditional' West Country building was constructed with stone walls and stone or slate roof coverings; the walls tended to be very thick to withstand the cold wet winters. The materials had to be relatively local due to the high cost of transport. In contrast, houses built in Hertfordshire at the same time were timber framed, using local oak and elm with **wattle and daub** or **clay brick infill**; the roof coverings were clay tile or thatch. In Essex, the walls were clay brick or timber frame, depending upon the wealth of the owner. Which building method is 'traditional' in the UK? Generally the term 'traditional' means that the walls would be load bearing, supporting upper floors and the roof.

Non-load bearing walls are mainly used as partitions in domestic dwellings to form rooms within the structure (Figure 5.1). However, if a **framed** structure has been used, the external walling could be non-load bearing **curtain walling** or **sheet cladding** as shown in Figures 5.2 and 5.3. Non-load bearing walls do not require a foundation, so the costs are reduced (see Chapter 8).

5.2 Wall design

The design of the wall is influenced by the following factors:

1 function of the wall
2 materials used
3 cost of construction.

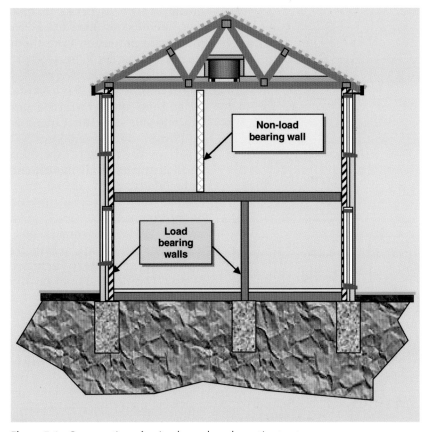

Figure 5.1 Cross section of a simple modern domestic structure.

Wall functions

The functions of the wall could include:

- **weather resistance** – protection against the elements: rain, sun, wind and snow
- **thermal insulation** – reducing the amount of energy required to heat or cool a building
- **sound insulation** – reducing the amount of sound leaving or entering a room
- **fire insulation** – providing a fire barrier and not adding fuel to the fire
- **load bearing** – transferring loads onto lower components
- **stability** – providing support to another wall
- **privacy** – providing areas that regulate or reduce vision
- **security** – preventing easy access or entry by people and animals.

Materials used

The choice of materials has a direct relationship to the functions of the wall. If the wall is to be load bearing, the material will need good compressive

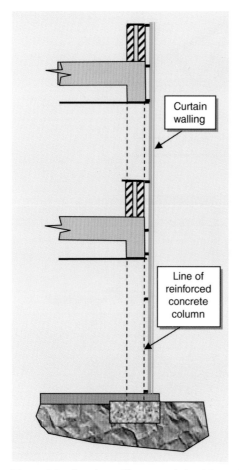

Curtain walling

Line of reinforced concrete column

Figure 5.2 Curtain walling on a skeletal framed building.

qualities, such as bricks or blocks. In contrast, the wall function may not be load bearing, but to provide protection from the weather so glass curtain walling could be selected.

The relationship between selected materials requires consideration. For instance, clay bricks will expand and contract with changes in temperature; concrete and steel also expand and contract, but at different rates or amounts. When several materials are used together the differing rates of expansion (**differential movement**) should be accommodated. As a general rule, clay bricks will expand when heated, so movement joints should be incorporated at every 12 m straight run (Figure 5.4). At returns (corners) of a wall, the movement joint should be 6 m from the end. The principle is based on ability to move. A straight run of wall 12 m long should be allowed **longitudinal movement** along its full length, whereas a wall that cannot move at one end, such as at the corner, will have about 50% less allowable movement. Figure 5.5 shows the comparative movement; at point A the stresses will be pulling at right angles.

In comparison, lightweight concrete (including aircrete) should have movement joints at 6.00 m on straight runs and at 3.00 m from returns (Figure 5.4). Aircrete, in common with most cementitious materials, shrinks as it dries out, so movement joints must allow for contraction. The design should allow materials to move **longitudinally** (along their length), yet must provide stability to both sections of wall. Connections with metal dowels or plates at every other block course allow longitudinal movement whilst preventing lateral movement (Figure 5.6).

Proprietary slip bars are available that consist of a bar within a plastic sheath. The bar *must not* be pushed fully into the sheath as it will prevent the piston like movement. Figure 5.7 shows a cavity wall detail including **compressible caulking**. The caulking strip should be installed as work progresses, ensuring that no mortar 'snots' from the bed joints bridge the gap. If the wall is to be left **face finished** internally (as is common in commercial, industrial and larger buildings), fire resistant mastic jointing should be applied to provide fire integrity to the wall at the joint.

Alternatively, a suitable gun grade mastic joint can be applied at the joint. Where the walls are plastered, **plaster stops** should be detailed on both wall edges and filled with gun grade mastic fronting the compressible caulking.

Movement joints also allow for **drying shrinkage**. An allowance for drying shrinkage should be detailed where concrete blocks have not dried out

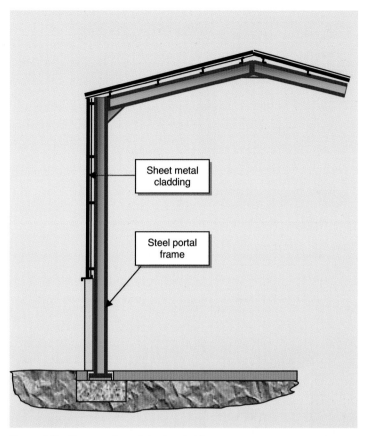

Figure 5.3 Cross section of steel portal frame.

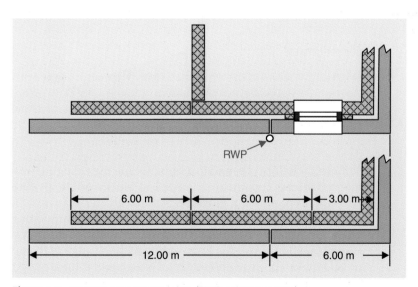

Figure 5.4 Masonry movement joints (RWP, rainwater pipe).

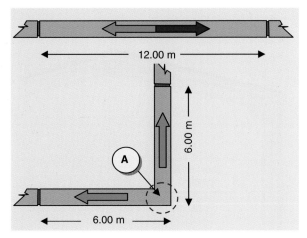

Figure 5.5 Lineal movement in masonry.

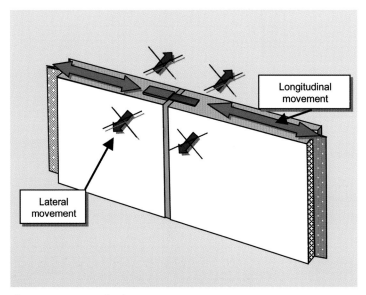

Figure 5.6 Longitudinal movement joint in masonry.

sufficiently before use, and to accommodate the changing water content of the mortar. During construction, inner leaf blockwork to an external cavity wall will commonly have a moisture content greater than 10%; this will drop to about 5% after a year and the wall will shrink as it dries out. Figure 5.8 shows an example of poor detailing. The designer specified the movement joint in the correct position above the lintel in line with the edge of the opening but made no allowance for lintel movement. Consequently the drying shrinkage of the blockwork has created extreme tensile forces that have pulled the lintel bearing and cracked the blockwork beneath. To prevent the problem a slip bearing is

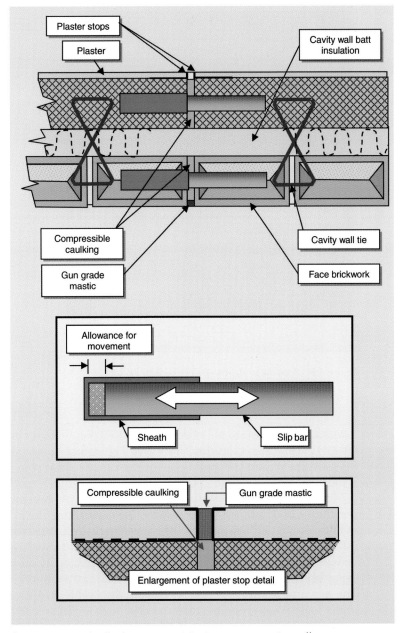

Figure 5.7 Longitudinal movement joint in masonry cavity wall.

detailed comprising two sheets of plastic laminate or similar non-porous dense material placed on the mortar beneath the lintel bearing as shown in Figure 5.9. A compressible caulking board or foam placed at the end of the lintel and finished off with gun grade mastic as shown in Figure 5.10 will allow longitudinal movement.

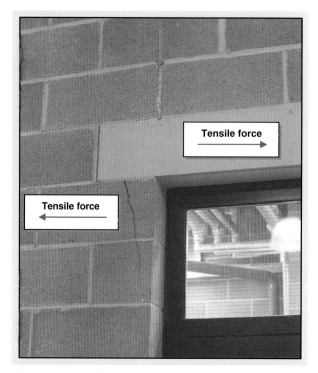

Figure 5.8   Poor detailing resulting in bearing failure.

Movement joints are not normally required below damp proof course (dpc) level as the masonry is less likely to move. Figure 5.11 shows poor detailing with the movement joint continuing below the dpc for three courses of brickwork. If the wall does move the dpc could fail at that point.

Materials such as brick will expand when heated. A movement joint correctly positioned and detailed will allow for the expansion. The bricks will expand in both length and height; therefore where panelled brickwork is designed as infills to skeletal framed buildings a movement joint should be detailed at the ends and top abutment. Figure 5.12 shows a suggested method to overcome the expansion problem at the top of a masonry wall and a reinforced concrete beam. The movement joint will also overcome the problem of the beam deflecting under load. A compressible filler material such as Rockwool will provide a fire stop. Other proprietary fixing methods are available.

Figure 5.13 shows what happens if the wall is built without any allowance for movement. The brick wall has expanded and cracked the reinforced concrete column. Figure 5.14 shows a new movement joint cut into every storey height of a high rise brick clad building. When the building was designed, no allowance for differential movement had been made. Decades later on a cold day the Sun heated one façade of the building causing it to expand. The butterfly wall ties could not hold the masonry skin and large areas of brickwork spalled from the concrete framed building, cascading to the ground.

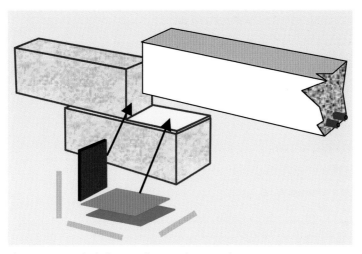

Figure 5.9 Exploded view of correct bearing slip joint.

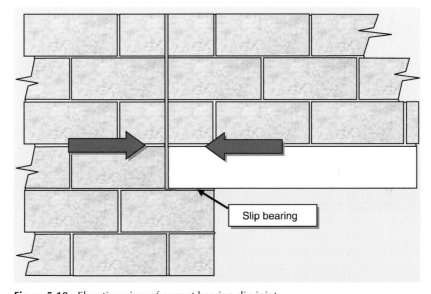

Slip bearing

Figure 5.10 Elevation view of correct bearing slip joint.

Where a wall significantly changes in height or material, a break should be detailed and the walls **tied** together with wall ties or expanded metal. The old method of **toothing** or **bonding** should not be used as there is a likelihood that the wall will crack at the stressed area and break the bond as shown in Figure 5.15. Figure 5.16 shows a movement joint positioned correctly at the point of greatest stress. If the change in height is all new build, expanded metal lathing (ideally in stainless steel) at 450 mm course heights at least 600 mm long should be used (Figure 5.17).

Several proprietary wall starter bars are available for where an extension to an existing wall requires tying. A stainless steel plate is fixed to the existing

Figure 5.11   Incorrect use of a movement joint.

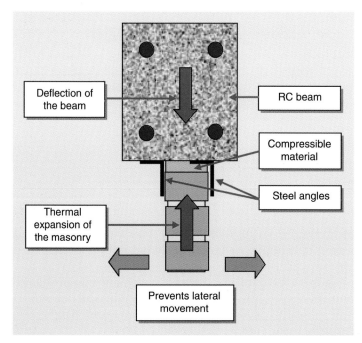

Figure 5.12   Section through a suitable vertical movement joint in masonry.

Figure 5.13 Column shear caused by thermal expansion.

wall using stainless steel coach screws or wood screws. One plate is required per wall leaf (Figure 5.18).

Movement joints can be positioned behind rainwater pipes which are used to mask the joint (Figure 5.19). Alternatively, the joints can be positioned at a reveal utilising the opening as a position for movement, as shown in Figure 5.8.

Designs and stability

Wall design has been a matter of trial and error over the centuries. However, technological advances and an understanding of structural principles now allow modern designers to confidently create fantastic structures.

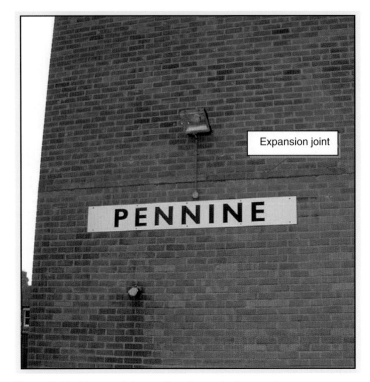

Figure 5.14 Masonry joint to allow for vertical expansion.

Figure 5.15 Differential movement in a masonry wall.

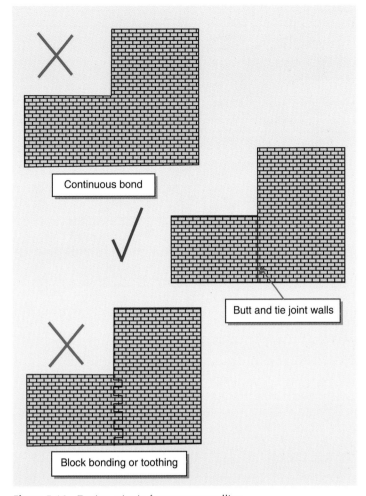

Figure 5.16 Design criteria for masonry walling.

Historic designs

Castle outer walls

These were made of stones of irregular shapes and sizes bedded in clay with straw. The walls tended to be very thick to provide security and stability.

Walls of grand houses, churches and cathedrals

Stone, either of irregular shapes and sizes bedded in clay or lime/sand mortar, or dressed and bonded was used. The walls tended to be thinner as security was no longer a major issue. Many church and cathedral towers fell down and were rebuilt, either thicker or using later techniques. The Normans were fine masons and there are many monuments to their skills still in use today. Buttressing enabled taller thinner walls to be built by stiffening them. Cathedral walls 20 m tall stabilised with solid and **flying buttresses** of limestone were built in Normandy during the 12th century. In the 14th century the limestone

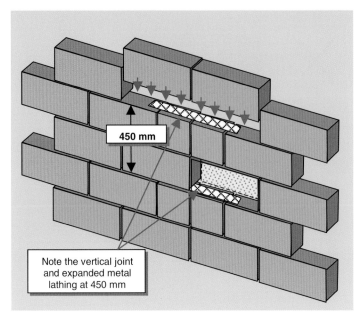

Figure 5.17 Tying a butt joint at change of height in walling.

nave of Westminster Abbey topped 31 m in height and was stiffened only by an array of flying buttresses. Figures 5.20 and 5.21 show the inside of the nave of a 13th century cathedral in Normandy and the flying buttresses that stabilise the walls.

If you try to balance a book on a single sheet of thin card it will collapse as in example A of Figure 5.22. Now if you score and fold a series of small returns as shown in example B the weight of the book can easily be supported, even though the actual material has not changed in thickness. In construction, this is referred to as the **buttressing technique**. Walls that may not directly support floor joists may be providing **lateral** or **buttress** support to the load bearing wall. If the support is removed the load bearing wall may collapse.

Modern designs

Building Regulations Approved Document A provides guidance for the stability of low rise buildings of 'traditional design'. i.e. load bearing domestic masonry walls. Where other forms of construction are being considered, the British Standards and Eurocodes should be used. A structural engineer would normally prepare calculations on behalf of the designer. The stability of a wall will be influenced by its height, length, thickness and how it is connected to other components. For guidance of minimum wall thicknesses look at Table 5.1.

Cavity walls comprise two leaves tied together with wall ties. The ties should be placed at 900 mm centres horizontally and 450 mm vertically at 450 mm offsets (Figure 5.23). Where openings occur, the ties should be spaced no more than 300 mm vertically (it is usual to place a tie between every course of blockwork,

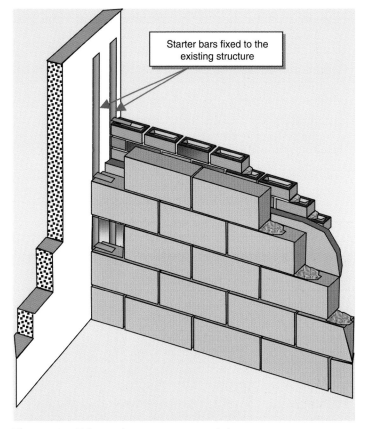

Starter bars fixed to the existing structure

Figure 5.18 Tying cavity masonry to an existing structure.

225 mm) and no more than 225 mm from the opening. The opening is referred to as the **reveal**. Figure 5.24 shows a typical brick/block cavity wall reveal. The wall tie shown is a special tie that holds the insulation in place as the outer leaf is brought up. Cavity walls should be built no more than 1 m in height before the other leaf is built. It is common practice to build the inner leaf first, and then install the cavity insulation which is referred to as **batts** and held in position with special flange washers (Figures 5.25 and 5.26). The outer leaf is then brought up using a gauging rod based on course heights of 75 mm. Figure 5.27 shows a plan view of a reveal with a thermal break cavity **closer** comprising a polypropylene box section filled with dense foam. Compare the detail with Figure 5.24. Note that when using masonry to close the cavity a vertical dpc is required with a 20–25 mm overlap to prevent moisture ingress. In Figure 5.27 the thermal cavity closer does not require an additional dpc.

Figures 5.24–5.26 show the stainless steel wall tie and flange washer. The latest Building Regulations Approved Document A states that zinc galvanised wall ties no longer comply: they rust away within the wall. Simple austenitic stainless steel wall ties should be used of which there are several designs. For high level construction over three storeys specialist fixings are required.

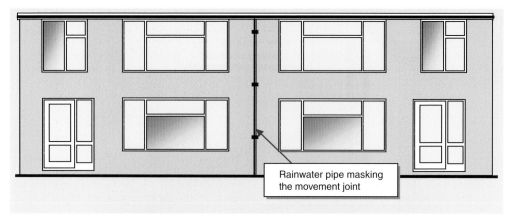

Rainwater pipe masking
the movement joint

Figure 5.19 Masking movement joints.

Figure 5.20 Norman cathedral nave, *ca* thirteenth century.

Figure 5.21 Norman limestone flying buttresses, *ca* thirteenth century.

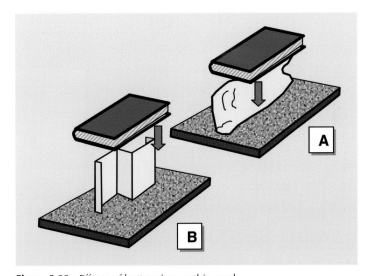

Figure 5.22 Effects of buttressing on thin card.

Mortar

Up to the early 1950s mortar comprised lime and sand. Quicklime would be slaked in a pit dug on site to produce lime putty. (Slaking lime is a process of hydration where water is added to the lime to produce a chemical reaction

Table 5.1 Minimum thickness of certain walls, compartments walls and separating walls.

Height of wall	Length of wall	Minimum thickness of wall
Not exceeding 3.5 m	Not exceeding 12 m	190 mm for the whole of its height
Exceeding 3.5 m but not exceeding 9 m	Not exceeding 9 m	190 mm for the whole of its height
	Exceeding 9 m but not exceeding 12 m	290 mm from the base for the height of one storey and 190 mm for the rest of its height
Exceeding 9 m but not exceeding 12 m	Not exceeding 9 m	290 mm from the base for the of one storey and 190 mm for the rest of its height
	Exceeding 9 m but not exceeding 12 m	290 mm from the base for the height of one storey and 190 mm for the rest of its height

Crown copyright: Building Regulations Approved Document A: 2004, Table 3.

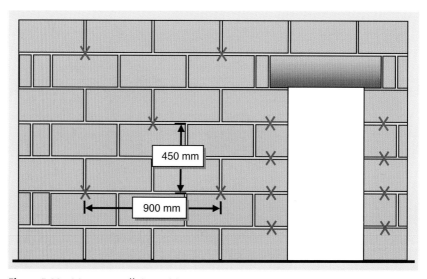

Figure 5.23 Masonry wall tie positions.

before it can be used safely – today **hydrated lime** is used instead.) The lime putty would be mixed with sand to produce a lime mortar. Lime mortar has been used for centuries, and there are millions of buildings still in use today built using lime mortar. During demolition, bricks in lime mortar can be easily cleaned off for salvage. In comparison, modern cement mortar is difficult to

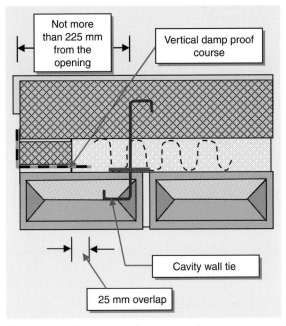

Figure 5.24 Plan section of a cavity wall reveal.

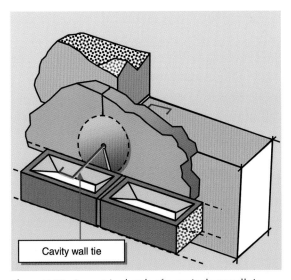

Figure 5.25 Isometric sketch of a cavity batt wall tie.

remove from bricks or blocks which therefore require crushing into masonry hardcore.

Modern mortar above dpc level comprises Portland cement and soft washed sand plus a **plasticiser** or **hydrated lime**. A useful mix would be specified as 1:1:6 mix above dpc level. The mix is a ratio of volumes starting with

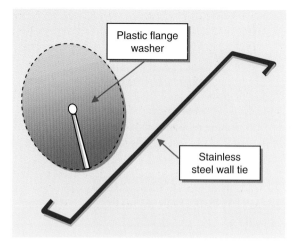

Figure 5.26 Cavity batt wall tie components.

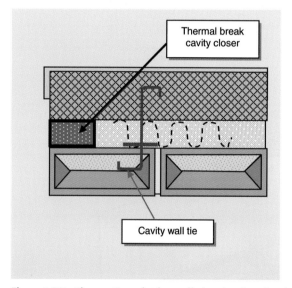

Figure 5.27 Plan section of a thermally insulated cavity closer.

Portland cement, then hydrated lime, followed by sand. Other mixes include 1:5 masonry, where the Portland cement type includes a powdered plasticiser, and five volumes of sand. Both these mixes are suitable above dpc for both bricks and blocks. Using hydrated lime produces a smooth mortar that retains enough water to make it flow and enough suction to help it bond to the masonry. This is known as a **fatty** mix. When set and cured the hydrated lime remains slightly soft in comparison to the Portland cement that once set is rigid. Buildings move as they change in temperature and the substratum compresses slightly: the term **settlement** is used for this period. Mortar comprising just

Portland cement and sand has less elasticity when set and can crack during the settlement period. To overcome some of the problems, masonry reinforcement should be used; a lightweight and heavy duty straight wire or rod versions are shown in Figure 5.28. On a typical dwelling the weakest sections of masonry are above and below openings; therefore continuous runs of reinforcement should be used every 450mm in height and continue past both ends of the openings by about 600 mm (Figure 5.29).

Mortar below dpc level is subjected to more water; therefore lime is not normally recommended. Thermal movement is minimal so a stronger mortar can be used such as 1:4 where there are four volumes of soft washed sand per volume of Portland cement. If there are known sulfates in the ground or groundwater, **sulfate resisting Portland cement** (SRPC) is recommended. Most brick and block manufacturers quote their recommended mortar mixes in their trade literature.

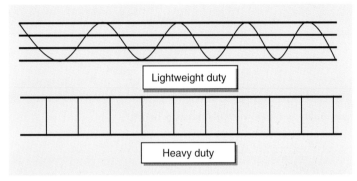

Figure 5.28 Metal masonry reinforcement.

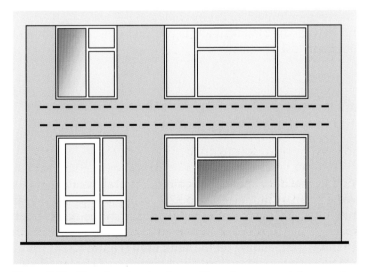

Figure 5.29 Elevation showing the position of masonry reinforcement.

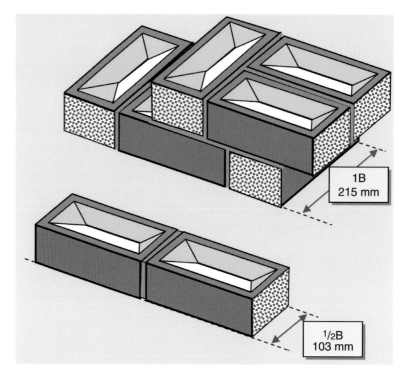

Figure 5.30 Isometric view of brickwork bonds.

Bricks are arranged in specific patterns termed **bonds** to give the wall stability. The main principle is to ensure the perpendicular (upright) joints (**perps)** are not directly above each other. There are several bonds; however, Flemish bond and English bond were more popular with solid walls, and stretcher bond was commonly used for cavity walls and ½B walls. (½B walls are 103 mm thick and 1B walls are 215 mm thick; Figure 5.30.) Figure 5.31 shows three of the most commonly used bonds.

5.3 Openings

Lintels

Window and door openings require support above them. The simplest method uses a beam or lintel (Figure 5.32). A simply supported beam has a bearing at each end that carries the entire load from the lintel. If a brickwork panel (spandrel panel) is being supported, the load will spread outwards between 45° and 60° depending on the bond used (Figure 5.33).

Modern lintels are available in standard modular lengths, usually in 150 mm increments.

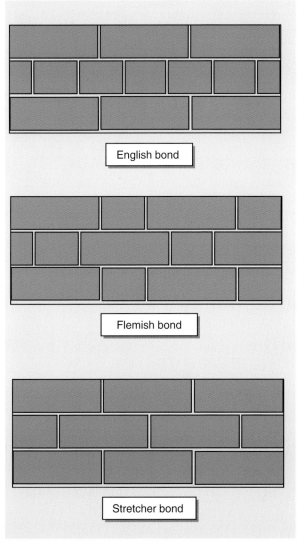

English bond

Flemish bond

Stretcher bond

Figure 5.31 Common brickwork bonds.

Pressed steel lintels

Figure 5.34 shows a range of steel lintels now available.

Inverted tee section lintels

These are available in pressed steel finished in epoxy resin paint to prevent rusting. Where the two sections meet a thermal break of plastic reduces the conduction of heat. Similar designs are also available in austenitic stainless steel for use in more exposed conditions (Figure 5.35).

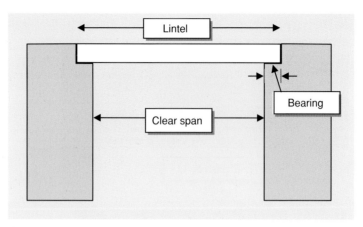

Figure 5.32 Masonry support at openings.

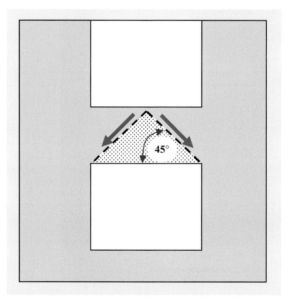

Figure 5.33 Forces acting at an opening.

Top hat section lintels

These are available in mild steel and austenitic stainless steel. Unlike the inverted tee section the top hat is one continuous folded sheet (Figure 5.36). It is debatable whether there is significant transfer of heat via conduction. Most lintels are supplied with a rigid insulation insert to provide additional thermal insulation. It is important to use an additional dpm (damp proof membrane) with this design of lintel. When specifying any type of lintel it is advisable to check that the product has a current British Board of Agrément (BBA) certificate. The certificates can be checked online at: http://www.bbacerts.co.uk/frames.html (Figure 5.37).

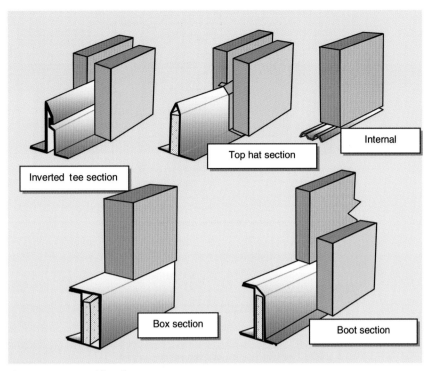

Figure 5.34 Metal lintels.

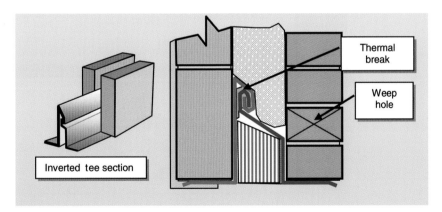

Figure 5.35 Inverted tee steel cavity lintel.

The British Board of Agrément is an independent organisation that carries out tests based on a manufacturer's claims for its products. If, say, a steel lintel manufacturer produced a new design of lintel they would have to comply with all the relevant British Standards and Eurocodes for that product. However, to ensure that the lintel does comply the manufacturer has it independently tested both as a product and how it is intended to be used. For example, the

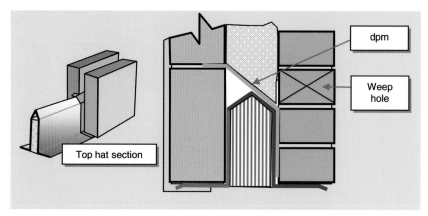

Figure 5.36 Top hat section steel cavity lintel.

Figure 5.37 Electronic version of a British Board of Agrément certificate.

company IG have been producing their famous L1/S range of top hat design lintels for many years. When using the lintel on an external cavity wall a dpm must be used. There should be an overlap of about 150 mm beyond each end of the lintel bearing to ensure the quality of workmanship as stated in the BBA certificate. It should be noted that a BBA certificate does not necessarily

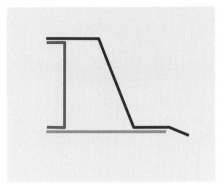

Figure 5.38 Boot style cavity steel lintel.

confirm compliance with all the regulations; for example, thermal insulation and sound insulation as required in the Building Regulations.

Internal lintels

These are intended to support load over internal doorways. The maximum length of lintel is 1·20 m which provides a clear span of 0·90 m having a **safe working load** (SWL) of 7 kN. This is adequate to support a panel of, say, masonry but not loadings from floors, etc.

Boot lintels

Boot lintels are available in both mild steel and stainless steel. Unlike the previously described lintel types, boot lintels are generally three separate steel plates spot welded together to form the boot. Steel lintels are then either hot zinc galvanised, or coated with epoxy paint or resin paint. In the 1970s, lintels required the additional protection of black bituminous paint to cover up scratches that occurred during transport and handling. The current protective finishes are scratch resistant and technically far more durable (Figure 5.38).

Box lintels

These have a similar construction and finish to boot lintels; they are used to support single leaf walls.

Concrete lintels (Figure 5.39)

There are two types:

- precast
- prestressed.

Precast concrete lintels

This type of lintel is factory made in a mould. A steel reinforcement cage is positioned using spacers, and concrete is placed and vibro-compacted to remove air pockets. After a setting period the concrete lintel is removed for further curing before dispatch. The design of the reinforcement varies amongst manufacturers, but generally lintels have a 'top' and a 'bottom' marked on them as the diameter of the rods may differ. Figure 5.40 shows a cage with red rods or bars at the top denoting they are for compressive forces, whereas the blue bars may be of larger diameter for tensile forces.

When a simply supported beam is under load the upper surface will be in compression and the lower surface under tension as shown in Figures 5.41 and 5.42. Through the depth of the lintel the compressive forces diminish to a point halfway down where they develop into tensile forces which increase through to the bottom edge. At the centre line neither compressive nor tensile forces

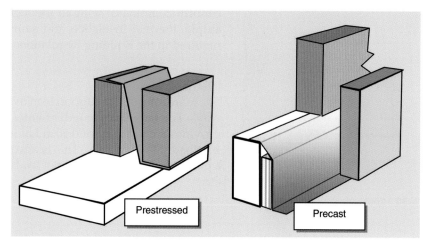

Figure 5.39   Concrete cavity lintels.

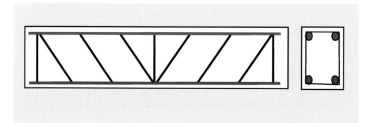

Figure 5.40   Precast concrete lintel steel reinforcement cage.

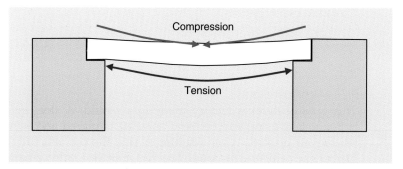

Figure 5.41   Structural behaviour of a lintel or beam.

act, so it is called the **neutral axis**. As previously described, the design of the reinforcement of some concrete lintels means they *should not be used on their side*.

Precast concrete lintels can be used with steel tray lintels for cavity walls. To meet the required U value of $0.35 \, W/m^2 \, K$, rigid insulation may be required between the steel and the dense concrete lintel and insulated plasterboard (Figure 5.43).

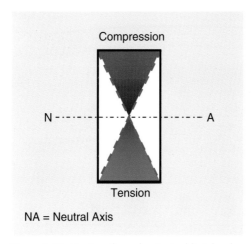

Figure 5.42 Section through a typical lintel or beam.

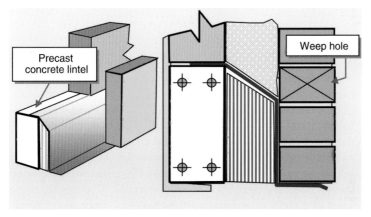

Figure 5.43 Section through a steel and concrete cavity lintel.

Prestressed concrete plank lintels

These lintels are shallow in depth (65 mm) to **course in** with face brickwork. The theory underlying the reinforcement and manufacturing process is very different from that of precast concrete. Reinforcement is by a series of wires (**tendons**) that are placed under very high tensile forces in very long moulds. Their position is in the neutral axis of the lintel. Fresh concrete is then placed and compacted into the mould surrounding the tendons. A low voltage current is then connected to the tendons, significantly speeding up curing of the concrete which is completed in hours instead of days. When the curing has finished the mould is struck (taken apart) and the continuous plank is then cut to standard stock lengths before removal. As the tendons have been stretched, cutting them releases some of the tensile force that becomes a compressive force bending (**deflecting**) the plank as shown in Figure 5.44.

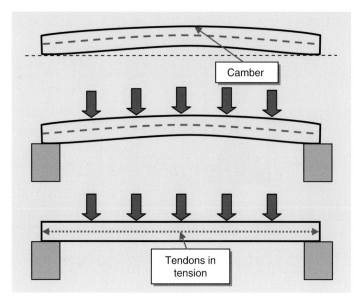

Figure 5.44 Behaviour of a prestressed concrete plank lintel.

The design principle of plank lintels can be demonstrated by the following experiment (Figure 5.45). Using a sheet of polycard (or similar), score a series of grooves to about one-third depth from both surfaces. Place a slightly smaller rubber band around the board and allow it to deflect. The deflection is termed the **camber** and it must always be placed upwards. Note that the grooves will have opened up, indicating the surface is in tension, whilst the other side will close up indicating compression. Place the polycard over two bearings as shown, with the deflection facing upwards. Now gently place a small weight, such as a piece of wood, to act as the uniformly distributed load (UDL) and note the camber is reduced to almost straight. (The UDL would now be creating increased tension on the tendons in a real plank lintel.) If, however, the load is greater than the design safe working load (SWL), the tendons may either slip within the concrete or deflect to a negative camber with subsequent failure of the lintel.

In practice, plank lintels should be supported in the centre before the load is applied. A minimum number of brick or block courses are required to spread the loading from any floor joists, converting the **point loads** into UDLs. A period of curing (normally several days) should be allowed before removing (**striking**) the support.

To comply with thermal regulations (Building Regulation Approved Document L1), thermally insulated plasterboard on battens or dabs is recommended. Dense concrete is a good conductor of heat so a potential **cold bridge** may occur (Figure 5.46; see also Chapter 8).

Cast in-situ lintels have generally been superseded by factory made units, although they may still be used where access is limited.

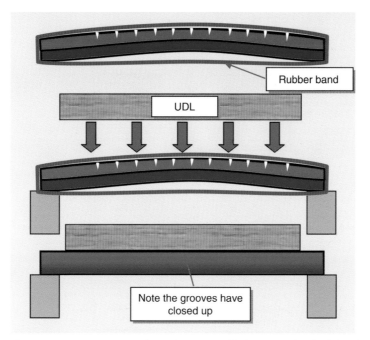

Figure 5.45   Experiment to show the structural behaviour of a plank lintel.

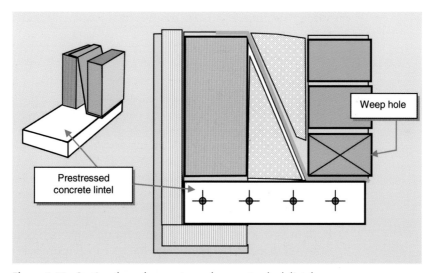

Figure 5.46   Section through a prestressed concrete plank lintel.

Windows

The following are the most frequently used designs:

- side hung casement
- top hung vent

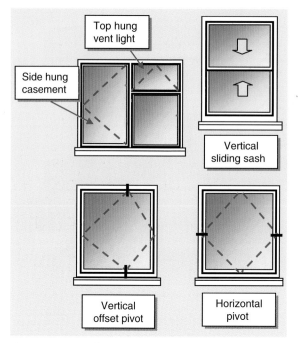

Figure 5.47 Window types and terms.

- fixed light
- vertical sliding sash
- horizontal pivot
- vertical offset pivot.

When preparing drawings a dashed line indicates the position of the hinges or pivots, as in Figure 5.47.

Fixings

There are three main methods of fixing windows:

1. Built in as the wall is erected using frame anchors (Figure 5.48)
2. Screwed through the frame into the reveals (Figure 5.49)
3. Screwed to the reveals using fixing lugs (Figure 5.50).

Materials

Window frames can be made in softwood timber, hardwood timber, extruded aluminium alloy, mild steel, PVCu, or composition. Composition window frames comprise softwood frames with an external covering of coloured aluminium, providing excellent thermal insulation and low maintenance.

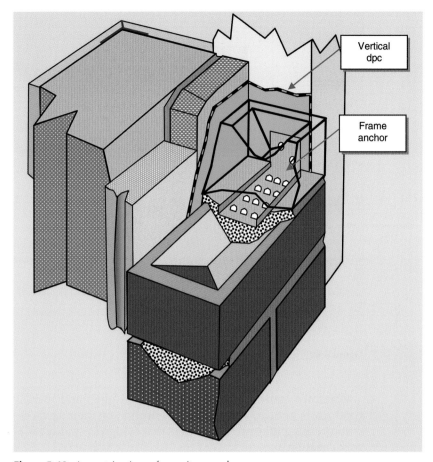

Figure 5.48 Isometric view of a cavity reveal.

Glazing

Single glazing, double glazing, triple glazing and sealed units are available.

U value issues

Glass is a very good conductor of heat energy, which means it is a poor insulator. It is an ideal glazing material in many ways because it will allow easy access of light, is weather resistant, and provides good insulation against fire and sound, plus it can provide security.

The effectiveness of thermal insulation is difficult to see unless X-ray film or special thermal imaging cameras are used; however, one easy tell-tale sign is condensation. Moisture in the air will condense into liquid when in contact with a cold surface. Figure 5.51 shows a damaged sealed unit with condensation between the glass sheets.

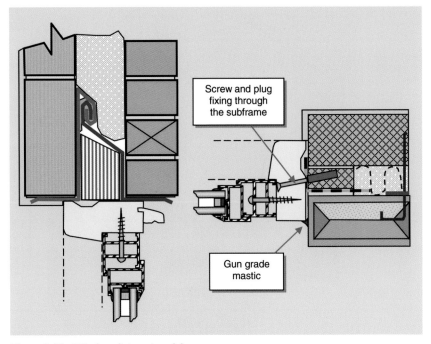

Figure 5.49 Window fixings in subframe.

Windows made from materials such as glass and metal, which are good conductors of heat, will form condensation on their surfaces when their temperature is significantly lower than that of high humidity air. Factors favouring condensation are:

- significant difference in temperature
- high humidity
- good thermal conduction (poor insulation).

Design can help overcome problems other than that of temperature difference which is determined by human comfort. Lowering the humidity levels can be achieved by ventilation. Unless it is raining outside, simply opening a window may help reduce the humidity inside. Venting water vapour from, say, a tumble drier directly outside the building, installing extractor fans in shower/bathrooms, and extraction hoods above cookers, will all reduce the humidity inside a building.

As humans, we produce water vapour all the time, so condensation will form in rooms such as bedrooms, where it is common for the door and windows to be closed for long periods and heating levels to be lower than in the rest of the building. If the walls, floor and ceiling have been adequately insulated to the requirements of Approved Document L1a, the windows must also comply. The choice of glazing and materials is important. Glass, as previously mentioned, is a good thermal conductor, so layers of insulation will need to be incorporated. Air, similar to most other gases, is difficult to heat; therefore trapping a layer of

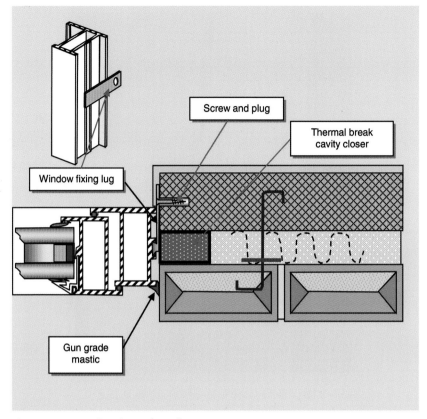

Screw and plug

Thermal break cavity closer

Window fixing lug

Gun grade mastic

Figure 5.50 Window fixing direct fix.

air between two sheets of glass will improve the thermal insulation by about 100% although the insulation is still very poor. Coating one side of the glass will reflect some of the heat: a layer of metal is atomised onto the surface of the glass which is so thin that vision through the glass is unaffected and light penetration is only slightly reduced (by an amount which is only detectable if clear glass and coated glass are placed side by side on white paper). The two sheets of glass can be made into a sealed unit with plastic spacer tube holding the panes of glass together (Figure 5.52). The air trapped between the glass should have a low humidity; silica gel crystals contained in the spacer tube will ensure the air inside the hermetically sealed unit is dry and cannot produce condensation. To improve the insulation qualities, the trapped air can be replaced by an inert gas such as argon. The thermal insulation (**U value**) of single glazing (irrespective of how thick) is $5.7 \, \text{W}/\text{m}^2 \, \text{K}$, double glazing $2.8 \, \text{W}/\text{m}^2 \, \text{K}$ and coated double glazing with argon gas is $1.8 \, \text{W}/\text{m}^2 \, \text{K}$. In comparison with $0.35 \, \text{W}/\text{m}^2 \, \text{K}$ for the wall, even the most efficient double glazing is poor insulation. For further technical information look at the *CIBSE Guide A* published by the Chartered Institution of Building Services Engineers in 2006.

Figure 5.51   Condensation in failed sealed glazing unit.

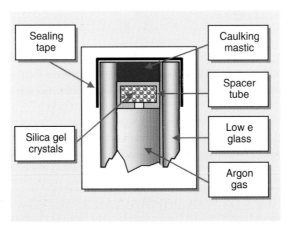

Figure 5.52   Section through a sealed glazing unit.

5.4   Sound insulation

Sound insulation of walls has become an important design factor. Unlike thermal insulation, the slightest fault can be very noticeable. There are two ways

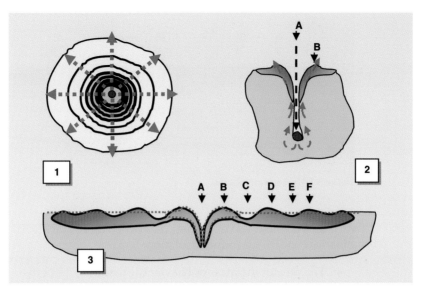

Figure 5.53 Sound waves.

that sound can travel:

1 through the air
2 through structural components (impact sound).

Airborne and structural sounds are interchangeable: for instance, slow heavy vehicles can generate airborne sound that can cause structures to vibrate; someone operating an electrical switch or plugging an electrical appliance into a socket can be heard through a solid masonry party wall which converts structural sound to airborne sound.

Sound can travel through liquids, solids or gases by vibration. Figure 5.53 shows a pebble dropped into water. Note that as water is very dense it does not compress; instead the force causes the displaced water to produce an upward wave before filling the void formed as the pebble sinks. As the action is fast, the overcompensation causes an upward pressure at the water surface point 2B where the water is now higher than that of the bulk water surface, so gravity will pull the water downwards (point 3C). The reaction continues to radiate from the source, diminishing in height as distance from the source increases and causing **waves**. As the forces try to stabilise the peaks of the waves B, D, and F become further apart until the forces are in equilibrium (Figure 5.54). The main forces acting are:

- The initial pebble being pulled downwards by gravity
- Hydraulic forces pushing the water upwards rather than compressing it
- Atmospheric pressure (air pressure)
- Cohesive forces between water molecules (formerly known as surface tension).

In a similar way, sound needs a medium through which it can travel. It cannot travel through a vacuum such as outer space. If the density of a material is not

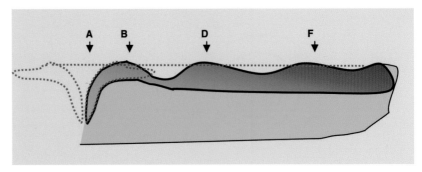

Figure 5.54 Wave amplitude.

regular, the sound waves become dampened more quickly. There are two main theories concerning sound insulation:

- The greater the mass or heavier the material, the better the sound insulation. This is only partly true; for example, a party wall made of dense concrete blocks will reduce airborne noise more effectively than lightweight aircrete blocks purely as a result of its greater mass.
- Soft materials will absorb vibration and therefore reduce airborne and structural noise. An example is when the noise bounces around a room after all the soft furnishings have been removed. Heavy curtains and carpet will reduce unwanted sound by absorption.

The amendments to Building Regulations Approved Document E 2004 provide guidance for maximum sound transmittance. For new build dwellings, airborne sound must not exceed 45 dB through walls and floors, and structural (impact) sound must not exceed 62 dB.

Sound insulation between dwellings should be measured through the party wall. (This is a wall that is shared between two separate dwellings.) The mass of dense concrete blockwork walling, including mortar and plaster, should not be less than 415 kg/m^2. To comply with health and safety requirements for lifting heavy weights, it is recommended that the blocks should be 440 × 215 × 100 mm laid flat to produce a 215 mm thick wall with 110 mm coursing. A block density of at least 1840 kg/m^3 must be used. There is a problem of connecting to other walls with 75 mm or 225 mm coursing (Figure 5.55). Either the mortar joints in the party wall will be thicker by about 3 mm making them 13 mm thick or the walls cannot be bonded together.

Failure resulting from bad workmanship is the most common cause of party walls failing to meet good standards of sound insulation. Where dense blocks have been specified, the bricklayer may **shell bed** the mortar. It enables quicker laying (and a better bonus) if a narrow bead of mortar is run along the edges of the block or brick: this means that the resistance will be far less when the bricklayer taps the brick or block to level it. Figures 5.55 to 5.57 show the effects of shell bedding on standard height and blockwork on its side. The defect can also be found on poor quality brickwork. Figure 5.58 shows a **buttered perpend** and an **empty perpend** where no mortar has been used. To overcome

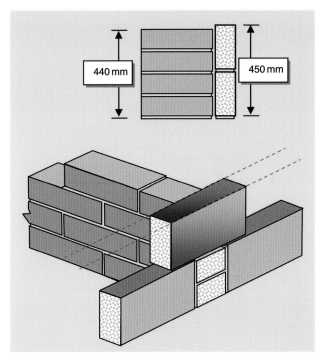

Figure 5.55 Masonry party wall junction.

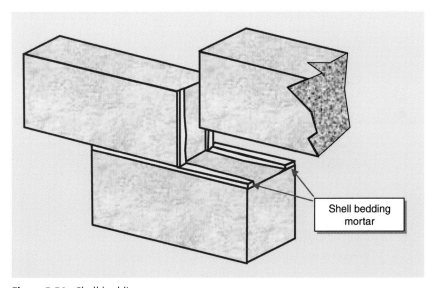

Figure 5.56 Shell bedding.

the problem of possible poor workmanship in cavity party walls, sound insulation **batts** can be used. The batts are commonly made from mineral wool or fibreglass with a greater mass than those used for thermal insulation, so it is important that the correct insulation is installed on site. The design is obviously more expensive, but ensures the structure will meet stringent new sound tests

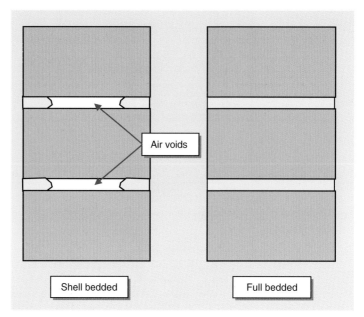

Figure 5.57 Shell bedding comparison.

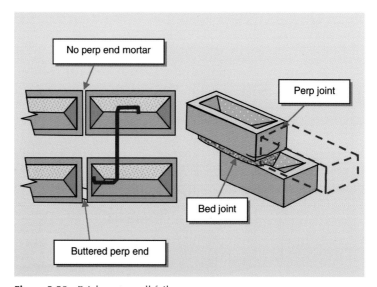

Figure 5.58 Brick party wall failure.

carried out on site. Where clay bricks are being used in a party wall, the sound insulation qualities are reduced if they are laid **frog down** because of the presence of air voids, a point noted in Building Regulations Approved Document E (Figure 5.59).

Poor sound insulation, unlike poor thermal insulation, can be easily detected without technical equipment. Sound travels through taut, dense and rigid ma-

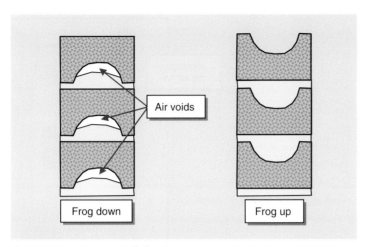

Figure 5.59 Masonry wall density.

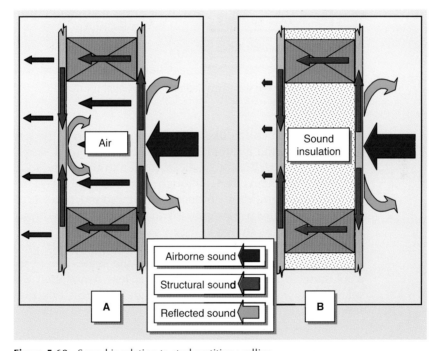

Figure 5.60 Sound insulation to stud partition walling.

terials easily, so using materials such as fibreglass or mineral wool will absorb the sound waves quickly without them bouncing off the wall. If the fibres are packed together relatively densely, the sound waves will be absorbed more effectively. Sound proofed studwork walls can be constructed using dense acoustic glass or mineral wool. Figure 5.60 A shows a timber studwork wall without sound insulation: airborne noise will bounce off the hollow skin of

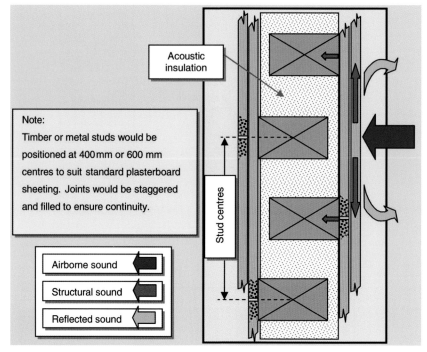

Figure 5.61 Improved sound insulation to stud partition walling.

plasterboard which acts like a diaphragm transferring the sound to the timber studs and into the air void between the plasterboard skins. In contrast, the wall in Figure 5.60 B has acoustic insulation to absorb the sound waves and reduce the diaphragm effect, although sound will still travel through the timber studwork. To significantly reduce structural sound, two sets of studs offset and clad with two skins of plasterboard on both outer faces with acoustic insulation will be required as shown in Figure 5.61.

It is important that the operative uses the correct grade of insulation and does not consider that they are all the same. The fibreglass or mineral wool suitable for thermal insulation is less dense than that for sound insulation, so traps more air within it and is a less effective barrier against sound.

Simple experiments

For the first experiment (Figure 5.62) you will need the following items:

- an old strong cardboard packaging box
- five old telephone directories (sound insulation)
- a wooden board with a hole the size of a keyhole near the middle
- roll of masking tape
- an alarm clock or something that makes a noise.

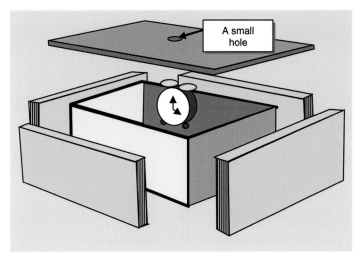

Figure 5.62 Sound testing experiment 1.

Proceed as follows:

1 Tape four of the telephone directories around the four sides of the box, ensuring there are no gaps (you might need to cut the directories to size with a saw).
2 Place the board over the fifth directory and drill a hole about 10 mm in diameter through both.
3 Activate the alarm clock and place it in the box.
4 Place the wooden lid over the box and note the sound.
5 Now plug the hole with your finger and note the reduction in noise.
6 Remove the wooden lid and replace it with the telephone directory with a hole in it. Note whether the sound is as loud in comparison with the board.
7 Now plug the hole and compare the difference.

To develop the equipment try inserting hard surfaces such as hardboard, or irregular soft surfaces like carpet. A hole the size of a keyhole lets sound through, so it is important that the workmanship on site is of suitable quality. Building Regulations Approved Document E: Sound highlights the importance of ensuring the work has been carried out correctly. Testing the construction is now a requirement. Robust is a non-profit making organisation that will advise on and test designs (their address is shown in Amended Document E) by using a sound source in one room and receivers in the others. They can monitor the frequency and pitch of the sound as well as the volume. Higher frequency sounds such as the treble range on home entertainment will travel through air, but are significantly reduced by solids and liquids. The bass sounds will travel through both air and the structure relatively easily. The energy of the waves can vibrate whole buildings: this has become a common cause of complaint from people who live in flats and terraced houses.

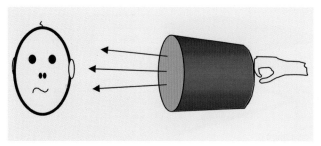

Figure 5.63 Sound testing experiment 2.

Another simple experiment is to aim a plastic wastepaper bin or cup in the direction of an assistant and flick the end of the bin/cup with your finger. The sound will travel mostly in one direction and can be physically felt as an airwave some distance away (Figure 5.63).

5.5 Thermal insulation

External walls should reduce the speed that heat can pass through them. The process is referred to as **thermal insulation** and is measured in **U values**. Generally, the more dense a material the more easily heat will transfer through it; this property is termed **conductivity**. The symbol for conductivity is a lower case letter k; however, manufacturers use the ancient Greek letter lambda (λ) when they quote conductivity. Different materials will conduct heat at different speeds, the slower the speed the more **resistance**, identified by the letter R. Heat energy will be conducted or travel through more dense material more easily than through air and most gases: all materials have an R value. The amount of material that the heat has to go through is termed the **thickness** (t); thickness is expressed in **metres**.

5.6 How to calculate U values

All the information can be entered in a table as shown in Table 5.2. To calculate R, divide the thickness t by the conductivity λ. When all the R values for the materials have been calculated enter the R values for the surface resistance for inside and outside the wall: the figures are a constant at ground level through external walls. To find the surface resistances for other applications look in the *CIBSE Guide A* published by the Chartered Institution of Building Services Engineers in 2006. Total the R values and enter in the box ΣR (the Greek symbol Σ means total or sum of). To calculate the U value, divide ΣR into 1 to find the **reciprocal**. The smaller the number, the better the insulation.

Generally, the denser a material, the more easily heat will transfer through it. However, there are other factors to be considered; for example, the colour of a material or how shiny it is. Heat energy can be reflected by materials such as highly polished aluminium or silvered material. The science becomes very

Table 5.2 U value calculation.

Materials	Thickness (t) (m)	Conductivity (k) or (λ)	Resistance (R)
Brick	0.103	0.84	0.123
Cavity insulation	0.050	0.03	1.667
LD aircrete block	0.100	0.11	0.909
LW plaster	0.013	0.16	0.081
Surface resistance (outer)			0.125
Surface resistance (inner)			0.055
			$\Sigma R = 2.960$
			$1/\Sigma R = 0.337$
			U value = 0.34 W/m^2 K

complex and is beyond the scope of this book. For more information about heat transfer, see Chapter 11.

5.7 Fire insulation

Fire is the process of gas burning (**combustion**). Depending on the temperature, most materials will give off gas if heated. When considering insulation against fire, the purpose or reason will dictate the materials and forms of construction. For example, timber will burn if the temperature is high enough to produce gas. To occur, combustion also needs oxygen. Starve the fire of oxygen and eventually it will stop burning. Some fire extinguishers use carbon dioxide gas which rapidly cools the surface of a material and reduces the oxygen content of the air, stopping the fire.

The reason for designing a building with **fire resistance** is to increase the time available for escape from the building. Walls, for example, may be load bearing; therefore the design will include a time period that the wall must support the structure if the building is on fire. If the fire continues, the wall will eventually fail; therefore domestic dwellings range from 30 minutes' fire resistance to 4 hours' fire resistance, depending on the number of floors and the use of the rooms. For guidance, Building Regulations Approved Document B states the length of time for which specific elements must resist failure: the requirements are the minimum acceptable. There are other factors to be considered, so fire officers should be consulted for their expertise and experience. The subject of fire resistance is very complex, so only the basics are included in this section.

- The Building Regulations are concerned with health and safety of people. Therefore the design of dwellings should enable people to escape from a burning building as safely as practicable, which means the structure must remain intact long enough for them to get out – the **time element**.
- The regulations require that a fire shall not spread to other property very quickly – the **time element**.
- The regulations require minimal emission of smoke, poisonous fumes and particulates – **health considerations**.

Walls that are built of brick will expand in a fire and can cause a structure to collapse, so even though the material is non-combustible the wall may fail. Timber, however, will burn easily. Floor joists, floors, roofs, stairs and timber studwork can all be structural materials; however, if correctly protected they will be excellent in a fire. Larger sections of timber, even if burnt, will form a charcoal layer preventing oxygen or combustible gases leaving the timber. Figure 5.64 shows the relatively small sections of timber that have survived a very severe commercial fire. Note that the steel cladding had melted.

Gypsum plasterboard has excellent thermal insulation and low combustibility; 13 mm thickness will normally provide 30 minutes of fire resistance if it is correctly fixed. All edges of the plasterboard must be supported. If fixed to

Figure 5.64 Fire damage.

ceiling joists or the underside of floor joists, **noggins** 32 mm × 32 mm minimum should be positioned behind all joints and edges to prevent hot gases pushing the boards upward (see Chapter 6, Figure 6.11). Joints between the plasterboard should have a gap of approximately 3 mm to enable plaster filler to seal between the boards. Gypsum plaster coving will add to the protection where the ceiling meets the top of the walls. Fire can travel through an element either by hot gases which will be under pressure finding any gaps or holes, or by heating the element so that the other side produces combustible gases that eventually ignite. Rooms that reach very high temperatures can produce combustible gases from most surfaces that ignite simultaneously causing a **flash over**. Buildings rarely catch fire; commonly it is the contents that ignite such as furnishings, wall coverings, possessions, liquids such as cooking oil, paraffin and alcoholic spirits, and most commonly, cigarettes. Electrical fires have become less common in recent years as technology and materials have improved (see Chapter 12).

Chapter 6

Floors and stairs

This chapter introduces floors and stairs as the third building element. Opening with the functions of a floor, the chapter identifies the types, structural composition and detail. The graphics show the more common solutions to detailing junctions with walls and in relation to the ground and/or foundation. Practical details and comment are included as appropriate, emphasising the need for the design to be practical. Stairs have been included in this section as they are inseparable from floors. This book is not intended as a carpentry textbook; therefore only the fundamentals of regulations, terminology and fixing are addressed.

6.1   Floors

The functions of a floor are to:

- provide a smooth level surface without the risk of condensation or damp penetration
- provide lateral stability to walls
- prevent poisonous gases, liquids or other contaminates from entering the building. This is particularly important where radon gas is known to exist, or the land or surrounding land has previously been contaminated by industrial use or landfill, for example.

There are two main types of floor:

- solid
- suspended.

6.2   Solid floors

Solid ground floors comprise a sub-base commonly formed in consolidated **hardcore** to provide a level drained layer of at least 150 mm thick (Figure 6.1). The hardcore can be broken bricks, concrete, stone or rock. To fill in the voids between the hardcore a 50 mm layer of sand or ash **blinding** should be well compacted and levelled off. The blinding provides a flat layer without sharp points or edges that may puncture the polythene **damp proof membrane** (dpm). It is important that the hardcore and blinding are well compacted to prevent future settlement and the possibility of subsidence of the concrete **oversite**.

The concrete oversite is laid onto the dpm. It should be a minimum of 100 mm thick or ideally 150 mm, and **tamped off** with the edge of a long board. The process of tamping knocks trapped air pockets free and consolidates the concrete. The surface can either be left as a tamped finish or **floated** smooth, as required for a garage or outbuilding.

Figure 6.1 shows a typical sectional detail at a cavity wall junction of a solid floor. Note the dpm has been taken over the edge of the inner leaf and an additional damp proof course (dpc) has been detailed above. When laying polythene dpms the corner detail becomes very awkward; therefore to ensure

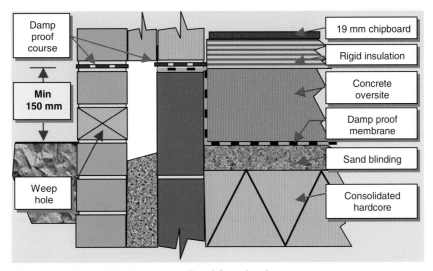

Labels on figure:
- Damp proof course
- Min 150 mm
- Weep hole
- 19 mm chipboard
- Rigid insulation
- Concrete oversite
- Damp proof membrane
- Sand blinding
- Consolidated hardcore

Figure 6.1 Thermal insulation to wall and floor detail.

there are no gaps in the damp barrier an extra dpc should be used above the dpm. When tamping the oversite, the finished level will be taken from the top edge of the block or brickwork. Note the dpc does *not* go across the cavity.

To comply with the current Building Regulations the floor will probably require thermal insulation. This can be achieved with the additional layer of rigid thermal insulation under sheet flooring. Flooring grade chipboard 2440 × 1220 × 19 mm tongue and grooved on all edges should be glued together with polyvinyl acetate (PVA) adhesive. A gap of 10 mm around the floor is required to enable the floor to expand. This is particularly important in room such as bathrooms, shower rooms and kitchens where water may overflow onto the floor. Chipboard expands when it becomes wet, and if saturated it will disintegrate. Therefore the National House Building Council (NHBC) requires all new buildings to have moisture resistant chipboard in rooms where water could come into contact with the floor. The code V313 is used for a flooring grade chipboard that is resistant to moisture. The boards are tinted green for easy identification. From a costing point of view they are more expensive than standard flooring grade chipboard so developers tend to use both board types. The advantage of detailing thermal insulation over the oversite is to reduce the heat loss into the concrete slab. Although relatively low by calculation it is physically noticeable, especially in the colder months of the year, because the concrete slab does not heat up. The insulation can be laid after the building has been **weathered.** (When a building has a covering over the roof and all the window and door openings have been covered, even temporarily with polythene, the building is said to be weathered. The work can proceed internally whatever the weather outside.)

In contrast, Figure 6.2 shows a similar detail with the insulation below the oversite slab and up all external edges. More insulation board is required, and more time is taken cutting it to size and ensuring it stays in place during pouring of the oversite concrete, especially in windy or gusty conditions. The oversite

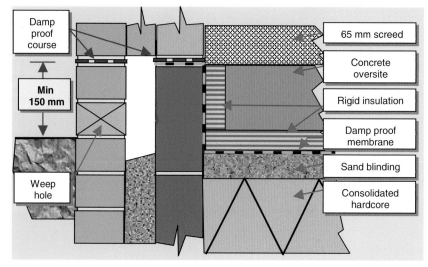

Figure 6.2 Alternative design of thermal insulation to wall and floor detail.

should be finished with a medium tamp which will provide a key for the sand and cement screed. The advantages of screeding the floor are that it is more robust, it can be finished with vinyl sheet flooring, carpet, or ceramic floor tiles; however, tiles have the disadvantage of being difficult to change in line with fashion over the longer period. The sand and cement screed plus the oversite act as a heat sink: in the warmer months this may be advantageous; however, in the cooler months the screed and oversite will absorb heat until they are in equilibrium with the room temperature, so it will feel cold underfoot.

The prevention of dampness from the ground can be achieved using the dpm and dpcs shown in the details. The dpm, where possible, should be in one sheet; however, if joins are required they must have a minimum 150 mm overlap and a self adhesive tape run along the whole join. The minimum gauge polythene is 1000 g; however, it is recommended that 1200 g be used as stated in Building Regulations Approved Document A. The dpc should be lapped at all corners and joints, and bedded on fresh mortar both sides. Where there are known gases such as methane or radon, specialist dpc and dpm products must be used. Manufacturers supply technical data for their installation.

Solid floors have no limit to their area or shape, and are suitable for relatively level ground that is free from ground swell. If pile and beam foundations have been considered to overcome possible ground heave, solid floors are probably unsuitable.

6.3 Suspended ground floors

There are three main types of suspended domestic ground floors:

1 concrete plank
2 block and beam

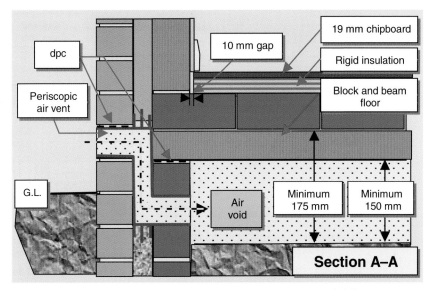

Figure 6.3 Thermal insulation to wall and block and beam suspended floor 1.

3 timber.

For concrete plank floors see Chapter 8, Sections 8.2 and 8.4.

Block and beam

Block and beam floors have the advantage that they can be laid in most weather conditions as, apart from the bedding of the edge blocks and the final grouting, all the materials are dry laid. A series of prestressed concrete 'tee' beams or joists are laid on a continuous dpc spanning between load bearing walls with a maximum span of 6.00 m (Figures 6.3–6.5). Concrete blocks span between the joists, forming a deck or floor. Sand and cement grouting is then brushed into the joints, locking the components together and improving the sound insulation where required (Figure 6.6). A layer of rigid thermal board is laid over the floor deck and topped with flooring grade 19 mm thick chipboard to meet Building Regulations thermal requirements.

To prevent a build up of unwanted gases, the void formed below the floor must be ventilated. Periscopic air vents with a minimum air flow of 1500 mm^2 per metre run of wall are required on opposing walls to allow adequate airflow ventilation. If soil gases are known, there are specialist gas proof membranes that must be fixed according to the manufacturer's recommendations.

In recent years there has been a change in the preparation of the ground beneath a suspended concrete floor. To prevent possible water retention after periods of flooding, the soil beneath the floor should be left uncovered and a weed killer applied. If in the event the area is flooded the water can percolate through the soil and eventually dry out. Previously it was considered that the ground beneath a suspended concrete floor should be concreted to prevent

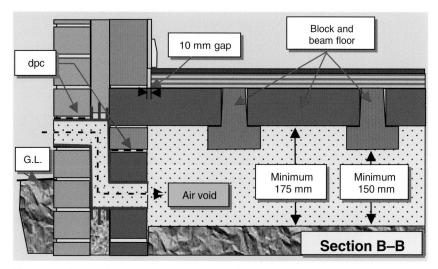

Figure 6.4 Thermal insulation to wall and block and beam suspended floor 2.

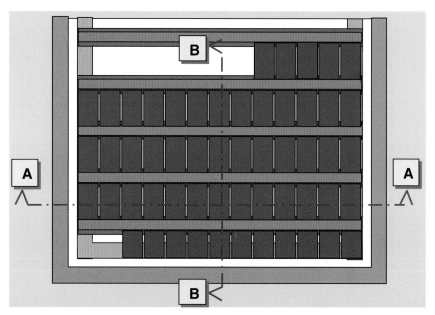

Figure 6.5 Plan view of a block and beam floor.

weed growth. Using periscopic air vents no light can enter below the floor therefore virtually no plant life could develop.

6.4 Suspended timber ground floors

The use of suspended ground floors has become less popular in recent years because of the cost. However, the introduction of composite joists has meant

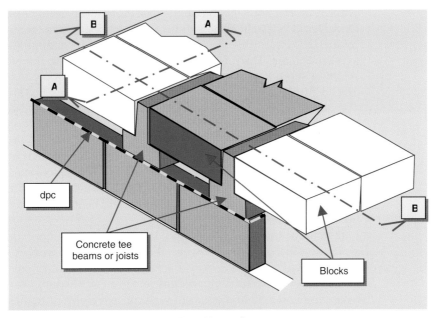

Figure 6.6 Isometric view of a block and beam floor.

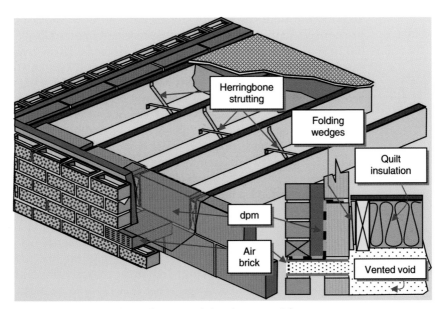

Figure 6.7 Isometric view of a suspended timber ground floor.

that the spans can be increased, therefore reducing installation costs. Figure 6.7 shows a typical suspended timber ground floor with a sectional detail. Unlike the floors of buildings up to the 1950s, modern floor joists are built into the inner leaf, sitting on a dpc. The ends of the joists should be cut square and not protrude into the cavity. Some textbooks suggest that the ends are splay cut to prevent mortar snots laying on the joist ends and possibly bridging the cavity. However, it is rarely done on site; with good site management ensuring the cavities are clean plus the use of full cavity fill insulation, there are unlikely to be problems. The joists may twist as they dry out; therefore herringbone strutting should be used centrally (or solid strutting if preferred) where spans exceed 2.5 m, with two rows equidistant if exceeding 4.5 m span.

When specifying floor joists, state that timbers are to be **regularised**. Where much of the timber is converted at source, it is possible to have lengths of timber that have been sawn on separate machines. Each machine has an upper and lower limit of tolerance, so it is possible to have lengths of timber with various depths and breadths within the permissible limits. Regularising the timber means the sawyer running all the timbers through and reducing them to the minimum depth timber. As the variation will be in millimetres it is normal to use a **thicknesser plane**.

To achieve the required thermal insulation for the floor, quilt insulation is **slung** on plastic netting between the joists. Mineral wool is very easy to install as it will fill any variations between the joists. To prevent the floor from lateral movement, **folding wedges** or tight cut blocking should be used between the end joist and adjacent walling as shown in the detail at the end of any strutting run. The floor deck is commonly 19 mm sheet chipboard tongue and grooved all round; the joints should be glued using a suitable adhesive such as PVA woodwork glue. A gap of 10 mm all round the edge, with the exception of door thresholds, makes allowance for moisture movement. As previously mentioned, V313 moisture resistant floor grade chipboard is essential in wet areas such as bathrooms and kitchens. However, for the minimal extra costs involved, the whole ground floor should be moisture resistant. Fixing sheet chipboard to the joists can be by **lost head nails** or **countersunk screws** at 300 mm centres. Where joints between boards occur, a noggin or strut should be located to provide the required edge support. Unlike traditional floor boards, chipboard flooring is virtually impossible to take up, so ensure all pipes and cables are clear from nails and leaks. On upper floors it is easier to carry out maintenance by accessing through the plasterboard ceiling.

Timber joists have a limited span determined by their depth; therefore **stub walls**, also known as sleeper walls, are required. Unlike suspended concrete floors, the ground beneath the suspended timber floor must be capped with concrete with a minimum thickness of 100 mm. Figure 6.8 shows a section through a typical suspended timber floor, highlighting the position of the dpcs. The ground capping must be higher than the adjacent finished ground, including patios, paths and driveways. Weep holes should be positioned in the base of both the outer leaf and the inner leaf to enable any possible flood water to drain out; to aid drainage, a slight fall to the capping should be introduced. The details are labour taking, so if carried out correctly increase the cost of the floor. There are weep hole inserts that have insect mesh to prevent ants and wasps from entering the floor void. It is recommended that all the floor joists are either pretreated with insect repellent preservatives or that these are applied on site.

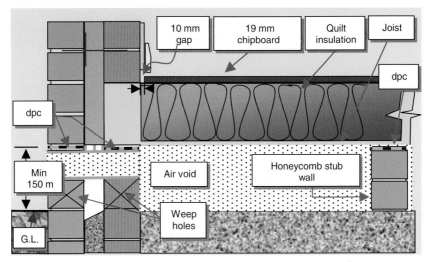

Figure 6.8 Sectional view of a suspended timber ground floor.

6.5 Suspended timber upper floors

Similar in detail to ground floors, the joists would be laid at 400 mm, 450 mm or 600 mm centres to match the sheet floor and ceiling boarding. Strutting requirements are the same, but additional stability is required to the adjacent walls. Galvanised metal straps or austenitic stainless steel straps are hooked over the inner leaf of the cavity wall as shown in Figures 6.9 and 6.10. The 30 mm × 5 mm straps should be long enough to span at least three adjacent joists, so are likely to be 1200 mm or longer. The Building Regulations state that straps must be used where the run of wall exceeds 2.00 m; therefore, where practical, position them over the runs of struts for maximum effect. The joists will need to be notched either top or bottom to enable the straps to remain flush, so if solid strutting has been specified reduce the depth of strut by at least 5 mm before fixing to save grooving later.

It is usual to span the joists onto the shortest span; however, if that means the joist ends will enter a party wall, joist hangers will be required. Ideally, the joists should not span onto a party wall because of structural sound transmittance. The sizing of floor joists can be by calculation or reference to the tables in Building Regulations Approved Document A (see http://www.planningportal.gov.uk/england/professionals/en/1115314110382.html). As a general rule of thumb, upper floor joist depths tend to be about 200 mm, or 225 mm if larger spans are needed. All the floor joists must be to the same depth, otherwise steps in the floor will occur. To overcome the problems of very long spans a steel joist is used as shown in Figure 6.11.

Where the joists are built into the wall as shown in Figure 6.11, air tightness has to be assured to comply with Approved Document L1A. The new

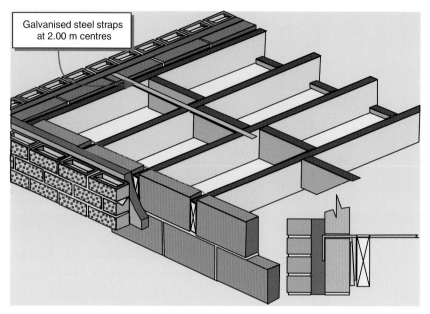

Figure 6.9 Isometric view of a suspended timber upper floor.

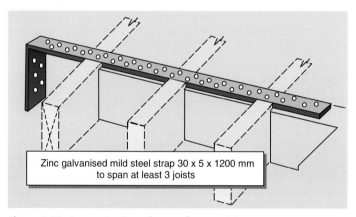

Figure 6.10 Isometric view of upper floor retaining strap.

legislation requires air permeability across the building envelope with a resistance of $50\,N/m^2$. One method to achieve resistance would be to apply flexible seals, such as gun grade mastic, as shown in Figure 6.12. The mortar used to make up course height must be well compacted to prevent shrinkage. The joists will shrink as they dry out during the first year, so a bead of silicone sealant applied when the joists and the mortar are dry should prevent possible gaps.

As an alternative, joist hangers may be used. Galvanised mild steel hangers cradle the end of the joist, and the tab either slots in at a mortar bed joint

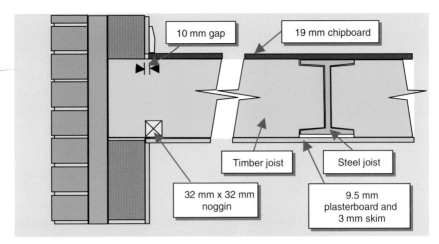

Figure 6.11 Sectional view of built in floor joist and double floor.

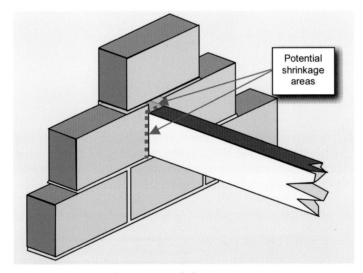

Figure 6.12 Joist ends must be sealed.

or mortise cut into the wall. Figure 6.13 shows a standard joist hanger with holes for location screws or nails to be driven into each side of the joist. Note the notch cut into the lower edge to enable the plasterboard ceiling to continue flush beneath the underside of the joist. Other designs are available that include hook tabs that provide longitudinal connection and stability to the wall. It is more difficult to cut the noggins between the joists when on hangers. Use of hanging joists is advantageous for party wall connection where joists must not penetrate the wall, and connection to an existing structure such as an extension (Figure 6.14).

6.6 Detailing for fire resistance

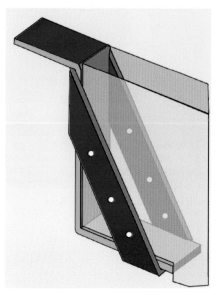

Figure 6.13 Joist hanger.

Timber floors must provide a minimum of 30 minutes' fire resistance: this is achieved by using plasterboard or fire boards at least 13 mm thick nailed or screwed to the underside of the joists (Figure 6.11). Fire causes an increase in pressure within the room; therefore it is important to support all edges of the plasterboard. Where the boards abut the walls, 32 mm square softwood noggins must be fixed. They will provide a suitable support for fixing and, more importantly, prevent pressurised hot gases bending the plasterboard upward at the edge (point A in Figure 7.21, Chapter 7).

6.7 Stairs

The function of stairs is to enable safe access to other levels or floors. Building Regulations Approved Document K provides the minimum statutory requirements regarding safety aspects of design. Basically, if there is a change in height between floors greater than 600 mm, handrails and edge barriers (**containment**)

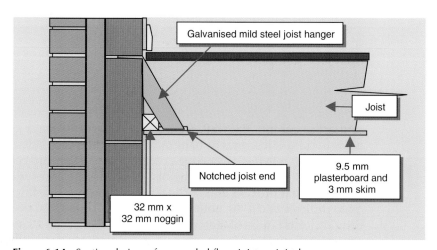

Figure 6.14 Sectional view of suspended floor joist on joist hanger.

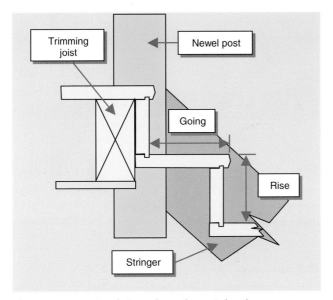

Figure 6.15 Sectional view of wooden stair head.

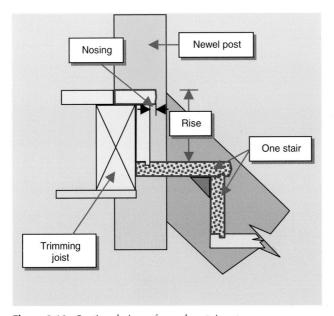

Figure 6.16 Sectional view of wooden stair.

are required to prevent falls. As there are several designs with specific safety criteria, this section has been simplified to the basics.

Ready made flights of stairs are available with standard components; however, if **specials** are required, several good books on joinery are available. The house designer can select a straight run of stairs termed a **flight** comprising twelve **stairs**. Each stair has a **rise** and a **going** (Figure 6.15). The thirteenth

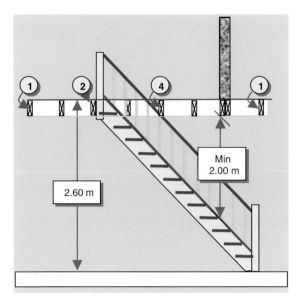

Figure 6.17   Section through timber floor at stairwell.

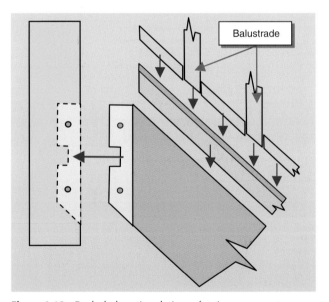

Figure 6.18   Exploded sectional view of stair components.

stair is formed by attaching a rise to the front of the trimmer joist and a nosing fronting the upper floor (Figure 6.16). The storey height is taken from the finished floor to finished floor and is commonly 2.60 m. If the floor joists are 200 mm deep with 19 mm floor decking and 13 mm plasterboard ceiling, the room height from finished floor to ceiling will be 2.37 m (Figure 6.17). To connect the flight of stairs, the stringer can be screwed to an adjacent wall and fixed

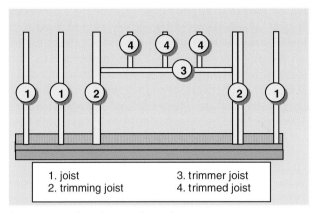

1. joist	3. trimmer joist
2. trimming joist	4. trimmed joist

Figure 6.19 Plan of stairwell joist layout.

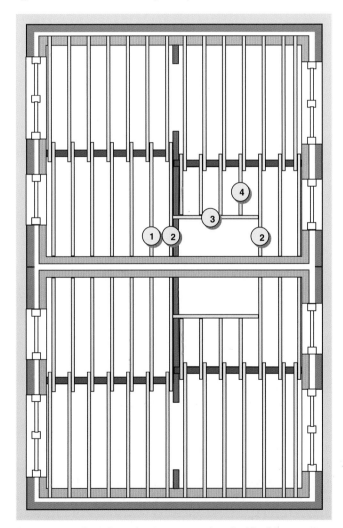

Figure 6.20 Plan of joist layout to semi-detached buildings.

to the trimmer joist and upper newel post using a dowelled double mortise and tenon joint. The newel post should be cut and fixed to the trimming and trimmer joist to provide stability for the handrail and balustrade. With modern precut kit handrails it is a simple task to assemble and fix on site (Figure 6.18).

The floor joist layout in Figures 6.19 and 6.20 shows the stair well cut out formed by the trimming, trimmer and trimmed joists. Floor joists span between load bearing walls and are either built in or on joist hangers as previously described. Where a load bearing partition wall is available, the joists bear down

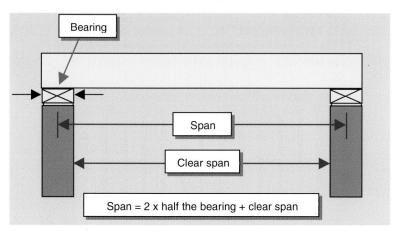

Figure 6.21 Beam and joist terminology.

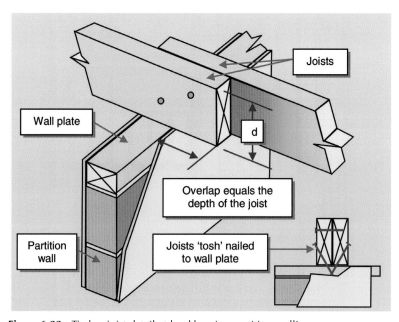

Figure 6.22 Timber joist detail at load bearing partition walling.

onto 100 × 50 mm sawn softwood wall plates bedded on fresh mortar and fixed to the top of the walls. The joists should overhang the wall plate by a minimum of the depth of the joist (at least 200 mm). The first joist nearest the wall should have a gap of about 25–50 mm to allow for any deviation or bowing of the wall and joist, and be packed out with folding wedges or fixing blocks to prevent movement. The joists can be centred at 400, 450 or 600 mm to enable standard sheet materials to be used for the floors and ceilings. If the space between the last joist and the other wall is greater than 50 mm, say 75 mm, another joist will be needed to support the plasterboard and flooring materials.

If the span of the joist is greater than 2.50 m, strutting will be required. Figure 6.20 shows the first floor joist layout for a pair of semi-detached houses. The blue coloured walls are obviously load bearing and require foundations beneath them. However, the green coloured walls are not directly carrying the loading but are acting as buttressing walls. They should not be removed unless they have been checked by a structural engineer who will then make alternative structural arrangements. (Note no internal door openings have been shown for clarity.)

There are two similar terms, **span** and **clear span**. Figure 6.21 shows the definition of clear span as being between the faces of opposing walls and span being half the bearing each end plus the clear span.

Where the first floor joists span onto load bearing partition walls they should be **tosh nailed** into a softwood wall plate as shown in Figure 6.22. Tosh nailing means knocking the nails in at an angle from both sides, as shown in the insert, thus providing a stronger fixing to the wall plate. Another nail is then driven through horizontally for further stability.

Chapter 7

Roofs

This final chapter on building elements describes the changes in roof design over the past few decades. It is important to have an appreciation of the older traditional roofs and materials, as there are millions of roofs that will require attention or replacement over the years.

The design concepts and requirements have changed from the need to provide basic shelter from the elements to include the conservation of heat and fuels. Roofing materials have changed significantly, especially for roof covering. For example, polyester in roofing felt can provide extended serviceable life to flat roofs. This chapter describes design details, including graphics, for both pitched and flat roofs, concentrating on the housing market. Design criteria for roofs include stability and fire related issues.

7.1 Roofs

There are three main groups of roofs:

1 Pitched
2 Flat
3 Others – dome, barrel, hyperbolic paraboloid.

Functions

The functions of the roof include:

- Protection from the elements – sun, rain, snow and wind
- Stability to the walls – without the roof the top of the wall will be weak
- Protection from fire – one of the main hazards during the Great Fire of London was the way in which the fire spread, consuming the roofs first
- Insulation – thermal and sound insulation.

Materials used

Mainly softwood timber is used for the main parts of the structure, with a covering of sheet metal, clay tiles, concrete tiles, stone (such as slate), mineral felt or polyester felt. Also, reeds and straw are used for thatching.

7.2 Pitched roofs

The **pitch** of a roof is the angle measured from the horizontal (Figure 7.1). Anything less than 10° is considered to be flat and anything greater than 80° is considered a wall. Historically, roofs have tended to be steep pitched (greater than 40°) because of the roof covering materials available. Since the 1970s, when Timber Research and Development Association (TRADA) roof trusses (see Figure 7.4) and interlocking tiles became available, the pitch has decreased

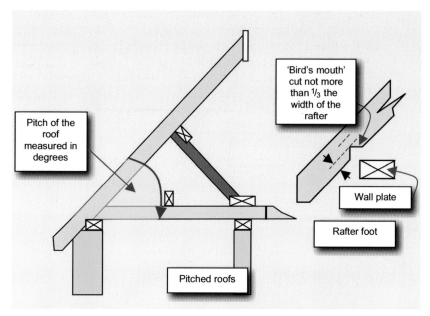

Figure 7.1 Sectional view and terms of a simple cut roof.

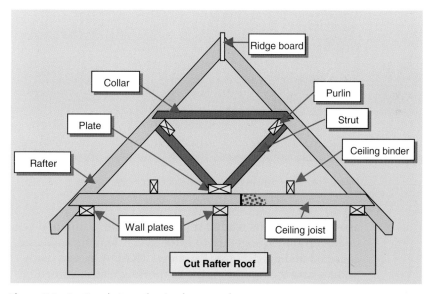

Figure 7.2 Sectional view of a simple cut roof.

to about 30°. The result has been a reduction in the height of roofs, reduced surface areas of roof covering and reduced weight.

Pitched roofs can be **cut**, meaning they are made from lengths of timber cut on site (Figure 7.2), or prefabricated with TRADA roof trusses, for example (Figure 7.3). Prefabricated roofs are factory made and delivered to site in

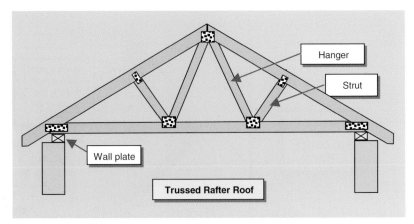

Figure 7.3 Sectional view of a prefabricated fink style truss.

Figure 7.4 Roof trusses banded and stacked on site.

assembled sections ready for erection. There is minimal waste as offcuts can be finger jointed together to form longer lengths and used. Sawdust is also collected and used. Figure 7.4 shows TRADA trusses ready to be erected. The point where the roof meets the supporting wall is termed the **eaves**. Figure 7.5 shows a typical eaves detail supported on a cavity wall construction.

7.3 Ventilation

Pitched roofs and cold deck roofs must be ventilated to remove water vapour in accordance with Building Regulations Approved Document C which requires adequate ventilation to roof voids to prevent interstitial condensation.

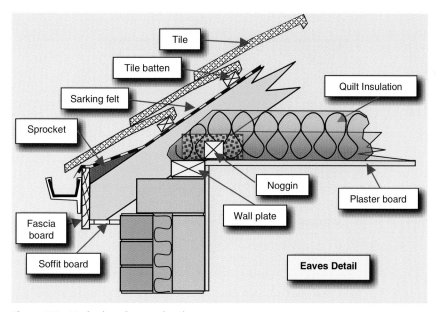

Figure 7.5 Pitched roof eaves detail.

Why do new roofs need ventilation? Starting with old roofs: unless they were on high quality work, roofs rarely had sarking. For example, the roof on a Victorian chapel, church or stately home would have been diagonally boarded with softwood before it was slated or tiled. In contrast, the majority of Victorian dwellings had no sarking beneath the roof covering and were therefore very draughty. However, buildings with diagonal sarking boards were more susceptible to insect attack and fungal growth as a result of lack of air movement. More modern buildings, say since the 1950s, started having sarking felt (thin bituminous roofing felt, lightly sanded to stop it sticking on the roll). The felt could be easily torn and in hot weather it sagged. The roof void would still be draughty enough to remove the water vapour. Loft insulation was rarely used, if at all (see Chapter 10). **Building paper** or **kraft paper**, comprising a brown paper sandwich containing bitumen and nylon strands, became popular for sarking as it was more difficult to tear, remained in place during hot periods and reduced draughts. Water storage tanks located in the loft evaporate large amounts of water vapour during warm periods. The gradual introduction of loft insulation in the mid 1970s reduced the heat loss via the ceiling so the roof void remained cold at night and water vapour condensed on metal gang nail plates and other metal surfaces, resulting in a build up of dampness through condensation.

Figure 7.5 shows a ventilated soffit board with either slots or holes backed with insect mesh to prevent wasps and bees from entering the roof void. As an alternative, the fascia board can be vented, or plastic inserts between the tiles and the fascia board installed. All variations must have insect mesh fitted. To enable the ventilation to work efficiently, high level vents, either at ridge level or via ventilation tiles, should be used. To maintain the 25 mm air passage

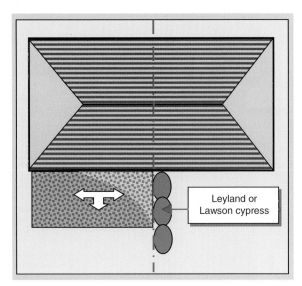

Figure 7.6 Potential tree problems.

between the sarking and quilt loft insulation, plywood boards or proprietary plastic ventilation boards can be used.

Cold deck flat roofs are more difficult to ventilate. Single storey extensions containing high humidity rooms, such as bathrooms and kitchens, are particularly vulnerable; if built near the site boundary the ventilation efficiency could be significantly reduced if cross ventilation has been used and the neighbour of a pair of semi-detached houses, for example, were to plant a row of coniferous trees to mask the single storey extension next door. Leyland and Lawson cypress trees have thick evergreen foliage; they are quick growing (up to 450 mm per year) and can grow up to 10 m tall when mature In Figure 7.6, if the roof design had been based on cross ventilation the shaded area would be susceptible to stagnant and potentially damp air because the trees block the through draught. To overcome the potential problem, a proprietary ventilation system is available for the cold deck roof wall abutment (Figure 7.7). Proprietary plastic vent inserts are available to ensure that efficient through ventilation is maintained. Figure 7.8 shows the eaves detail with soffit ventilation over the quilt insulation on the cold deck roof. Note that if ventilation is used at the wall abutment, the verges do not require ventilation (Figure 7.9). Warm deck roofs do not require ventilation.

7.4   Flat roofs

Any roof that has a pitch under 10° is considered flat. Most flat roofs have a slight fall, which means they are not completely horizontal, to enable rain and thawing snow to run off the roof. There are exceptions; however, they are

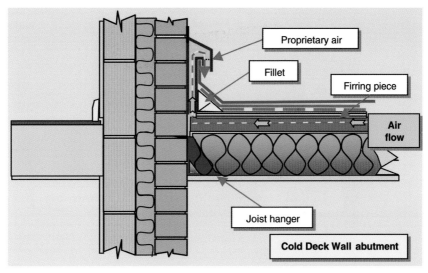

Figure 7.7 New cold deck flat roof abutment to existing cavity wall.

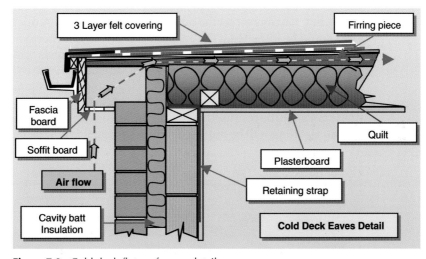

Figure 7.8 Cold deck flat roof eaves detail.

mainly commercial or industrial roofs that use stored water on the roof as a form of protection against thermal movement and ultraviolet (UV) degradation.

Construction details

Softwood timber joists span the shortest direction bearing on softwood timber wall plates 100 × 50 mm, either built into the cavity walling or supported on joist hangers, depending on the application. The details are similar to those shown in Figure 6.14 for suspended timber floor joists. To provide a soffit to

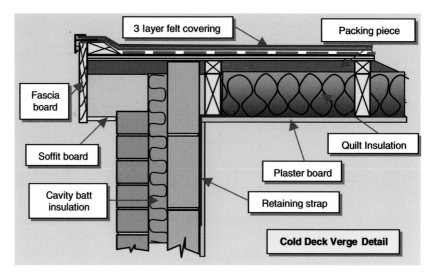

Figure 7.9 Cold deck flat roof verge detail.

adjacent wall junctions, short joists (**stub joists**) are fixed to the side of the end joist as shown in Figures 7.10 and 7.11. The stub joists do not bear onto wall plates, but usually rest on the outer leaf of the masonry. A dpc beneath the stub joists could be included if the location is prone to long periods of dampness or exposure. It is good policy to pre-paint all the stub joists with a minimum of two coats of timber preservative to deter fungal or insect attack. The ends of the exposed joists will also benefit from the same application.

To support the corners of the roof and fascias, **dragon** or corner joists are used (these are not to be confused with dragon beams). If the roof span is greater than 2.50 m, strutting will be required; however, noggins will be needed to support

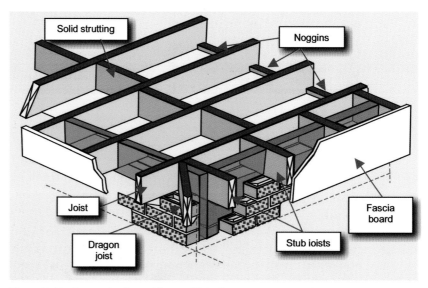

Figure 7.10 Isometric view of flat roof joist layout.

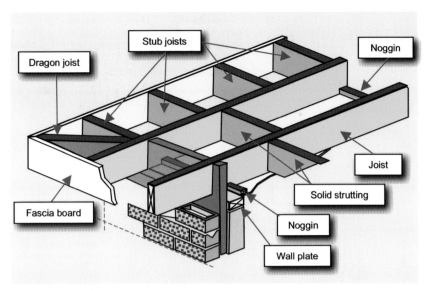

Figure 7.11 Isometric view of flat roof stub joist layout.

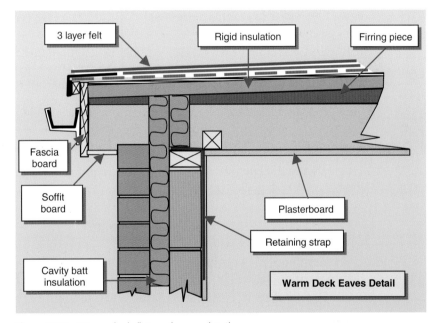

Figure 7.12 Warm deck flat roof eaves detail.

the edges of any roof insulation and decking. When the joists have been fixed, retaining straps should be positioned at not less than 2.00 m centres along the wall plate as shown in Figure 7.12 and to the length of the edge joists as shown in Figure 7.13. Alternative designs of straps are shown in Figure 7.14 to suit different conditions. At this stage the masonry can be built up around the joists to increase the stability of the walls; then additional batt insulation can be installed before the roof decking is fixed. To provide the fall to the roof,

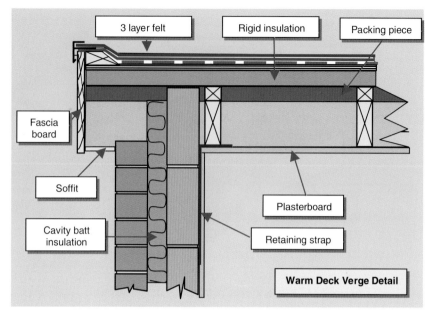

Figure 7.13 Warm deck flat roof verge detail.

firring pieces should be nailed either along the length of the joists or battens at 600 mm centres, or at right angles across the joists. The pitch ratio should be about 1 in 80 which means that for every 80 mm in length the height increases by 1 mm.

Example

Calculate the thickness of the firring pieces required for a joist that is 3.50 m long.

$$\frac{3500\,\text{mm}}{80} = 44\,\text{mm}$$

The minimum thickness for the firring piece is 12 mm; therefore the overall depth of the firring will be 56 mm down to 12 mm (Figure 7.15).

The roof decking is then fixed to the firrings as shown in Figure 7.16.

7.5 Roof terminology

Cut roofs

These comprise rafters as the main framing of the roof. They are cut to size on site and assembled in-situ. Disadvantages include offcuts of timber, and variable quality of timber and workmanship.

Trussed roofs (new type)

These are factory made frames delivered to site ready for erection. There is no waste, they are relatively quick to erect and are quality assured.

Cold deck flat roofs

These have quilt type insulation supported by the ceiling boards. An air gap and ventilation system remove moisture and moisture vapour from the void. Ventilation problems occur, especially when they are installed next to a boundary.

Warm deck flat roofs

These incorporate rigid insulation above the roof deck. A vapour barrier must be detailed to prevent vapour permeating through the insulation and condensing behind the felt covering. The designer may choose a rigid insulation board with an integral vapour barrier comprising silvered plastic foil bonded between the plywood sheathing and the rigid foam layer. A second vapour check is bonded to the underside of the board. However, the manufacturer recommends a 3 mm gap between the board and sealing with a continuous bead of silicone mastic or similar sealant. In practice, many roofing contractors use 1000 g polythene vapour barrier beneath the insulation and butt joint the insulation boards dry with no sealant. The designer may specify foil backed plasterboard; however, the recessed lighting fitments and cabling for switches are not sealed and thus breach the barrier of warm deck details.

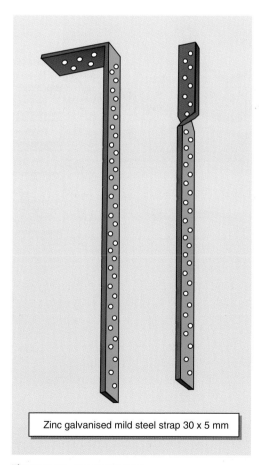

Zinc galvanised mild steel strap 30 x 5 mm

Figure 7.14 Retaining straps.

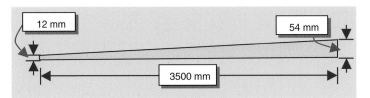

Figure 7.15 Firring pieces.

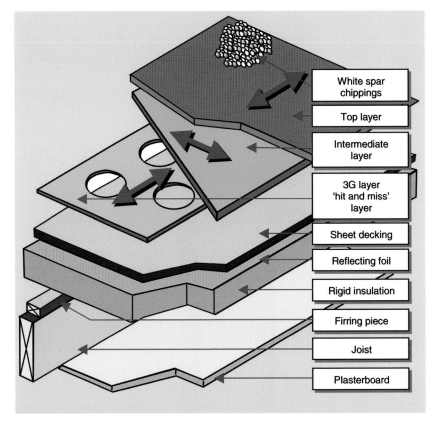

Figure 7.16 Exploded view of a warm deck.

Covering

The most common covering to flat roofs is three layer roofing felt. In recent years the technology has developed longer lasting materials such as polyester. Roofing felt has traditionally been based on thin layers of bitumen bonded at right angles to the layer below with hot applied bitumen. The base layer is fully bonded to the roof decking to prevent the wind lifting it off the roof. To reduce the harmful effects of the sun's UV rays hardening the felt, it is covered with reflective white mineral chips. The other benefit is to reduce the noise from heavy rain by increasing the mass on the roof. During prolonged periods of direct sunlight the roof will expand and therefore the covering will stretch. When the roof cools down the materials should return to their original dimensions. However, after many cycles and the effects of the UV from the Sun the three layer felt will begin to become brittle and crack. If moisture, such as melting frost, thawing snow or rain, enters the crack by capillary action it will eventually penetrate the roof covering and soak into the roof structure. It is particularly difficult to find the origin of a leak on a flat roof as the moisture may travel horizontally some distance before dropping onto the ceiling.

Modern three layer flat roofing systems start with an underlay or first layer of felt that has a series of large holes in it as shown in Figure 7.16. The hot bitumen is applied over the felt, enabling the intermediate layer to be bonded to the underlay and onto the roof deck where the holes are. The underlay is technically referred to as 3G; it works on a basis of 'hit and miss', hence the other name.

Application of 3G flat roof covering

The roof deck should be clean, dry and flat. Drips or aprons should be tacked into position using galvanised clout nails ready for dressing (Figure 7.17). The

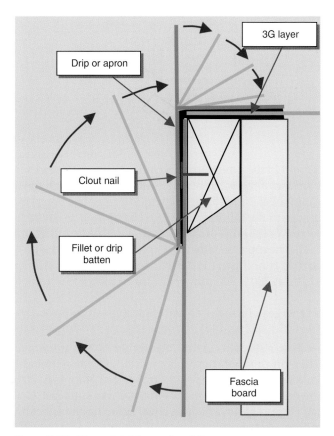

Figure 7.17 Verge detail for roofing felt.

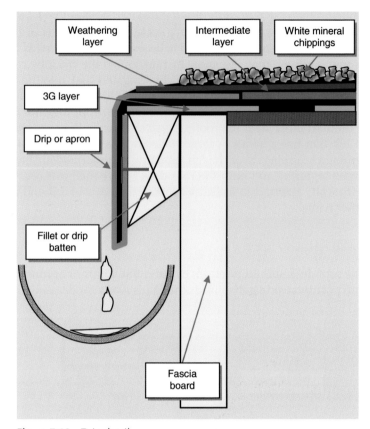

Figure 7.18 Drip detail.

3G is rolled out in the direction of the fall of the roof *dry*. Hot bitumen is applied over the surface of the 3G allowing contact only through the holes – hence the name **hit and miss** – and the next layer (intermediate layer) is applied running at right angles to the fall (Figure 7.16). The aprons are then dressed back with hot bitumen over the edge of the 3G and square butted to the intermediate layer (Figure 7.18). The top or weathering layer is then bonded with hot bitumen over the whole roof, running in the direction of the fall. Hot bitumen is then applied to a band about 200 – 300 mm wide into which white mineral chippings are bedded.

The advantages of using mineral chippings are as follows:

- there is a reduction in the possibility of the mineral chippings being washed off the roof during heavy rainfall; the remainder of the roof can be covered with dry chippings or bonded chippings as required
- white chippings reflect the UV from direct sunlight
- white chippings act as a heat sink, reducing the effect on the roof deck
- chippings reduce the drumming effect of heavy rain by increasing the mass of the covering.

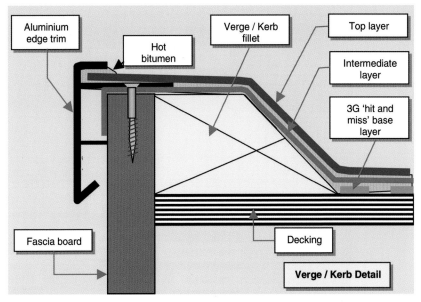

Figure 7.19 Ideal verge detail.

The disadvantages are:

- in certain locations moss will grow, benefiting from the retained moisture held in the limestone chippings
- the chippings can puncture the weathering coat if walked on, by the window cleaner for example, during very hot weather; in that case either leave a bare band next to the window area or provide spreader boards to walk on.

To finish off the edges of the roof (the **verge**) a kerb is formed in softwood as shown in Figure 7.19.

Inverted flat roofs

These roofs have the insulation above the waterproof layer. The principle is to protect the waterproof layers from UV rays from the Sun.

7.6 Stability

In recent decades roof pitches have reduced from 40–45° to 30° or less. Roof technology has enabled lighter roof construction; therefore roofs are now more vulnerable to wind pressure. Roofs with a pitch greater than 15° do not require strapping; however, the extra cost provides peace of mind in storm conditions. Before the requirement for roof restraints, a complete pitched truss roof was blown off a terrace of houses in Milton Keynes during a freak wind.

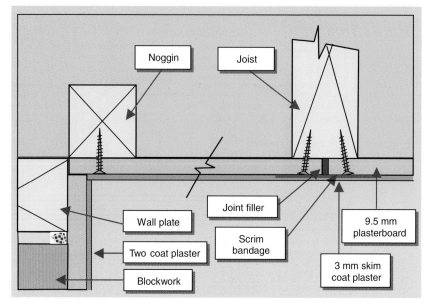

Figure 7.20 Plasterboard fixing detail.

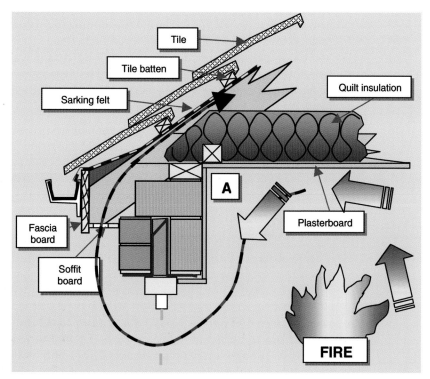

Figure 7.21 Soffit detail fire resistance.

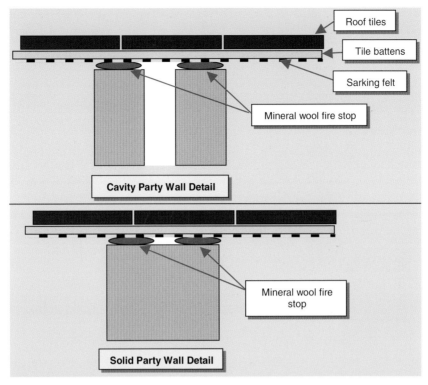

Figure 7.22 Party wall fire stops.

7.7 Fire insulation

The first regulations or building bye-laws involved insulation against the spread of fire. Specifying fire resistant materials in construction that work effectively is difficult; therefore consideration should be given to preventing fire from spreading. For example, if a terraced house is on fire, the main source of combustible material is likely to be the contents. Protecting the combustible structural materials of the building, such as timber joists and roof members, will slow the spread of the fire and destruction of the structural integrity of the building. A layer of 13 mm plasterboard or fire board sheeting to ceilings will provide 30 minutes' fire resistance (Figure 7.20). In most cases the fire brigade will be in attendance within minutes therefore half an hour is usually sufficient resistance. However, pressure from the fire may blow the glass out of the windows and the flames and hot gases can enter the ventilated roof void via the eaves (Figure 7.21). The timbers of a typical modern roof are relatively thin and therefore combust easily. To prevent the hot gases entering the neighbouring loft/roof voids, fire breaks should be built in beneath the sarking felt and the roof battens as shown in Figure 7.22. Proprietary sausage like tubes containing mineral wool should be attached to the top of the party wall, one per leaf or one per side in the case of solid walls, before the sarking felt and battens are fixed. The fire stops will slow the hot gases entering the neighbouring roof.

Chapter 8

Construction techniques

Following the elemental approach in Chapters 4–7, this chapter describes construction techniques used to build the structures. The fundamental concepts of structure are discussed, with relevant explanations of the history and reasons for the design. To help with selection decisions, advantages and disadvantages of the various designs are also listed. Further details relating to the science of specific materials employed with the construction techniques are included. On-site issues are addressed, with practical discussion on the construction process.

8.1 Traditional masonry

This term is used for structures incorporating load bearing masonry walls such as stone, clay bricks or concrete blocks. For more detail see Chapters 4–7.

8.2 Cross wall construction

The original concept started in Saxon times when structures were divided by a masonry cross wall. Eventually, an additional cross wall formed a masonry corridor across the structure where open fires were lit on large **flagstones** (large flat pieces of stone) with masonry around the fire. The new innovation had two main advantages:

- The building was less likely to catch fire as there was stone all round the **fireplace**.
- As the masonry heated from the open fire, it convected heat into the two main rooms. It took about another four centuries before the chimney was developed during the early Tudor period. The Romans used chimneys; however, the technology was lost after they left 'Briton' in 200AD.

Cross wall construction required a **foundation** beneath the two heavy walls. In contrast, the framed panels that made up the main parts of the structure had baulks of elm or oak known as **plates**, either as buried sleepers or placed directly onto the ground, with a layer of stones beneath to aid drainage.

Modern cross wall structures were redeveloped during the 1970s to save on materials and site labour. The principle was to minimise the amount of work on site and reduce the amount of face brickwork. The concept was to build terraced housing utilising factory made concrete plank flooring units spanning up to 6.00 m between load bearing walls (Figure 8.1). The front and rear elevations were basically infills made of softwood frameworks that could be factory made and sent to site as full width panels. The roof design utilised either **boxed beams** or **girder beams** spanning between the load bearing walls and carrying all the weight of the roof (Figure 8.2). Variations on the design included prestressed concrete guttering and eaves beams which were reinforced with steel tendons that have corroded over the years and eventually failed, allowing part of the weight of the roof to bear onto the window frames below.

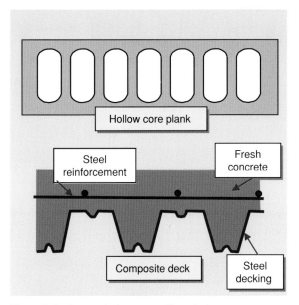

Figure 8.1 Suspended concrete floor design.

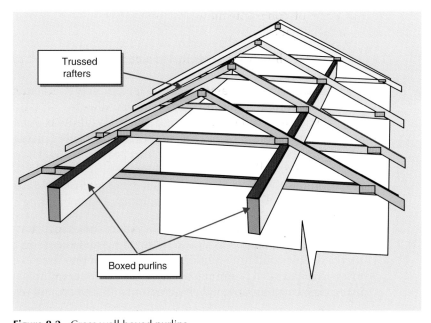

Figure 8.2 Cross wall boxed purlins.

The advantages of cross wall construction are as follows (Figure 8.3):

- Reduced runnage of foundations – foundations are only required beneath the cross walls.

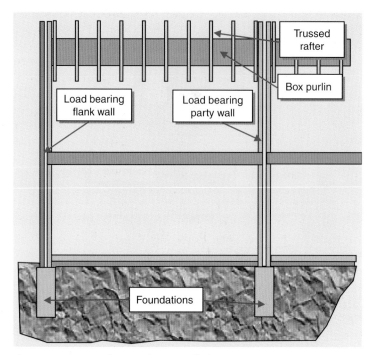

Figure 8.3 Sectional view of cross wall domestic construction.

- Only the cross walls need to be load bearing. All other walls can be non-load bearing infill panels that can be factory made and finished ready for site erection using cranes.
- Modular plank floor units can be factory made and craned into position, spanning between the load bearing cross walls.
- The need for skilled labour and materials on site is reduced.
- Factory made roof trusses supported on factory made box purlins or girder beams span between the load bearing walls. The purlins would be slotted into shoes bolted onto the load bearing walls.
- Window units can be factory glazed and ready to fit directly into the infill panels to the front and rear elevations of each building.

Essentially, cross wall construction is ideally suited to factory processed components, having the advantage of minimal waste generation, as offcuts can be recycled. Softwood timber infill panels can be factory finished, fully insulated, pre-wired and with plumbing as required. The external finish can be tile hung using clay, concrete or wood shingles, or sand and cement rendered.

8.3 Modern timber frame

Timber frame housing has been available on a commercial basis for several decades in Britain. At the height of its popularity, about 21% of the new build housing market became timber frame. Perhaps the most popular multiple house builder was Barratt Homes who built estates of timber frame housing. The structural engineers designed a standard foundation that enabled several alternative house types to be built. The builder carried out the groundwork and constructed a show home where potential buyers could choose their own cosmetic details, such as face brickwork, timber cladding, rendered work, Georgian windows and leaded lights, bathroom colours and kitchen units, etc. The positions of the walls and the roof were standard and were factory made to order. The cosmetics of each house unit would be tailored to the house buyer and could be ready for **hand over** in less than six weeks from the official order. Upon hand over, the contractor transfers ownership to the client. It is important that the client takes out an insurance policy from the agreed hand over date, as the contractor ceases to be responsible for the building from that point.

As the popularity of timber frame increased, vested interests from the masonry companies, especially the concrete block manufacturers, launched campaigns such as 'Be Wise – Build Traditional'. Rumours about timber houses being potential fire hazards were spread in an effort to force the decline of the ever increasing change to timber frame. At the height of the campaign, a television documentary showed bad building practice and how timber frame housing would be difficult to sell and not worth much when the houses were sold to the next buyer. Barratt shares plummeted overnight and the company changed back to traditional building techniques. Decades later, timber frame houses are beginning to increase in popularity again. The technical aspect has changed significantly. Many companies market their developments as being eco friendly – something that is discussed in Chapter 10, Section 10.12.

Advantages are as follows:

- Components are factory made, so there is minimal waste.
- Quality assurance is superior.
- Modular units enable a small range of components to suffice, making factory production more competitive and more mechanised, thus reducing costs.
- Minimal site labour is required as the panels are craned into position and bolted together: typically, the house unit can be weathered in two or three days.
- External finishes can be built onto the timber frame after the roof has been weathered, so internal trades can operate whilst external trades finish.
- The method is ideal for fast track building and not subject to the restraints of inclement weather.
- Thermal insulation can be superior to masonry for the same wall thickness (Figure 8.4).
- Sound insulation can be superior to masonry for the same wall thickness.
- Fire insulation and resistance has to reach the same minimal standards as masonry structures.

Disadvantages of modern timber frame are as follows:

- Significant damage will result if buildings are in a flood area.
- Structures are vulnerable to insect infestation if not correctly sealed.
- Components are susceptible to drying shrinkage if not stored correctly.

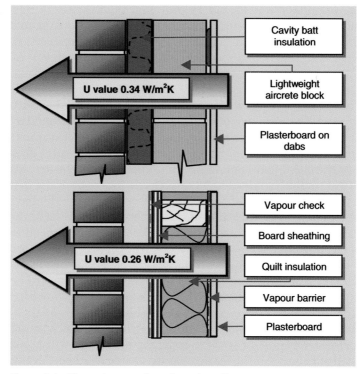

Figure 8.4 Thermal comparison through walls.

- Buildings are more susceptible to fire damage if holes are cut through the walls, e.g. tumble dryer outlet, kitchen hood extractor, air conditioning equipment, after they have been occupied. This is not a problem if the work is carried out correctly using appropriate fire checks and stops.
- Installation of additional power sockets needs specialist knowledge because of the possibilities of vapour barrier damage and insulation disturbance, or they tend to be face fixed.

Modern panel type

Component manufacture

Modern timber frame housing can be made in panel form in a factory environment. The timber can be stored under cover to ensure that the correct moisture content is maintained. Computer guided mills feed straight lengths of timber through multisider machines where the timber is planed smooth on all edges at the same time. This ensures the timbers are all of exactly the same dimensions (**regularised**). The timber is cut to length by computer guided crosscut saws and any small offcuts are sent to the finger cutting machines. The ends of the offcuts are jointed and glued together to produce long lengths of usable timber (Figure 8.5). The sawdust is collected in suction silos for animal use or board manufacture so there is no waste problem.

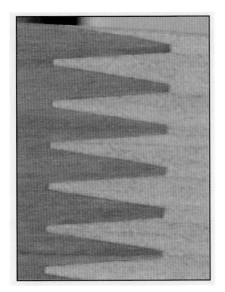

Figure 8.5 Finger joint.

The machine cut component parts are laid out on large benches and clamped into position by pneumatic jaws before the ends are machine nailed together. The board sheathing is then placed over the frames and machine nailed onto the timbers. Some companies use a tough fabric-like material to cover the sheathing. All joints are overlapped with the top lapping over the bottom sheet. This is the **vapour check**. The material is microporous which prevents water droplets passing through whilst permitting water vapour to pass easily; it works in a similar way to the plastic used to cover freshly baked bread through which the steam or vapour can escape easily via minute holes which also prevent any water from passing back onto the loaf.

Some manufacturers install electrical cables and plastic water pipes before filling the void with quilt insulation. Plastic water pipes are ideal for timber frame housing. The pipes are suitable for drinking water (cold) and special layered pipes are used for hot water and central heating runs. They are available in coil form similar to a hosepipe, and can be threaded through joists and studs in a similar way to cables. They have push fit joints which can be opened and re-jointed easily without the need for any heat or special sealants (Figure 8.6).

25 mm water pipes

22 mm water pipes

15 mm water pipes

Electrical cables

Figure 8.6 Timber frame services – water.

Any noggins for cupboard fixings will have been positioned during the initial stage of assembly. Thick polythene sheeting is stapled through material tape to prevent the plastic tearing and forms the **vapour barrier**. (Unlike vapour checks, the barrier is present to prevent vapour passing through and into the timber panels.) The panels are date stamped and coded for easy assembly on site. The floor panels comprise joists and trimmers forming frames. Floor grade chipboard is machine nailed to the floor joists. All the panels are flat stacked and stored under cover ready for delivery.

Roof trusses are made in a similar way to the panels. The joints, however, are held together with two galvanised steel gang nailed plates (Figure 8.7). The truss members are computer guided cut and placed on adjustable benches. An overhead pneumatic jaw press pushes the timber components together and presses the two gang nailed plates into the timbers simultaneously. When all the joints have been finished, the truss is lifted by crane, ensuring that it is not **racked** (twisted), and taken to the storage area. When several trusses have been finished they are banded together to provide greater stability during transport and site offloading (see Chapter 7, Figure 7.4). Larger trusses are made in sections for on-site assembly. During transport the panels and trusses are supposed to be sheeted to keep the timber dry and clean in inclement

Figure 8.7 Timber frame roof connections.

weather. Road spray and incorrect on-site storage can significantly wet the timber, causing it to swell.

Internal wall panels that are load bearing have the studs at close centres. They do not usually need sheathing or vapour checks, etc. Exceptions are bathrooms or en-suites with showers, where vapour barriers are required on the **humid room side only**. The room will require sealing at all the panel joints to prevent moisture vapour accessing the timber. Moisture resistant flooring grade chipboard should also be used in all wet area rooms (this is also a requirement of the National House Building Council: NHBC). Vapour barriers must not be used on the other side of the internal walls in case any moisture does penetrate the timber as the result of damage or accidents. The timber will be able to dry out naturally.

Site preparation

The foundations will be purpose designed to take the timber frame assembly. Some designs incorporate a concrete upstand with cast in steel bolts ready to receive the softwood timber floor plate (Figure 8.8). The timber plate would have a plastic damp proof course (dpc) beneath it to prevent any moisture ingress into the wood. Some designs require the dpc to step down onto a lower level of external cladding to ensure that any water penetration through the external cladding would be directed away from the bottom of the timber frame (Figure 8.9). During erection the ground floor panels are bolted or cleated to a floor plate and at all corners.

The following are some of the problems with earlier timber frames:

- incorrect detailing
- poor assembly on site
- inferior materials

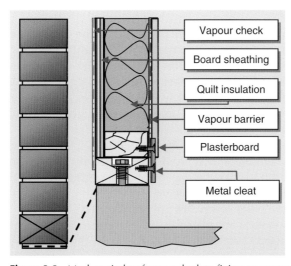

| Vapour check |
| Board sheathing |
| Quilt insulation |
| Vapour barrier |
| Plasterboard |
| Metal cleat |

Figure 8.8 Modern timber frame soleplate fixing on upstand.

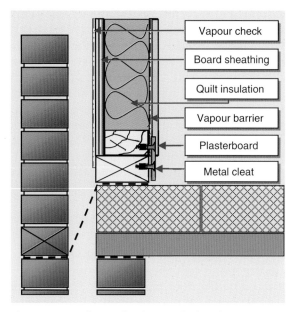

Figure 8.9 Modern timber frame soleplate fixing to concrete block.

- poor quality control and understanding by site erectors and associated tradespeople.

There are still some textbooks showing old detailing, such as a **vapour barrier** on the **cavity side** of the panel. Any moisture vapour from within the structure cannot escape and therefore builds up behind the barrier. In the cooler months the vapour will condense into fluid that will soak into the timber. Any fungal spores present in or on the timber will grow with the added moisture content and, in extreme cases, rot the timber. In contrast, modern systems use impregnated timber that has been chemically poisoned preventing fungi from living on or in the timber. **Vapour checks** are used on the **cavity side** of the panel, allowing vapour to permeate into the cavity. In reality, there should be virtually no water vapour entering the timber frame. The vapour barrier behind the plasterboard should prevent most of the vapour permeating through; however, holes in the barrier for light switches, sockets, water pipes, etc. breach the barrier. From a practical aspect, unless there is high humidity in the room, the amount of vapour breaching the barrier will be minimal.

Plasterboard with a mist coat of plastic emulsion followed by two good coats of full strength emulsion will provide a reasonably effective vapour barrier. A mist coat is a diluted coat of emulsion one part paint to about ten parts water. It soaks into the card facing of the board and when dry provides a thin plastic coat. Plastic emulsion paint comprises polyvinyl acetate (PVA), water and pigment. The ceiling materials are plasterboard, similar to that of the walls; however, there is no vapour barrier behind them, with the exceptions of bathrooms, en-suite showers and kitchens. Vapour can permeate through the holes in the ceiling where cut for electric cables for lights and switches.

Some designers show polythene patches bonded to the plasterboard and sealed with gun grade adhesives. They may reduce or stop vapour from escaping through the hole; however, the practicability of the process being carried out every time on site is doubtful.

Condensation

Water vapour (a gas) is less dense than liquid water because the molecules are further apart: the water molecules comprising gas and liquid are exactly the same. To convert liquid water into water vapour, it is generally boiled or heated to a higher temperature than the air around it. There are exceptions: rain, for example, will evaporate into vapour without boiling or heating; plants evaporate water vapour from their leaves. In neither case has heat been used to change the state of the water from liquid to gas. Saucepans, kettles and other cooking utensils produce water vapour (steam); showers and baths produce large volumes of water vapour which commonly condenses on colder materials such as ceramic tiles, glass and metals. Humans produce approximately 750 ml of water vapour per adult per night; if the room is cool but not ventilated, the vapour will condense and soak into soft materials or form condensation on glass and cold dense materials, leading to problems including fungal growth on wallpaper and soft furnishings, such as curtains and beds, chairs and sofas. If the room is ventilated the vapour will be removed so there are unlikely to be problems.

The following two scenarios illustrate social issues involving water vapour.

Scenario 1

A couple of adults live in a dwelling. Each day they both shower before going to work. The weather is cold and wet. They have the heating timed to go off whilst they are at work and switch on again 30 minutes before they return home. They are security minded and ensure all windows are closed and locked before leaving their dwelling.

The problem

The bath towels they have used are damp and even if they are left to dry over a radiator will only form water vapour which will go into the rooms. The heating will only be on for a short time after they leave for work so the ambient temperature will fall, causing the water vapour to condense. The ceramic tiles in the bathroom are likely to be covered in condensation and the humidity level in the bedroom will still be reasonably high.

The solution

- Humid air should be extracted from the bathroom mechanically using an electric fan whilst the shower is in operation.
- There should be background trickle vents in the heads of the window. If no windows are in the room, there should be a trickle vent through the wall via either a plastic or an aluminium tube; both materials will prevent the

vapour entering the timber framed wall or cavity. All joints must be sealed with aluminium tape, foil tape or silicone mastic. There should be similar trickle vents in the bedroom windows; the vents must be left open. The pressure in the room will be greater than that outside, so the humid vapour should be pushed out of the dwelling. A correctly ventilated dwelling should not need a dehumidifier as there should always be some moisture in the dwelling.

Scenario 2

A young couple have two very young children; one parent stays at home to look after them. The weather is cold and wet. The income for the household is very low and in an effort to keep warm and keep fuel bills to a minimum they have closed all the trickle vents. They cannot afford a tumble drier and dry the washing by hanging it over the radiators. Even when cooking, they never use the extractor fan. The WC is in a separate room so there is no extractor in the bathroom. The temperature in the dwelling tends to be quite high as they feel their home is damp.

The problem

There is a significant volume of water vapour and no ventilation. Increasing the temperature in the dwelling will enable the air to hold more water vapour. The drop in temperature during the evening and overnight will cause large volumes of water to condense on windows and other colder surfaces, and to soak into carpets, wall coverings, curtains and other similar furnishings.

The solution

Look at the cost of drying clothes by other methods. Young children do tend to significantly increase the washing loads.

If the timber framed dwelling has been correctly erected and finished, neither scenario should result in any structural problems. However, in Scenario 2 there is a strong possibility of the high humidity soaking the plasterboard, causing it to swell and possibly become softened. If carrying out a survey on a timber framed house, there is virtually no way of seeing into the timber frame panels and an endoscope would breach the vapour barrier if used. Therefore you should look for social signs when carrying out building surveys; they can often give clues as to whether there is a problem with the building or whether it is a social problem.

Finishing timber framed buildings

Terminology

- plasterboard – gypsum sandwiched between layers of cardboard
- studs – vertical members
- rails – horizontal members
- muntins – small vertical members between horizontals
- noggins – small horizontal members between studs
- plates – full length horizontal members at the top and bottom of a wall

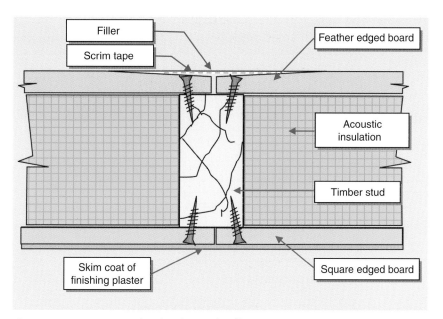

Figure 8.10   Acoustic insulated timber stud walling.

- transoms – members over doors between studs
- jambs – end studs either side of a door or window opening
- tacking – formerly galvanised plasterboard nails; now more commonly screws at 100 mm centres on all edges (Figure 8.10)

Plasterboard is the most common finishing material used internally. The rigid sheets are available in a range of thicknesses and stock sizes that coincide with the centres of studs and rails, etc., which should be based on 400 mm or 600 mm centres (Figure 8.11). Walls generally require 12.5 mm thick boards tacked to the timber studs and rails. All edges of the boards should be supported (Figure 8.10). Although the *ivory* cardboard surface is suitable for applying wall coverings or paint, many designers prefer the plasterboard to be skim coated with a gypsum finishing plaster. The finish is applied to the *grey* side of the board which provides a better bond for the plaster. Plasterboard fixing (**tacking**) is commonly screwed to the stud using self tapping pozidrive screws which commercially come in boxes of 1000 screws or cartridges for use with electric battery operated screwdrivers for speed fixing. The screw system can be used on both metal and timber studwork. Other methods of fixing include electrically operated nailing machines, although these can only be used on timber studwork.

Fixing the boards is dependent on the final finish. Square edge boards are butt jointed and a 3mm thick skim coat of plasterboard finishing plaster is applied. The finish enables wall coverings to be more easily removed, especially when painted high relief papers such as anaglypta or wood chip with several coats of emulsion on them have been applied.

Other designers prefer virtually dry application (the system is known as 'dri-wall' in the USA). Feather edged boards are fixed to the studs with a 3 mm

Figure 8.11   Timber frame floor and walling junction.

gap between them to allow a good key for the filler (Figure 8.10). When they have been **tacked**, the joint is filled with a thick creamy plaster before a **scrim** bandage is applied. The scrim reinforces the wet plaster over the joint and helps prevent cracking between the boards. When the filler has dried, a final filler coat is applied and smoothed level with the faces of the two boards. If applied correctly, there should be a minimum of light sanding before the wall is sealed off as previously described. An alternative scrim is more commonly used. A plastic self adhesive fine mesh is bonded over the joint before any fillers are used. The filler is then applied and finished in one process, saving time. The finished board surface must be sealed with either a polyvinyl acetate (PVA) wash or mist coat and emulsion because the cardboard finish is porous.

Door openings have door linings fixed to the timber jambs and transom; then the skirting boards and architraves are fixed after the plasterboard has been finished. This is referred to as **second fixing** (Figure 8.12). In Building Regulations Approved Document E, internal partitions have been categorised as types A and B. To improve the sound insulation acoustic grade mineral wool or equivalent material should be used. This should not be mistaken for thermal insulation grade mineral wool (see Chapter 5). For further detail the Approved Document can be freely viewed on line at www.planningportal.gov.uk.

The timber frame wall panels carry the imposed loads from the floors and roof plus live and wind loads, so calculations by a structural engineer are *essential* and no modifications should be made without written permission. Openings over windows and doors can be supported by timber joists bolted together (Figure 8.13), or where the spans are wider, steel joists are used

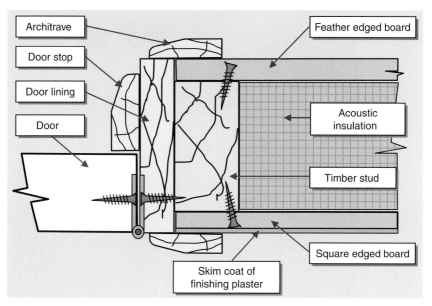

Figure 8.12 Plan section of door lining in a timber stud wall.

(Figure 8.14). New buildings now require a log book providing the user with full information regarding the structure, services and maintenance requirements. Any approved modifications must be included in the official log book.

External finishes are required for weathering purposes. They can take the form of a veneer of brickwork tied to the timber frame with special metal wall ties. The centres will be the same as those used in masonry cavity walls; however, they should coincide with the studs or rails for maximum fixing. Various types are available, including flat stainless steel ties mechanically fixed during the assembly process. The ties are pulled out from the timber stud as the external brickwork is brought up. Other versions include L shaped stainless steel timber frame wall ties that the bricklayer fixes to the panel as work proceeds.

Steel lintels specifically designed for use on timber frame houses are fixed to the timber sheathing and joists; in addition, a plastic dpc should be positioned above the lintel to ensure conformity with the Agrément certificate and manufacturer's instructions, to prevent any water that penetrates the outer leaf bridging onto the timber panels.

When the brick outer veneer is complete, most buyers would not realise that the building was timber frame (Figure 8.15). Alternative outer skins can include rendered blockwork and timber boarded cladding.

Fire precautions

Timber frame housing should have clear cavities on all external walls. The materials are mainly combustible, so the plasterboard covering the inside of every room is not only decorative, but the main fire resistant cladding. If the cladding

Figure 8.13 Timber frame roof and walling junction.

Figure 8.14 Steel lintel and joist in modern timber frame housing.

is breached and hot gases are able to enter the external cavities, a chimney effect will help the fire to spread. Fire stops can contain hot gases in a localised area; stopping the air flow reduces the supply of oxygen which is essential for the fire to continue. Fire stops are available as special non-combustible rolls containing

Figure 8.15 Outer cavity leaf brick veneer to modern timber frame housing.

mineral wool. They should be fixed at all floor levels, ideally below the floor line, at the roof level, and vertically between dwellings at party walls. They are required around all openings such as windows and doors, plus around any pipes or flues that breach the wall: this is especially important around gas appliances, tumble dryer outlets and sanitation outlets. Some designers detail battens with mineral wool to fill any air voids. There are specialist fire collars for large diameter pipes, such as soil pipes and gas appliance flues. In the event of a fire, the collar contents expand and totally constrict the aperture.

The important point made above is that it is the **designer** who has specified precautions to be taken. Even though the designer may have specified fire stops, the on-site workforce must be aware of why they are important and how to fix them properly. Several years ago a pair of semi-detached timber framed houses were seriously damaged by a chip pan fire. The kitchen, obviously, was engulfed in flames and plenty of smoke. The fire brigade took only minutes to arrive, but the fire had already taken significant hold on one house, so sledge hammers were used to break open the cavity walls of the adjoining house in an effort to pump water into the cavities. Hours later the home with the original fire had been very badly damaged; after the fire brigade's report it was demolished and a brick/block masonry house built on the original foundations. Repairs were made to the adjoining property which is still being used. The report concluded that the main reason for such a devastating fire was the absence of fire stops. They had either not been installed or had been installed incorrectly. The fire exploded the glass in the UPVC kitchen window and breached into the clear unprotected cavity. The hot gases produced the necessary heat for the upper floor, second storey walls and roof to burn without the fire brigade being able to access the fire. The house contents provided additional fuel and the heat generated many metres away was significant. *Fire stops are essential and must be fitted correctly.*

Stick type

Commercially, timber frame housing is also available in **stick** form. Frames can either be factory made and delivered with lengths of timber for assembly on site. The method is very common in parts of the USA, although less popular in the UK. The principle involves studs being fixed to the floor plate and a plate nailed to the top to complete the frame (Figure 8.16). Noggins are cut and nailed between the studs providing fixing rails for the plasterboard finish and outer sheathing. Noggins are also positioned ready for wall hanging units such as in the kitchen areas. When the studs (sticks) have been fixed either orientated stranded board (OSB) or plywood boarding is nailed to the outer face of the external walls, providing structural rigidity. When completed and the internal load bearing walls have been fixed, floor joists are fixed to the wall heads adding to the rigidity of the structure. When complete, the next floor is virtually a repeat of the ground floor. American versions observed under construction in California had two storey studs and the floor joists bolted to the inner face. The roof assembly is the same as previously described.

Problems with stick frame construction

Problems observed include the following:

- poor on-site quality control
- more on-site carpenters and machinery, and slower construction time than with panel construction
- saturated timber having thermal insulation pushed in apertures in an irregular way so that some places had no insulation whilst others were highly compressed and therefore less effective
- notches in studs to allow rigid waste pipes to be recessed into the wall substantially weakened localised wall strength.

Infestation

If the structure has not been completely sealed, insects such as wood wasps, wasps, bees and ants can nest in the cavity or within the loft area. This is true

Figure 8.16   American 'stick' timber frame housing partitions.

of masonry structures also. However, an insect new to the UK could arrive within the next decade or so – the termite. As climate change takes place, termite infestation is heading further north from hotter countries. The climate in the UK is becoming annually warmer with fewer hard frosts or long cold spells, thus allowing the insects to survive. According to insect exterminators, in southern France termites are becoming as common as the common furniture beetle: they are, however, far more destructive than common furniture beetles. The questions are, how effective against termites are the chemicals currently used for timber frames and roofs, and for how long do they remain active?

Historical perspective

Softwood timber frames have been used in England for well over five centuries. Figure 8.17 shows a timber framed house in Essex dating back to the early 17th century. There are many examples in Essex and the south eastern counties. Traditionally, the wooden planks nailed to the softwood frames had no insulation and lath and plaster internal finish. The dwellings are termed 'Essex boarded' and could be considered 'traditional' to the county. They should not

Figure 8.17 Early seventeenth century timber framed housing.

be confused with medieval timber framed buildings. During the 15th and 16th centuries, Tudor and Jacobean manorial homes were built using large sections of green oak and elm (Figure 8.18). Many of the buildings have survived the centuries and are now open to the public. As this book is aimed at building in the 21st century, the techniques of the medieval carpenter can only be noted; however, comment and reflection are made in Chapter 10.

8.4   Steel frame

Steel frame is also termed **skeletal frame** and comprises many long steel components bolted together to form a load bearing frame. The upright members are **stanchions** and horizontal members are **beams** or **joists** onto which the floors, roof and walls are supported. The loadings are directed through the stanchions onto isolated foundations, such as pads or piles and caps. The walls are commonly curtain walling literally hanging on the main steel frame, and therefore in many cases negating the need for foundations.

The method of attaching the steel components has changed from riveting to welding or using nuts, bolts and washers. Rolled steel, formerly known as rolled steel joists (RSJs) and now British steel beams (BSBs), is used. The profile has changed, enabling greater strength to be achieved whilst keeping the same

Figure 8.18   Timber framed manor house, *ca* fifteenth century.

overall dimensions. Rolled steel is a more ductile metal than cast steel, having the advantage of bending rather than breaking.

As molten iron cools it forms crystals which tightly pack together in characteristic order. The result is a brittle metal that will react with moisture and oxygen to form oxidation products commonly known as **rust** only on the surface. The crystals are so tightly packed that oxygen and moisture cannot enter between them. Rolled steel is a modified metal termed an **alloy**, meaning that it contains more than one metal, compound or element. For example, steel is formed by adding carbon to iron; however, if the steel is allowed to cool without being worked it will have similar properties to cast iron, as follows:

- it is easily cast into intricate shapes using moulds (commonly made of sand)
- it will only rust on the surface
- it will not bend
- it is brittle
- it can be sounded for cracks by ringing (railway engine wheels were cast steel and would be tapped to check for cracks, similarly to a cup not ringing if it is cracked)
- it can be brazed using another metal.

If the metal is **worked**, such as by hot rolling or cold rolling, the crystalline structure can be modified. Figure 8.19 shows a very simplistic comparison between cast and worked metals or alloy. The cast crystals will not allow

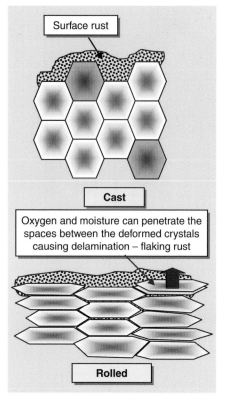

Figure 8.19 Diagrammatic difference between cast and rolled ferrous metal.

moisture or oxygen to enter due to the tightness of the packing. However, as the crystals are aligned in a similar way to holding knuckles together any movement will cause the bond between the crystals to break. If the metal or alloy is hot rolled, the crystals are deformed and elongate in a similar way to grain in timber; this will confer flexibility without fracturing. The metal can be stretched and bent and is therefore ideal for structural components; it is said to be **malleable**. (It should be emphasised that Figure 8.19 is purely diagrammatic; in reality the crystals are very complex.)

Now that the reasons for using rolled steel have been explained, we shall look at how to connect the components. The stanchions are commonly H in section with two flanges and one web. To connect a stanchion to the foundation a base plate or cleats must be prepared either by welding or by bolting angle cleats (Figure 8.20). To connect the horizontal members to the stanchions, angle cleats can be welded or bolted to the stanchions, and the beam or joist bolted through the flange into the cleat, or bolted directly through the end cleat into the stanchion flange or web (Figure 8.21).

There are several methods of constructing floors on steel framed structures. Profiled steel sections formed in thin steel sheet protected either by zinc galvanisation or epoxy paint finish (**decking**) are connected to the steel beams using **shear studs** (Figure 8.22). The steel decking can be used as formwork for cast in-situ concrete where it is left in place as part of the structure. Alternatively, a thick layer of acoustic grade mineral wool with 19mm thick tongue and groove flooring grade chipboard rests on the decking. There are no mechanical fixings between the chipboard and the decking so the floor is termed a **floating floor**. The advantages are a reduction in structural sound transmission. Another alternative used with steel framed structures is precast concrete plank floor panels as shown in Figure 8.1.

Advantages of steel framed structures

These are as follows:

- They can be factory made – all waste can be recycled.
- They can be made whilst site work is carried out, so are ideal for **fast track** building.
- A minimal number of site operatives are required for assembly.

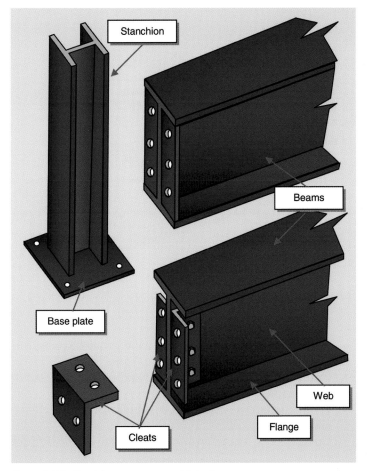

Figure 8.20 Steel frame component junctions.

Figure 8.21 Steel frame office tower under construction.

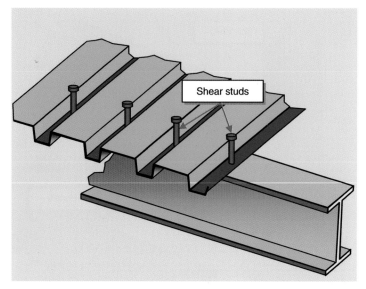

Figure 8.22 Connection between a steel frame joist and steel decking.

- Structures can be assembled quickly – there is no waiting period for materials to gain strength.
- Components can be shot blasted and primed in clean dry factory conditions before they are sent to site.
- Components can be factory bent or fabricated in sections to suit particular shapes.
- Components can form part of the aesthetic design when left exposed.
- Structures can be 100% recycled or possibly reused on another site.

Disadvantages of steel framed structures

There are two disadvantages:

- Steelwork must be protected from rusting.
- Steelwork must be protected from fire.

When steel members cool after milling (the process that forms the profile/shape of the steel (Figure 8.21) a thin film of rust forms known as **mill scale**. The steel must either be oiled to prevent the film forming or be cleaned back to bare steel before a primer paint is applied. If the steel is to be encased in concrete, the rust film can be left; however, if the steel is to be exposed to the elements, a weatherproof paint should be applied.

Steel used internally and exposed can be made fire resistant using specialist intumescent paints that expand to several times their original thickness when heated during a fire. The expanding paint film insulates the steel from further heat, so that it retains strength and stability for 90 minutes in accordance with BS 476.

8.5 Skeletal frame cast in-situ reinforced concrete

Cast in-situ means that the concrete is poured into moulds at the position in which the component will remain – **in-situ**ation. The moulds have different names for different applications:

- Upright concrete members are termed **columns** and the moulds are called **column casings**.
- Horizontal concrete members are termed **beams** and the moulds are called **shuttering** or **formwork**.
- Large flat areas such as floors and the roof are termed **slabs** and would be supported on **decking** with **formwork** around the edges. The edges may be thicker, forming an integral beam. If the beam is above the floor it is termed an **upstand beam** and if below the floor a **downstand beam**. Where the slab has a thicker section it is termed a **rib**.
- **Formwork** and **shuttering** are normally assembled by a carpenter who specialises in concrete moulds, known as a **shuttering carpenter** or **formwork carpenter**.

General requirements of formwork are as follows:

1 It must be strong enough to take the forces exerted by wet concrete, all working loads and vibration.
2 Joints must be tight to prevent loss of fines from the concrete.
3 Where possible, standard forms should be used.
4 It should be designed and constructed so that erection, **stripping** and **striking** are as simple as possible, and all units are of sizes that are easy to handle. (**Stripping** and **striking** are terms used when the formwork or shuttering is taken apart. Where tower cranes on large sites are used, the formwork must be strong enough to be lifted.)
   - It should be designed to a sufficient standard that it can be re-used where necessary.
   - All side forms should be able to be struck before the soffit/decking is removed.
   - It must be able to be removed without damage or waste to the formwork or the hardened concrete.
   - Nailing should be kept to a minimum, but where employed the heads should be left proud or better still double headed nails should be used.

Where multi-storey structures are built, metal formwork is commonly used. The initial cost of the formwork is very high, so many companies hire the components and thus reduce their capital plant costs.

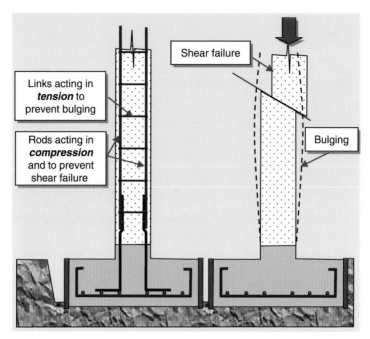

Figure 8.23 RC column design criteria to prevent failure.

Concrete columns comprise mild steel cages of vertical rods held in place by links or hoops tied with wire encased in dense concrete. Concrete is excellent in compression; however, in small section it is vulnerable to shear so the rods provide tensile strength and prevent shear failure or bulging (Figure 8.23).

To connect the columns to the foundations modified pad or pile caps are required. The foundation will require starter bars cast into the concrete onto which the column cages can be tied (Figure 8.24). It is common for the pad to be reinforced with steel rods bent to prevent slippage within the concrete. To ensure the column casings remain in the correct position during the concrete pour, a starter block (**kicker**) is cast as part of the foundation. When cured, the column casing is assembled around the kicker to ensure correct location and prevent **fines** (cement paste and smaller particles of sand) leaking out from the concrete.

Other materials used for moulds include:

- shuttering plywood
- tongue and groove boards
- plastic faced plywood
- steel shutters
- glass reinforced plastic (GRP) shutters
- plastic sheet strengthened with steel mesh.

All joints of the column casing should be either square edged or splay cut so that they are watertight. The surfaces must be kept clean and well oiled to

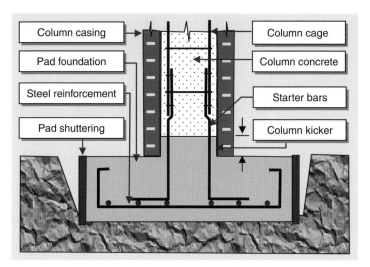

Figure 8.24 RC column to RC pad connection.

facilitate easy striking. This is particularly important where the concrete is to be **fair faced**, meaning that it will not receive any other finish, such as in a multi-storey car park.

For rectangular columns, the forms consist of four panels of plywood, boards or steel, nailed or cleated together. Panels are given strength by vertical studs and held together with column clamps or yokes (Figure 8.25). They should be more closely positioned nearer the base as there are greater forces/pressures. To ensure the columns remain vertical, adjustable steel raking props (**rakers**) can be positioned on at least two sides (Figure 8.26). Note the double handrail scaffold platform providing a safe working area for the concrete workers – not just a ladder which was common practice on site in the past.

Where the column is taller than 3·00 m, it is usual to erect only three sides to the full height and the fourth in 1·20 m stages as the work proceeds to allow access to consolidate the concrete and prevent segregation which may occur when concrete is dropped from a height into a mould. Circular column casings are made in two halves and yokes hold the sections together. Some steel column casings have a flange along both long edges enabling them to be bolted together (Figure 8.27; note the PPE being worn). If the workers should slip out of the working platform they are both wearing full harnesses clipped to the rails. On this particular project the reinforcement bars have fixed threaded sleeves (**female bars**) mechanically fixed to them. The next bar will have a threaded end and will be screwed into the socket, enabling the bar to continue in a straight line. This is in contrast to the more common approach of using cranked bars, overlapping them and tying them with wire (Figure 8.28).

Lapping bars can present problems with the concrete cover or the positioning of the reinforcement in relation to the imposed stresses. For example,

Rebar column reinforcement

Steel bar and wedges yokes

Wooden column casings

Figure 8.25   Timber column casing with steel adjustable yokes.

Steel rod yokes

Adjustable raking prop

Figure 8.26   Timber column casing with steel rod yokes.

Figure 8.27 Steel column casing with access platform.

if the design requirement was for the compressive rods or bars to be, say, 250 mm apart and, say, of 25 mm diameter, lapping would reduce the spacing to 225 mm. To overcome the problem the bars are normally **cranked** as shown in Figure 8.28. The minimum lap should be not less than 15 times the bar diameter or 300 mm, whichever is the greater. In the example the bar is 25 mm diameter so the minimum lap should be 375 mm. The lap dimensions vary according to the location and the imposed stresses; therefore the structural engineer would consider many factors for the design. BS 8110 provides guidance regarding lap requirements.

Where beams are connected to the columns **tie bars** and **continuity bars** are cast into the columns. When the concrete has set the next section of formwork can be attached using the shape of the previous casting as a starter for the next section. Figure 8.29 shows steel formwork being placed against a previously cast reinforced concrete wall. The reinforcement has been tied to the continuity bars. Other rods are bent to enable the floor reinforcement to be tied to the walls. There should be a minimum of 20 mm **nominal** edge cover to the reinforcement bars.

Nominal cover is the amount of concrete that covers the steel reinforcement (Figure 8.30). The reason for the cover is to prevent the steel from corroding

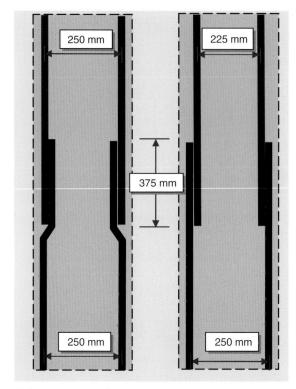

Figure 8.28 RC column continuity bars.

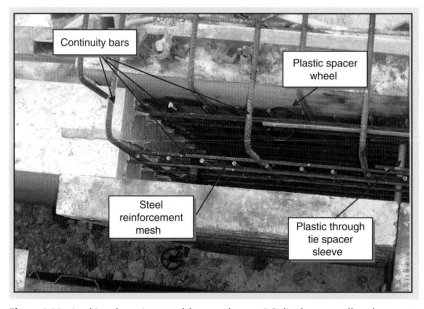

Figure 8.29 Looking down into steel formwork to an RC diaphragm wall under construction.

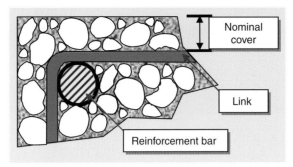

Figure 8.30 Reinforced concrete edge cover.

(rusting) within the concrete and to provide fire protection to the steel. Edge cover or nominal cover is specified in millimetres. During the concrete pour it is possible for the reinforcement to move slightly; therefore a **minimum cover** should be the nominal cover less 5 mm. (For example if the nominal cover is, say, 20 mm, then 15 mm should be the minimum cover, and plaster spacer clips should be used to hold the reinforcement away from the shuttering.) In practice, it is better to err on the safe side. High quality workmanship is essential. In Figure 8.31 the reinforcement is too close to the surface and has rusted. During the process of rusting the metal expands, spalling the surface concrete.

The concrete is usually compacted by vibrating poker; by the very nature of the job steel reinforcement can be moved during the pour. As the steel reinforcement is covered by concrete, it is possible that the problem of movement could be masked. Technicians can use electrical instruments to determine how much cover has been given to the reinforcement after the work has cured, eliminating the possible risks of lessened fire protection. For example, in the case of a reinforced concrete column where the cover has been reduced by poor workmanship, the structural engineer can specify that a cementitious render be applied to increase the cover. In some cases, where the tolerances do not allow the extra layer, this would mean that the concrete would have to be removed and the column re-poured, a very costly operation both in terms of extra work but more so in delaying the following work.

The maximum size of the coarse aggregate should not exceed the nominal cover. For most work 20 mm diameter ($\emptyset$) is suitable. End cover is only required to straight bars used in reinforced concrete floors or roof units where they are exposed to the weather or condensation.

The normal minimum cover is 20 mm in mild conditions, increasing to 50 mm in the most severe conditions. The density of the concrete also has a bearing on the nominal cover. The maximum free water (water/cement ratio – see Chapter 4, Section 4.9) of 0.65 with a lower cement content increases the minimum nominal cover by 5 mm. This means that a similar mix of concrete with a lower water/cement ratio and higher cement content can result in a reduced *nominal* cover as the concrete is more dense.

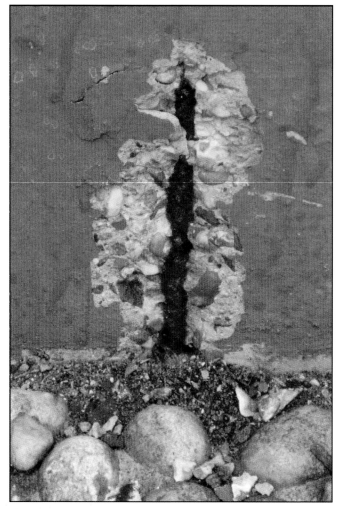

Figure 8.31 Minimal edge cover resulting in steel corrosion and concrete spalling.

BS 8110 Structural Use of Concrete states the minimum or lowest grade of concrete for reinforcement as C30.

Fire resistance

Fire resistance is an important issue. When heated during a fire steel will expand whilst losing its tensile strength. Concrete columns are generally in **compression** so concrete is the significant material, whereas floors, ribs and beams rely on the steel reinforcement for their **tensile** strength. The minimum concrete cover of 20 mm provides 30 minutes' fire resistance. BS 8110 provides a range of covers for specific elements such as beams, floors, ribs and columns against a range of fire resistances from 30 minutes to 4 hours.

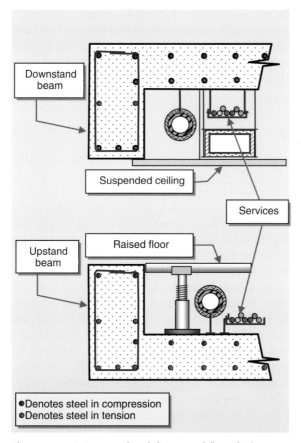

Figure 8.32 RC upstand and downstand floor decks.

As previously mentioned, the cover to the columns increases from 20 mm to 25 mm to improve fire resistance across the range, whereas beams, floors and ribs require a significant cover of up to 70 mm concrete to achieve 4 hours' fire resistance.

Cast in-situ beams

These can be square edged or rebated ready to receive precast concrete plank floors, or cast in with the in-situ floors. Alternatively, where the floor or roof is to be cast in with the supporting beams two main methods are used (Figure 8.32):

- downstand beams
- upstand beams.

Downstand beams

These stiffen the edge of the floor slab, enabling greater loading without increasing the floor thickness. An advantage of using a downstand is that the services can be run within the depth of the beam (Figure 8.32). Holes can be cast into the downstands, enabling rigid services such as large diameter pipes to pass through without the need for bends or cutting holes after the concrete has set. Fire is an issue where services run through compartments; however, specialist intumescent collars, mastic or bags can be used to overcome the problem.

Upstand beams

Where a flat soffit (underside) is required, such as in a shop unit, use of upstand beams should be considered. The advantages of upstand beams are that the floor can be stiffened and raised. Raised floors are ideal for modern offices because they enable most services to run beneath the floor, eliminating the unsightly cables required for communications and power leads being dropped from the ceiling. Special metal panels lock onto adjustable trees that are permanently fixed to the floor. Nuts and bolts connect the metal panels preventing them from rocking and providing rigidity to the floor. Access panels allow maintenance or major equipment renewal without the need for mobile or tower scaffolding. Power sockets and communication sockets with or without cover flaps can be fixed into special floor units, enabling communication terminals to be installed anywhere on the floor area. Additional floor surfaces are normally carpet or carpet tiles that can easily be removed when required for under floor access. If designed as part of the initial concept, the heating and ventilating ducts can be run below the floor, providing easier installation and maintenance. The added flooring and floor covering provide additional sound insulation for the air conditioning and ventilation units.

The main disadvantage of using upstand beams is cost. Raised floor systems are more expensive and in most cases a suspended ceiling is still required, so there is no initial cost saving using the raised floor to house the services.

The second disadvantage is that the location of specialist equipment such as motors, filters and valves must be carefully considered and noted to prevent store cupboards or shelves being placed over them. It may seem obvious not to place heavy storage cupboards over access to specialist equipment; however, the building user will want to locate their office furniture for maximum efficiency. It is therefore important that the **services engineers** and/or **heating and ventilating engineers** are consulted at an early stage of the design, perhaps during the **feasibility stage.** The initial extra costs up front of the contract can save many hours and thus money later on during the design stage.

Pouring concrete

To enable large areas of concrete to be placed, concrete pumps are used. The pumps can be mobile with a telescopic boom mounted on a lorry. A concrete transit vehicle feeds the mobile pump and the concrete can be placed on a continuous pour (Figure 8.33). During a $42\,\mathrm{m}^3$ pour for a basement slab, a mobile

Figure 8.33 Fresh concrete placement using a mobile concrete pump.

pump was used to place the concrete (Figure 8.34). The boom is directed by the pump driver/technician by remote control (Figure 8.35). To ensure the surface is flat and level the concrete workers spot level the surface using a laser level receiver (Figure 8.36) before tamping and hand floating the concrete (note the laser level in Figure 8.34). The tamping knocks the trapped air pockets in the concrete, bringing them to the surface. The pole float allows the surface to be smoothed without anyone having to walk over the fresh concrete (Figure 8.37). Alternatively, the whole floor can be poured and tamped and allowed to partially set, and then smoothed over using a power float (Figure 8.38).

To provide support to the upper concrete floors, metal multi-use decking can be used (Figure 8.39) which is primarily aluminium alloy sections with plastic coated plywood decks supported by adjustable props and stands (**tables**). The decks are supplied in standard modular forms and slotted into the horizontal rails either side of the column. When the main floor area is complete smaller panels are used to fill the space between the column sides and the decking. Note the use of the scissor jack working platform to provide a safe working area. The red edge barrier is erected with the floor deck and provides edge support for the concrete and safety during the pour.

Where multi-storey floors and columns are required, special concreting towers are beneficial. The concrete transit vehicles supply a continuous flow of ready mixed concrete into the site pump where large steel pipes transfer it to the towers. The continuous pressure pushes the concrete up the centre of the tower into an articulated extendable boom (Figure 8.40).

Figure 8.34 A 42 m³ fresh concrete continuous pour using a mobile concrete pump.

Lift shafts

Many modern multi-storey structures have cast in-situ concrete lift shafts (Figure 8.41). The formwork will be repetitious for each floor, so there are several methods of building the shafts. Many new skeletal framed office buildings have the lift shaft built either as a continual pour known as **slip forming** where the formwork is continually jacked up the outside of the shaft as the concrete is poured and cured. It is a *very* specialised technique as fine accuracy

Contiguous piles

Shutter sides for upstand

Cradles

Fabric

Steel chairs

Edge shuttering

Figure 8.35 Concrete pump technician operator remotely controlling the pour.

Figure 8.36 Laser levelling concrete before hand floating concrete slab.

Figure 8.37 Hand float finishing the concrete slab.

Figure 8.38 Hand operated power float.

Figure 8.39 Aluminium alloy and plastic coated decking panels under assembly.

is required throughout the entire pour. Concrete is pumped into the formwork as the shaft slowly creeps up in height.

Another method is to pour the concrete one floor per day and allow it to set. After 24 hours the enormous weight of steel formwork is jacked up on the inside of the shaft using the previous day's pour. The technique is similar to that used by tower cranes when they jack the mast up inside a lift shaft. Details of the techniques are outside the scope of this book.

Another method for providing a suitable lift shaft is to build storey height dense concrete blockwork walls around the lift well. A ring of cast in-situ concrete edge beams stiffen and strengthen the edge of the concrete floors onto which the dense concrete blockwork is erected.

Static concrete pump

Figure 8.40 Static concrete pump ideal for high rise concrete placement.

Measures include the following:

- Scaffold decks filling in the lift shaft are used to provide working platforms for the block layers at every level.
- Typically, as the stairwell goes around the lift shaft, edge barriers are essential to prevent falls down the lift shaft, off the edge of the stair well or edge of the stairs. The most common method is to have scaffold tubes clamped to columns, floor thicknesses, or onto adjustable steel props where other methods are unsuitable. There should be two horizontal rails as with all scaffolding: one should be high enough to prevent accidental falls over the rail at between 900 mm and 1100 mm from the finished floor surface; a second rail is required to prevent anyone falling through the gap beneath the top rail. The gap should not exceed 400 mm either side of the second rail, so it is normal practice to stand a scaffold board on edge and clamp it to the rail posts. The board is known as a **toe board** and prevents smaller items being knocked over the edge. Any ends of the scaffold tubes should have fluorescent end caps to protect against impalement.
- All stairwells should be well lit with low voltage lighting units and the power cables away from the walking area. This is particularly important where the stairwell is the emergency evacuation route.
- On larger sites, emergency evacuation routes are directed down stairwells so battery backup lighting and green signage with white motifs should be displayed on every floor and half landing.
- Every floor should have a set of fire extinguishers, emergency fire bells (Figure 8.42), excellent clear fire routes and trained fire fighters on site during working hours. The cost of providing the necessary equipment is often considered as extravagant; however, apart from being good building practice it is part of the legal requirement of the **employers' duty of care** under the Health and Safety at Work Act 1974.

8.6 Skeletal frame precast reinforced concrete

Precast concrete framed buildings tend to be low rise, although multi-storey structures can be built using precast units.

Figure 8.41 Cast in-situ concrete lift shaft surrounded by steel floor decking prior to receiving fresh concrete.

Figure 8.42 Essential fire fighting equipment ready for use.

The advantages include the following:

- Components are made in factories in steel purpose made moulds, ensuring accurate and consistent dimensions and finishes.
- Applied finishes such as exposed aggregates can be produced, as the moulds are laid flat during manufacture.
- Fixing lugs can be cast in to enable easy site handling (Figure 8.43).
- Quality control can be guaranteed and consistent.
- No on-site formwork is required other than localised shuttering around joints.
- Speed of erection can be achieved as no curing time or formwork positioning or striking is required.
- Precast floors and roof panels can be easily incorporated, including the precast stair units. The concrete steps should be protected during construction (Figure 8.44). In this example OSB has been used. Note the bolt head used to place the precast stairs during construction. The bolt will be removed and the hole filled

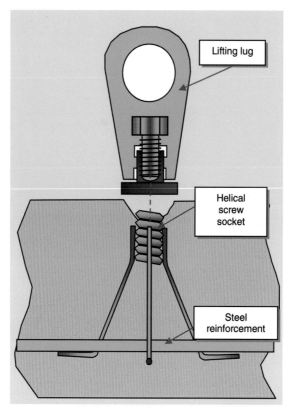

Figure 8.43 Lifting eye and lug used for moving precast concrete components.

with a cementitious grouting to prevent water penetration to the reinforcement.

- Fixing lugs and ports or sockets can be cast in for ease of lifting on and off transport and on site (Figure 8.43). Also cladding panels can be precast and bolted to the main precast skeletal frame.
- There are no on-site waste or pollution issues as it is essentially a clean dry process.
- No scaffolding is required other than edge protection to prevent falls. All fixers require full harnesses attached to specialist clamps or adjustable steel props strategically placed.
- Insulation can be placed when the building is completely weathered with either lightweight concrete blockwork inner leaves, or metal studwork and plasterboard finish.

Assembly

- Columns can be positioned into sockets cast in pad foundations or tied and cast onto steel reinforced pile caps.
- Precast concrete beams can be tied to pile caps and the junction shuttered and cast in fresh concrete.

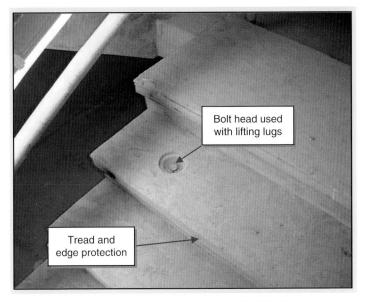

Figure 8.44 Precast concrete stairs protected with OSB sheeting.

Low rise commercial buildings

Factories, workshops, warehouses and farm buildings commonly require large uninterrupted floor areas. The building costs need to be relatively low and the finishes are mainly **utilitarian** (not too worried about what it looks like, as much as whether it performs well).

8.7 Portal frame

A popular system termed **portal frame** can be produced in precast concrete, steel, or laminated timber.

Precast concrete portal frames are single spanning frames comprising:

- **portal frame** – four precast concrete sections that are bolted together on site to form hockey stick profiles (Figure 8.45)
- **purlins** – precast sections that support the roof by spanning between the portal frames.

The portals (meaning **gateways**) comprise precast columns that fit into sockets within the foundation or into metal shoes bolted onto the foundations. When erecting the columns they are checked for verticality and suitably propped. Wedges are placed at the base or foot of the column to prevent movement within the socket. Another column is then erected opposite in a similar way. Due to the enormous weight of the sections, a mobile crane is used to lift the columns into position. When erected, the portal arm is lifted

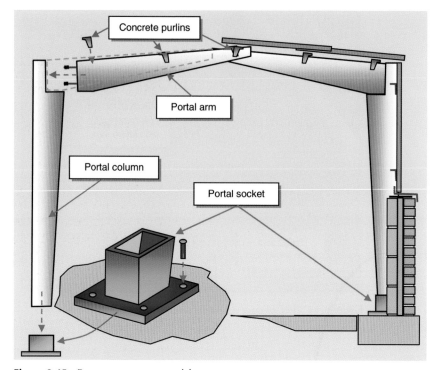

Figure 8.45 Precast concrete portal frame components.

into position and temporarily fixed to the column head. The arm at this stage needs to be fully supported using adjustable metal props to prevent the leverage pulling the column over or the arm drooping. Alternatively, two mobile cranes can be used. The second portal arm is then craned into position and temporarily fixed to the column and the apex. Both columns are further checked for verticality before the fixings are tightened to the designed torque.

When all the portals have been erected the structure looks like a giant toast rack; the purlins can then be fixed onto the arm using long bolts, washers and nuts. The whole structure is now finished and ready for cladding.

Cladding

Roof

There are various cladding techniques to suit customers' requirements, ranging from lightweight corrugated cement sheeting (formerly made of asbestos) to highly insulated plastic coated steel with flat cement sheet or plasterboard under-cloaking. The most popular method of fixing the roofing sheets is by zinc coated steel hook bolts, flange washers and nuts, hooked either onto the concrete or steel purlins (Figure 8.46). The first sheets are placed in position along the verge and marked out for drilling the holes.

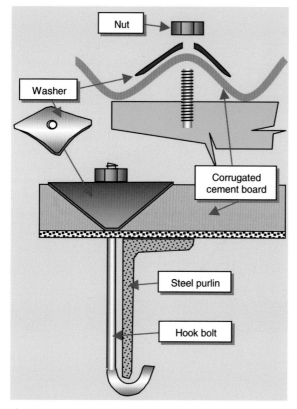

Figure 8.46 Simple corrugated sheet material fixing to steel purlin.

The fixers gain access either by using mobile tower scaffolds, or from hydraulically operated platform scissor lifts or single arm **cherry pickers** (Figure 8.47). The use of extending ladders is no longer acceptable, other than for access to the tower platforms. Gloves and eye protection are essential, as the edges of the cement sheets can be sharp and the dust from drilling may blow into the fixers' eyes.

Roof ladders or crawl boards *must* be used when moving over the cement sheets. Even when the sheets are new, they are brittle and not designed to take point loads such as the weight of a man. The next sheet overlaps the first sheet to ensure the rain runs over the joint in a similar way to roof tiles. The procedure is repeated on both sides of the roof until the top sheets meet at the apex. A capping piece covers the ends of both sides of the roof, providing a good weathering detail.

Walls

Barns and similar farm buildings, and also some factory units may be clad on the upper section of the wall using a similar method to the roof. Metal fixing rails are bolted to the portal columns and the cladding is hook bolted

Figure 8.47   Scissor lift access platform.

over the flange (Figure 8.46). The lower part from slab level is commonly masonry, either dense concrete block walls where the building will be used unheated such as a warehouse, or cavity walling with face brick outer leaf and dense concrete or aircrete block inner leaf with cavity insulation. Tying the masonry to the precast portal columns is normally achieved by **shot nailing** the hoop ties into the concrete at 450 mm course heights (Figure 8.48). Alternatively, the inner leaf wall is bonded around the inside of the column.

As commercial loadings tend to be high, the floor is commonly a solid concrete slab with a power float finish. Some buildings require floor insulation, so a thick layer of insulation may be positioned below the dense concrete oversite slab. Power floated finishes are produced after the concrete has started to set. The initial finish of the concrete is achieved by **tamping** (knocks by a series of frequent chopping actions releasing any trapped air within the concrete) to improve the compactability and therefore the density of the concrete. The concrete is left for a short period before a machine about the size of a shopping trolley comprising four large flat blades below a motor is swept over the surface in a side-to-side action pulled behind the operator (Figure 8.38). The blades sweep the tamped peaks and troughs into one flat surface which is ideal as a finished floor, or for application of vinyl sheet covering for shop floors, or a carpet finish. Power floating a floor is only suitable for larger areas; for smaller areas such as rooms, a hand trowelled screeded finish is more suitable.

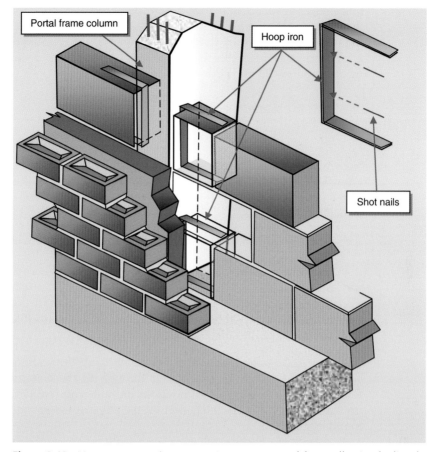

Figure 8.48 Masonry connection to precast concrete portal frame allowing for lineal movement only.

8.8   Stairs

Many cast in-situ buildings use precast concrete stairs. The cost of shuttering up flights of stairs and half landings is costly in terms of both time and skilled labour. Precast straight flights of stairs can be craned into position, or if access is difficult, taken to the fixing area on a trolley and jacked into position. Steel angles are required to be cast into the floors with either concrete blockwork or a concrete panel to support the half landing and upper flight. The precast flight has a rebate that sits on the steel angle, providing suitable bearing for the top of the stairs. Permanent fixing can be achieved using steel bolts or nuts and washers onto the cast-in bolts. Exact details vary with different manufacturers. As previously mentioned, when casting floors and roofs it is necessary to provide edge barriers to prevent personnel falling over the edge during the concrete pour. That also includes all openings known as wells or shafts: these include open lift shafts and rising main shafts where the main services such as water, electricity and sanitation pipes will be installed. Rainwater pipes on taller buildings tend to be run within the structure and can be large enough for a man to fall down, as can stair wells (large holes in the floor ready to take flights of stairs). Falls are the most common type of accident on site. Specialist edge barriers are available that fix directly into the metal decking used for cast in-situ concrete floors and roofs (Figure 8.39).

8.9   Walling

Skeletal framed buildings usually have walls that hang from the structure known as **curtain walls**. There are several different types of wall:

- **All glass** – toughened glass assemblies suspended from the structure and stiffened with glass fins
- **Glass and metal curtain walling** forming frames into which glazing units are fixed
- **Concrete panels** with openings for windows and doors
- **Sheet metal cladding** attached to light framing members fixed to the structure
- **Stone cladding** – thin layers of stone cladding fixed to blockwork backup walls attached to the main frame.

All the techniques mentioned rely on hanging the walling from the main structure. The weight is transferred through to the columns and onto the foundations, so additional foundations are not required. Ground level back-up walling may require thickening of the edge of the ground slab, but not a foundation.

Attaching the backup walls or infill walls requires special detailing. The objective is to ensure that there is a good mechanical fixing between the walls and the main skeletal frame. However, the walls must be able to move independently of the main frame to allow for differential movement, and the fixings must not corrode or deteriorate.

The objectives may appear simple, but were mostly ignored in designing the high rise concrete framed buildings of the 1960s. Everything is easy in hindsight though.

Masonry wall to column detail

Cast in-situ concrete columns are unlikely to move; however, the infill walling or cladding will be subjected to thermal movement and masonry will shrink as it dries out. A method of mechanically tying the wall to the column must allow for both longitudinal and vertical movement whilst efficiently restraining the wall panel. The system must be corrosion resistant, so austenitic stainless steel fixings are ideal. The alloy has good tensile strength and is therefore unlikely to stretch or fail in high wind loads. This is particularly important when considering taller buildings as the higher they are the greater the wind loads, both positive pressures and negative (suction). Pumping action is a combination of

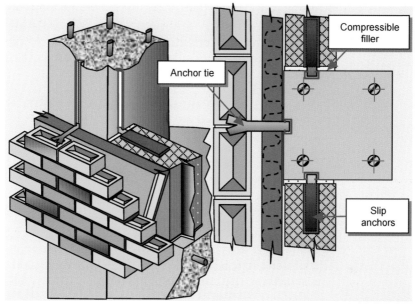

Figure 8.49 Masonry connection to a cast in-situ concrete column allowing for vertical and lineal movement only.

alternating positive and negative wind pressures which can cause more damage than continual pressure (see also Chapter 5 for further details of structural movement).

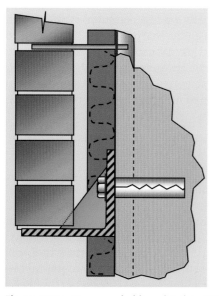

Figure 8.50 Masonry cladding detail to RC frame or floor.

To enable longitudinal movement to the walls, slip anchors should be used every other course (450 mm centres in height). The slip anchors locate into a cast in-situ channel. Different manufacturers have their own method of fixing the channels: for example the original Abbey anchor had an expanded polystyrene insert that temporarily held the channel in position using small nails through the wooden column casing. When the casing was struck the channels remained in place and the expanded polystyrene could be dug out to expose a clean vertical dovetail channel. Other manufacturers use mechanically fastened lugs that are held in the concrete, enabling square edged channels to be used. The anchors are sometimes referred to as fixing keys. They are twisted into position and can travel vertically throughout the channel. The grooves or dovetail shape prevent the key from pulling out or moving horizontally (Figure 8.49). To support panels of brickwork, specialist strengthened coated steel (or austenitic stainless steel) angles can be bolted into the concrete beams and columns at every floor level (Figure 8.50).

- Cement is alkaline and can cause burns to the skin if not washed off. Legs and arms are particularly vulnerable, so ideally all concrete workers should wear long sleeved clothes and trousers. In hot weather it is common for workers to wear shorts and no shirts, so clean water should always be available, either in shower units or, if the site is smaller, an easily accessible hose pipe. Under the duty of care, the site management should insist on all operatives be covered up whilst working to prevent splashes of cement on their bodies; however, on smaller sites, especially with subcontract workers, the rules in hot weather are commonly flouted.
- All concrete workers should wear protective gloves whilst working with fresh concrete. With several of the larger construction companies, such as Bovis Lend Lease and Mace, it is company policy for all personnel including visitors to wear eye protection and hand protection at all times whilst on site; this is in addition to the normal PPE. Industrial dermatitis is something that can and should be avoided. It is not macho to suffer with damaged skin or possibly blood problems as the result of working in the construction industry.
- Timber shuttering and formwork should always be neatly stacked and denailed when not in use. The writer has been noted in the site accident book on more than one occasion after stepping on a nail left protruding from shuttering ties. Modern site footwear should have steel toe protectors and, ideally, steel or strengthened soles. Mud on site can easily cover protruding nails; therefore all nails should be pulled or, if awaiting pulling after the formwork has been struck, they should be kept together. Larger sites use steel formwork, so nails are less commonly used. If you do receive a puncture with a nail or other sharp surface, it is useful to be protected by an anti-tetanus injection. Washing the wound may not cleanse it sufficiently to stop a condition more commonly called **lock jaw**. The first aid person should advise you to visit a hospital or doctor to receive an anti-tetanus injection immediately after the wound has been inflicted.

Toughened glass assemblies

Glass in the normal state is said to be **annealed**, meaning that it has been cooled gently allowing no stresses to be built up. Glass is technically a supercooled liquid at room temperature and is **amorphous,** meaning that, like golden syrup or treacle, it has no crystalline structure. If the surface of glass is scratched deeply, it will break along the weakened line. Glass is also brittle and will break into shards of razor sharp pieces. Thermally, glass is very conductive irrespective of its thickness and is therefore a poor thermal insulator. Therefore, why is it used to make a curtain wall?

To **toughen** glass, it is cut to the required size, and holes and notches are cut into the **plates** (a term from when large pieces of glass were hand ground);

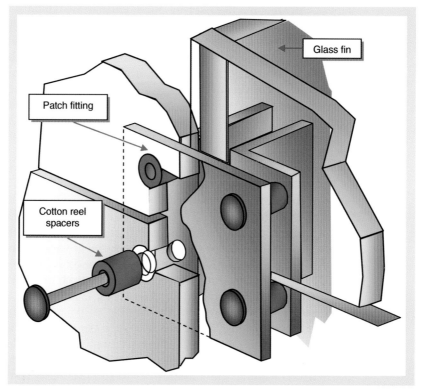

Figure 8.51 Toughened glass assembly curtain wall.

it is then heated until it becomes **plastic** (soft, just before it reaches melting point). The glass is then rapidly cooled, creating layers of stress within the plate: alternating layers of tension and compression through to the core where it is neutral, and then mirror image of the compression and tension layers. The outside layer is in tension, similar to an inflated balloon. The pressurised gas inside the balloon is trying to escape, whilst the rubber of the balloon is in tension. In comparison, the pressurised layers within the toughened glass are held in by the outside layer, thus increasing the impact strength about fivefold. Toughened glass is also known as **tempered glass** and has an increased resistance to applied heat. Glass 12 mm thick, with the edges ground and notches cut in the corners can be held together with metal plates known as **patch fittings**, to form one large area of glass. To weather the joints, silicon mastic or poly-sulfide mastic is applied which, when cured, forms a stiff rubbery bond that is impervious to water. The mastic also prevents the glass touching metal or other glass (Figure 8.51). Glass fins are attached to special patch fittings to stiffen up the thin glass wall. The fins are also bolted back to the structure to provide the necessary resistance to wind pressures on the main wall of glass.

Sections or plates can be left clear for door panels to be located. Like the main plates, the top of the door is pivoted in a patch fitting, whilst the heel of the door pivots on a floor spring. Floor springs are contained in metal boxes recessed into the floor. Two strong springs are attached to the pivot, providing enough resistance to prevent the wind from opening the door but still allowing a person to push or pull the door open. To slow the spring action, oil filled piston dampers add resistance to the spring action. The dampers can be adjusted where necessary to allow for windy locations. Fingers, especially children's fingers, can be badly crushed in all glass doors; therefore if the building is likely to be used primarily by children, there are either special fittings that reduce the gap on the pivot side of the door, or a plastic trim can be fitted.

In recent years toughened glass assemblies have improved. Double glazed sealed units can be used, significantly improving the thermal insulation properties of the glass wall. Glass can be modified to improve its thermal resistance by applying additional coatings of micrometre-thin metal during the manufacturing process. Inert gas such as argon can also improve the thermal resistance between the two plates of glass: the U value is about $5.6\,W/m^2\,K$ for single glazing but as low as $2.0\,W/m^2\,K$ for argon gas filled double glazed sealed units with low emissivity glass. At the time of writing the author is unaware of any double glazed all glass doors.

All glass assemblies require the facility to expand when heated. Where solar or tinted glasses have been used, the energy absorption increases so greater emphasis on thermal clearances is required. The William Faber building in Ipswich, Suffolk (the 'black piano') is an excellent example of all glass curtain walling. The **monolithic** (single) glazing is 12 mm antisun grey, with an allowance of 150 mm increase in height. The design is exceptional in allowing differential movement, especially in the cold yet bright winter months, where parts of the structure can be below zero in the shade with other large areas in direct sunlight having significantly expanded glass plates.

Design

Skeletal framed buildings require minimal materials to form the structure, a point we shall return to in Chapter 10. Occasionally, designers can forget that the structure should be functional and used by people, and is not just an architectural monument; for example, a high rise office block in London incorporated storey height glazing with mastic joints and no frames or handrails. Internally, it appeared that the floor finished and gave an uninterrupted view of the City of London. However, most people do not like standing next to the edge, especially when high up. From the outside, the building looks like a rectangular block of glass, although in recent years with terrorist activities, antiblast net curtains have been installed which has changed the crystal look. When the building was handed over to the client three decades ago, many of the office workers would not work next to the external wall until a continuous deep band handrail had been installed to provide a visual and physical barrier. The additional cost was considerable due to the size and number of floors requiring the handrail.

Metal and glass curtain walling

Metal and glass curtain walling has several advantages over all glass walling:

- It is easier to install in most cases using fixing lugs or cleats mechanically connected to the framing.
- Opening windows can be incorporated both as fire/emergency escape windows and for ventilation.
- It provides visual security for a building's users – human comfort.
- Sealed units can be replaced if they fail without the need for major plant use.
- Cosmetic changes can be achieved by replacing the curtain walling without structural modification – an important issue with sustainable construction.

Disadvantages include the following:

- Maintenance, such as window cleaning, is needed, although a glass is available that is self cleaning. If framed windows are used, cleaning will still be necessary.
- Frames need to be thermally efficient.

Metal cladding

Parts of a structure may not require light or vision through, such as plant rooms, store rooms, retail outlets, etc.

The advantages of using metal cladding include the following:

- There is a significant weight advantage compared with masonry cladding and therefore savings on the structural requirements of the frame and foundations.
- Metal cladding can be fixed quickly by a minimal number of workers in most weather conditions except high winds. There are no worries about waiting for mortars to cure, or freezing or wet conditions.
- At the end of the building's life the materials can be recycled or salvaged, which is environmentally advantageous.
- Bright colours and bold textures can be utilised where buildings are to be prominent, or light greys and greens can camouflage and blend buildings into the surrounding area if required in sensitive areas of natural beauty.

Metal cladding can be formed into sheet panels for roofing and curtain walling. Colour can be added by applying plastic coatings that also provide protection against corrosion or rusting. Embossing rollers can produce sculptured sheets with or without corrugations, therefore providing the designer with more scope than just flat sheets of metal.

Use of metal cladding has some disadvantages. Originally, large stores constructed using skeletal frames with metal clad curtain walls were easy targets for criminals. Because the wall panels were screwed onto horizontal rails from the outside it was relatively simple for thieves to undo the fixings outside opening hours and gain easy access to the contents of stores and warehouses. The alarm systems were typically attached to the entrance doors and side windows and therefore were not activated. To overcome the problem, secret fixing methods that cannot be easily accessed are now used on cladding systems. Masonry sub-walls are built behind the cladding, providing a more robust deterrent and surfaces onto which shelves and racking can be attached. Finally, detection using sophisticated passive infra red light sensors monitor any movement within the building, thus raising the alarm.

8.10   Laminated timber skeletal framed building

Commercial skeletal framed buildings can also be timber frame. The technique is based on forming larger sections of timber from many strips glued together. The market leaders are Glulam who have been forming architectural beams and stair components for several decades.

Timber compares favourably to steel as a structural material:

- Weight for weight timber is as strong as steel.
- Timber is combustible, whereas steel is virtually non-combustible.
- Timber can be farmed over a 35 year cycle – steel can be recycled.

Figure 8.52   Softwood timber girder frame supporting sheet metal roofing.

- Large sections of timber have a natural resistance to fire. When timber has charred to about 25 mm depth, it forms a thermal barrier which protects the timber beneath it. Therefore timber frames should be designed with a 25 mm sacrificial layer as part of the structure: this means that a beam 300 mm deep × 175 mm thick should be designed as 350 mm × 225 mm to incorporate 25 mm of exposed timber that will combust.
- Steel expands when heated and therefore produces unstable supported elements. (This is one reason why fire fighters play cold water on apparently non-combustible steel and/or brickwork.)
- Steel loses about 50% of its load bearing strength at 472°C and expands more than timber which retains its original dimensions under the same conditions.
- Steel can rust if not protected – timber also requires protection, but mostly from the sun's ultraviolet radiation and insect attack.
- Timber is aesthetically pleasing to most people – steel requires additional decoration or specialist materials such as stainless steel.
- Timber flexes without failure, a feature that the Japanese are investigating for their high rise buildings in earthquake zones; it has been proven to be superior to reinforced concrete in this respect.

In northern France large superstores are being built using softwood timber instead of steel. The uprights are termed **posts** and fit either into metal shoes

Figure 8.53 Softwood skeletal framed hypermarket under construction.

bolted to the foundation pad or into sockets. The beams are bolted to the posts, either directly through the timber or into metal angle cleats. Wind bracing beams also bolt into the main beams supporting the roof. The walls and roof are clad with rolled plastic coated steel with insulation and inner cladding to provide fire resistance and a decorative finish. Services are supported from the metal purlins on hangers. Large span beams are girders with either a composite of metal latticework or timber struts and hangers held in position with gang nailed plates. Figure 8.52 shows part of a timber framed shopping mall in Florida, USA. All of the buildings and covered walkways have been built entirely of softwood timber skeletal frames.

In France, laminated timber designs are more popular (Figure 8.53). Where it is very hot (in excess of 35°C air temperature), the timber is unaffected as long as it is protected from direct sunlight. Intumescent paints are available in clear finishes similar to varnish. In the event of a serious fire the paint expands to about eight times its original thickness, forming a barrier against the passage of heat. Similar types of paint are available for exposed steelwork. The paint was originally developed for the oil and gas industry to protect steelwork in the event of a major fire. The paint is very effective, although very expensive.

Chapter 9

Site Issues

This chapter describes some of the issues of working on-site and the practicalities of working to a contract. Site planning is often ignored in construction textbooks, and left to specialist volumes. The outset of a contract is vitally important. The chapter starts by considering pre- and post-contract site planning, with general considerations of what to look for and what should be planned. The work of the contracts manager and site manager is included, to give a complete picture of the construction process. The good neighbours policy is not only beneficial to all concerned, but is a legal requirement under duty of care. The chapter also considers health and safety, site security and concludes with selection of heavy plant.

9.1   Site planning

There are two stages to site planning:

1   Pre-contract – considerations that should be made before going in to tender for the contract, and
2   Post-contract – considerations when the contract has been signed.

At both stages the following issues should be considered:

- health and safety
- security
- accessibility
- plant
- materials
- workforce
- accommodation/welfare.

Pre-contract site planning

Before tendering for a contract, it is good policy to visit the proposed site. This will enable the estimators and contracts managers to form an appreciation of possible issues related to the above list. The initial site survey should include taking notes and photographing the following:

1   all boundaries and adjacent property
2   all services, both entering the site and running by
3   the road layouts and street furniture.

Why take photographs? Before tendering for a contract a **pre-tender plan** should be drawn up by the management and tendering team. Photographs, and possibly video, can help when deciding the type of plant required; for example, if a tower crane is to be considered, whether a rigid arm or a derricking jib will be most suitable (Figure 9.1). Accessibility to the site for plant and materials must be considered: whether there are any low bridges, weak bridges, narrow roads or acute road junctions that could prevent heavy plant or long vehicles access. If the site works require closing public footpaths or suspending

Figure 9.1 Derricking jib and fixed counterbalanced jib tower cranes.

parking or waiting bays, applications for a licence to the local authority or local police may be required (Figure 9.2).

Services are an important issue. The existing availability of prime services (**utilities**), such as water, sewage and electricity, will influence whether portable services will be required until temporary mains can be established. If live services pass over, alongside or through the site, protection will be required and certain plant cannot be used. For example, high voltage electricity cables suspended from pylons will prevent high masted plant from being used, whereas underground high voltage electricity cables require protection from excavating plant. All these factors will have an effect on the type and size of plant that can be used on the site and affect the tender pricing.

Post-contract site planning

After winning the contract and before any work commences on site, photographs of all boundaries, adjacent footpaths and roads should be taken. They may be required as evidence in any future dispute regarding damage and should therefore be date stamped. For example, it may be claimed that heavy lorries used for delivery of plant and equipment have damaged neighbouring property. Photographs can be used to confirm whether there was damage before works commenced. In most cases where the Party Wall Act 1996 comes into play, photographs will be part of the 'award'.

Figure 9.2 Highway notices.

The Health and Safety Executive (HSE) must be informed of the site workings using an **F10 Notification of construction project**. The notification is required for all construction projects that will last more than 30 days or 500 person days as required by the Construction (Design and Management) Regulations 2007. The forms can be downloaded from https://www.hse.gov.uk/forms/notification/index.htm. The HSE has an excellent website giving more information: http://www.hse.gov.uk/construction/information.htm#law.

The site must comply with all the current legislation for construction industry sites in the UK including the following:

- Health and Safety (First Aid) Regulations 1981
- Construction (Design and Management) Regulations 2007
- Health and Safety (Consultation with Employees) Regulations 1996
- Construction (Health, Safety and Welfare) Regulations 1996
- Party Wall Act 1996
- Lifting Operations and Lifting Equipment Regulations 1998 and amendment 26 January 2000
- Management of Health and Safety at Work Regulations 1999
- Control of Substances Hazardous to Health (Amendment) Regulations 2004
- Health and Safety (Control of Noise at Work) Regulations 2005.

Planning is essential. Poor or inadequate planning can cost a company a lot more than any time it has saved by not planning professionally. Much of the planning should have been carried out during the pre-contract stage, as it has a significant bearing on the tender pricing. However at post-contract stage the planning will need further development.

For larger sites a site management team would draw up a programme before the site is taken over. If the contract is based on a **bill of quantities**, much of the information will be contained in the **preambles** section. However, additional information will be required such as the names, addresses and telephone numbers of:

- suppliers
- subcontractors
- consultants for the project
- building control
- nearest accident and emergency (A&E) hospital
- the utilities companies (gas, electric, water, sewage/drainage, telephone, etc.).

A list of suppliers' telephone numbers on site is essential. A phone call from site to a supplier to change a delivery time can save abortive delivery charges if, say, the vehicle cannot be offloaded.

The location of the nearest A&E hospital to the site, including a map of how to get there, is essential. An accident on site may need medical attention but not an ambulance; therefore it is important for the site management to have the information quickly to hand.

The Fire Authority should also be informed of the site works. In the event of an emergency they will need to know where the site is and what access will be available.

When the contract has been signed, the **contractor** becomes legally responsible for the security and health and safety issues of the site from the agreed date. Public liability insurance should already be part of the company's policies. However, insurance of the site and its contents would usually be specific to the date the contractor takes possession through to either partial or full handover.

9.2 Health and safety

Health and safety has a direct link to security. Therefore it is important to make the site secure at the boundary; if this is impractical, any designated risk areas should be secured and appropriate signage erected (see also Section 9.3).

If existing structures are to be demolished they should be inspected for contents (especially people or animals) before being made secure. During inspection you should be watchful for used hypodermic needles, especially if the building has been empty for some time. The building may have been surveyed some time ago and it is possible that drug users have used the place since. If there is any risk, professional advice should be sought as to their safe disposal. Any casualties suffering **stick injuries** from used hypodermic needles should be immediately taken to the nearest hospital emergency room where emergency tests will be carried out and barrier drugs used to reduce possible infection. It is not only AIDS that can be fatal; hepatitis B and C are more

Figure 9.3 Asbestos clad factory under demolition.

common. To reduce the risk of unauthorised entry, it is usual to board up all window and door openings.

From an environmental point of view, the whole site and anything on it should have been surveyed at an early stage, such as the **feasibility stage**, of the project. An environmental audit should be undertaken before any site work is carried out to enable projected changes to be monitored. If any hazardous materials had been noted, they should have been targeted for risk assessment and programmed accordingly: for example, if there are asbestos roofing sheets (Figure 9.3), or chemicals have been stored or used and contaminated the ground or flooring.

If hazardous waste is on site warning signage must be displayed on the security barrier where the risk is. The risk may be present where a worker is in continual contact, such as during demolition or during the removal and transport procedure. Specific extra personal protective equipment (PPE), such as respirators and specialist boiler suits and footware, would be required. All vehicles transporting the waste would need their undersides and wheels washed in an on-site station that recycles the water which will eventually be disposed of as **low hazard liquid waste**.

All services, both running onto and adjacent to the site, must be located. The utilities companies will provide markers identifying their services; this should have been arranged during the pre-contract stage before any works on site had taken place. Actual flagging or marking is normally carried out when the site is ready to start. Sewers, water mains and gas mains are particularly vulnerable, especially if they are close to the surface. Protection in the form of large steel plates should be used to enable heavy vehicles to drive over them

Figure 9.4 Steel plate being moved using a web and shackles.

without causing damage. This is particularly important when first setting up site, and demolition plant and vehicles are required. Figure 9.4 shows a steel plate being moved to cover a services trench to enable vehicular access to adjacent property.

Goal posts should be set up where overhead obstruction or power cables are present. High voltage cables require significant gaps below them to prevent electricity from **jumping/earthing** via tall vehicles or those with crane offload facilities. Figure 9.5 shows goal posts used to warn delivery drivers of national grid cables carrying 132 000 V. The posts are painted red and white for easy identification. In this example a taut line with material flags has been used. All delivery vehicles were escorted through the goal posts and visually checked by a **banksman** (banksmen control all vehicle movements, such as reversing vehicles, guidance for unloading, crane use, etc. – they are basically a trained set of eyes that aid the driver). On other sites where cables are remote, goal post portals may have a rigid metal crossbar suspended on chains from the main crossbar. Bells or even old tin cans are tied to the suspended bar so that if the lorry knocks it the driver will hear the bells/cans clatter and be warned that the load is too high. Overhead cables carrying domestic voltage are much nearer the ground. Although there is no likelihood of voltage jump, they can still kill if broken. The most likely damage is from lorries with crane offloading equipment when the operator lifts the boom/jib.

Access for long vehicles such as low loaders used to transport crawling plant requires wide turning areas. In congested areas, such as towns and cities, Sunday or night deliveries are commonly the option used by site management. A very congested site in Lime Street in the City of London required careful planning for all deliveries: a special slip road was constructed to enable the delivery vehicles to leave the public highway to offload within strict time limits.

Figure 9.5 Goal post warning of very high voltage overhead cables.

9.3 Site security

Security

The contractor has **a duty of care** towards anybody entering, working or leaving the site, including authorised visitors, the workforce and unauthorised visitors. For example, during the working day access to the site must be restricted: some people are naturally inquisitive and may wander onto the site to see what is going on; sales representatives may try to gain access and wander around the site looking for the site manager; people may enter the site looking for work; the client may want to see how the work is progressing and visit the site; potential thieves may visit the site looking for materials or equipment. Under English law the contractor has a duty of care even toward thieves and vandals; it is not a defence to say that they should not have been on the site. The Home Office Report on Plant Theft (available at http://www.homeoffice.gov.uk/rds/prgpdfs/brf117.pdf and http://www.homeoffice.gov.uk/rds/prgpdfs/fprs117.pdf) estimates that stolen plant costs the construction industry between £600 million and £1 billion per year. Site security can take the form of close boarded panels termed **hoarding**. Figure 9.6 shows a good example of a 2.40 m high sheet hoarding secured to posts concreted into the ground. Note the extra height plastic sheeting to prevent dust escaping from the site during the demolition, and white lights to illuminate the footpath. In Figure 9.7 the footpath has been closed and pedestrians can walk safely in the road protected by timber baulks and a wooden

Figure 9.6   Excellent protection for the public during demolition.

Figure 9.7   Timber baulks and handrail to protect pedestrian access off the public footpath.

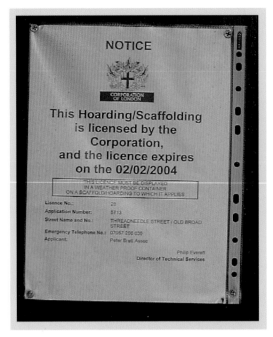

Figure 9.8 An official licence must be displayed at all times.

barrier. Red lights have been fixed to the baulks, highlighting the area for motorists. It is common practice to paint the hoarding in the company colours and display company logos and signage. Permission from the local authority planning department will be required if the hoarding is to be on the footpath or carriageway; a licence is issued and should be displayed on the hoarding. Figure 9.8 shows an example of a hoarding licence for the City of London, including the licence expiry date. Access gates should open onto the site or, where possible, a splayed entrance/exit where the gates can be opened out.

In contrast, Figure 9.9 shows an example of a poorly erected hoarding. Note that the car parking spaces have not been suspended and there are no protective baulks to stop vehicles knocking the scaffolding; the site office is directly above the footpath supported by the unprotected scaffolding; the end of the handrail has no protective cap and could impale a motorcyclist or motorist; the site entrance door has no security and there was a basement excavation less than 1 metre from the open door.

Close boarded hoarding has the following advantages:

- It will reduce dust and small debris from leaving the site. This is especially important during demolition.
- Vision panels either glazed in polycarbonate or steel meshed will help with security.

The disadvantages of close boarded hoarding are that:

- At night it can help thieves by preventing passers-by seeing what is happening.
- It is costly to erect.

Figure 9.9 Example of a dangerous site set up.

An alternative type of barrier to close boarded hoarding is grille fence panels or gates fixed into sockets in concrete or plastic footing modules (Figure 9.10), which has the following advantages:

- The wind can pass easily through the panels so there are minimal stability problems.
- Access is easier because individual panels can be opened or removed.
- Constant visibility is ensured which enables any out of hour attendance to be seen.

The disadvantages are:

- It enables criminals to see what on site is worth stealing.
- The panels can be easily levered out of their sockets, enabling large sections to be laid flat for easier access by vandals or thieves. To overcome the problem metal rods can be driven into the ground on the site side of the barrier and tied to the panels using wire, thus preventing them from being easily lifted.
- Wire panel barriers will not prevent flying debris during demolition or dust from site working, such as cement dust, cutting dust or simple sand dust from leaving the site. For example, if a worker is using a petrol disc cutter to cut steel reinforcement bars, the red hot sparks will not stop at the wire barrier. In contrast, a close boarded hoarding will stop sparks and dust.

Figure 9.10 Heras type wire mesh site barrier and shoe.

Chestnut paling fences are still used as site barriers on small sites. Wooden posts are driven into the ground and the chestnut palings might perhaps deter small children, but at 1.2 m tall they are hardly a deterrent to unauthorised entry. As the contractor you have a duty of care to reasonably prevent entry. Barbed wire or razor barb may be a solution in certain parts of a site, such as around a storage compound; however, it should not be used adjacent to the public highway or as the first line of a barrier. If an intruder is injured whilst trying to enter the site you may find that they can sue for compensation for their injuries. A recent example of unauthorised entry to a site was a drunken person at night who tried to reclaim an object that had entered the site by climbing over the hoarding, fell several metres and received extensive injuries. In this case the HSE agreed that the contractor had taken reasonable steps to prevent unauthorised entry; however, a determined person may still try.

Good neighbour policy

When setting up a site in a residential or commercial area, a **good neighbour policy** will help with security and should be introduced before the site starts.

Figure 9.11 Considerate neighbours and contractors schemes.

A letter and possible visit to let the neighbours know what will be going on enables better and safer working for everyone. The neighbours will possibly know who the local criminals are and potential problems. In cities, developers are encouraged to promote clean and considerate working (Figure 9.11).

On large housing sites it is not practical to erect close boarded hoarding all the way around the site. Developers often try to sell their houses 'off the drawings' which means that people move into their new homes as soon as they are built, whilst other houses are still under construction. Sites like these have other security issues such as **the weekend builder**. These are people who have moved into their new home and just help themselves to materials such as bricks, sand, ballast and cement from the houses under construction; they do not consider themselves to be thieves. Some even use the cement mixer to mix the mortars or concrete. Therefore site security is essential. Individual buildings that are still under construction must be protected.

Out of working hours, all scaffolding should be blanked off where practical, such as by tying a scaffold board over the rungs of all pole ladders to prevent use. Scaffold alarms are useful when the work is on residential sites such as for roof retiling. The scaffold looks interesting for children to play on. A scaffold alarm will let the neighbours know of the problem and hopefully they will call the police.

For services and utilities recognition see Chapter 12.

9.4   Site plant

Plant is categorised as being either heavy or light. Heavy plant includes cranes, diggers, haulers, etc., and light plant include barriers, ladders, scaffolding and hand held tools.

The choice of plant for site clearance is dependent on the following factors:

- Size of site – area covered
- What is on the site – trees, existing buildings
- Site topography – flat, hilly, marshy, etc.
- Type of soil/stratum – dense rock, soft rock, friable soil, cohesive soil (see Chapter 4)
- General weather conditions – seasonal weather conditions
- Time constraints – how soon must the operation be completed?

General heavy construction plant can be grouped under the following main headings:

- Lifting – cranes, hoists, fork lifts, loading shovels
- Digging – crawler backactors, piling rigs
- Hauling – dumpers, articulated haulers (dumpsters), rigid back tippers
- Earthworks – dozers, scrapers, graders, compactors, loading shovels, rollers
- Other – multi-purpose excavators, pumps, concrete crushers, concrete pumps.

Dozers are used for scraping to a depth of up to 300 mm depending upon the type of soil. Figure 9.12 shows the adjusting pistons enabling the blade to be angled a few degrees to one side pushing the spoil away from the scrape. The blade can be raised at one end enabling limited grading to take place. Modern

Figure 9.12   Tracked bulldozer.

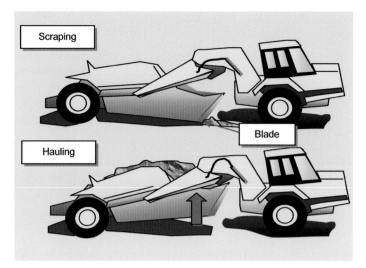

Figure 9.13 Earth scrapers.

techniques incorporate laser levelling for driver accuracy. An onboard laser sensor enables an accuracy of ±5 mm in height, speeding up site work. Dozers can also be used for towing earth scrapers, utilising the greater traction of caterpillar tracks. They are normally used on large area earthworks.

Earth scrapers are self loading tipping vehicles that scrape layers off the surface of the ground using a blade beneath the hauling body. They can be powered by tracked dozers where greater traction is required, or pneumatic tyre power units for faster coverage over rough terrain (Figure 9.13). They are only suitable for large area earthworks.

Graders enable greater gradients to be achieved over longer distances. They are typically used for landscape work, such as embankments and major road schemes, where broken soils or aggregates require angles from the horizontal. They cannot replace dozers. Figure 9.14 shows the fully adjustable mouldboard

Figure 9.14 Grader.

Figure 9.15 360° crawler backactor excavator loading an articulated site-use-only dumpster.

attached to the mould ring, enabling angles from the horizontal and from side to side to be achieved.

Articulated haulers, also known as **dumpsters** enable fast off-road hauling of spoil (Figure 9.15). They range in size from a carrying capacity of 11 m^3 to nearly 17 m^3 with a gross weight of up to 70 tonnes. Their large pneumatic tyres and very low tyre pressures enable them to travel over most terrain quickly. They are probably the most dangerous of all large plant on site due to their speed of operation. The driver has very limited visibility as the cab is very high above the front drive wheels. They have been known to reverse over small vehicles such as surveyor's trucks without the dumpster driver feeling anything in the cab. It is especially important to have strict site security where articulated or large off road tippers operate. The second biggest cause of fatalities on site is from vehicles on site.

According to the HSE about 70 people are killed and 1200 people injured each year from transport accidents. For further information visit the HSE websites at http://www.hsedirect.com/search/SQL/dataitem.asp?id=18941, http://www.hse.gov.uk/aboutus/hsc/iacs/coniac/201103/201110.pdf and http://www.hse.gov.uk/LAU/lacs/90-3.htm#appendixi, and a search engine for HSG 144, or E199:02 Safer workplaces.

Loading shovels vary in size, capacity and operation. They range from a small rigid based wheeled loader with a 1.2 m^3 loading shovel to the large artic-ulated wheeled loaders with a nearly 14 m^3 loading bucket. They can be used to pile site spoil in spoil heaps, although they are limited due to stability. Their pur-pose is mainly to load hoppers for on-site concrete batching or movement of ag-gregates for road construction, and to load other hauling vehicles (Figure 9.16).

Crawling backactor excavators are also referred to as **360s** as the body can rotate a full circle enabling it to excavate and place the spoil either side of the machine or directly into a tipper or dumpster. As for cranes, the horizontal rotation is termed the **slewing circle** or the **slewing ring**. The size of excava-tion bucket can be changed from 300 mm wide for services trenching and up

Figure 9.16   Pneumatic tyre loading shovel.

Figure 9.17   360° crawler backactor excavator loading rigid back road use tipper lorry for 'muck' clearance.

to 1.5 m³ for large volume excavations including basements with a maximum digging depth of up to 6 m. The power units are hydraulic and manufacturers such as Volvo use biodegradable hydraulic fluids so that if a hydraulic hose should fail the oil spillage will not harm the natural environment. Figure 9.15 shows a 360 loading an articulated dumpster for site hauling, and Figure 9.17 shows another 360 loading a rigid back tipper for clearance off site. Other attachments can be used with the power unit, such as nibblers used for demolition work, points used to break up old concrete and lifting hooks used with webs or chains for low lifting tasks (Figure 9.18).

Figure 9.18 Pneumatic and hydraulic tool attachments.

Compactors are used on large sites where large areas of fill are required. The steel wheels have knurled/toothed rims that compress the ground due to the heavy loading of the machine which vibro-compacts as it drives (Figure 9.19).

Figure 9.19 Soil vibro-compactor.

Rollers are used to compact loose soils and aggregates. Where roads or large flat areas are being constructed, heavy metal rollers compress the substrates. Figure 9.20 shows an articulated roller used to compress lime and clay as part of a soil stabilisation process. Figure 9.21 shows a smaller tandem roller used to compact hardcore over a new warehouse floor. Both machines have rollers that vibrate as they roll, increasing the compaction performance; this is known as vibro-compaction.

Concrete crushers are now common on demolition sites. The machine is fed broken reinforced concrete by loading shovel or, as in Figure 9.22, a 360 backactor. A series of rotating metal hammers crush the materials into various sized pieces that can be used for hardcore or aggregates. The steel reinforcement is removed from the machine and placed in a reclamation skip and sent for steel recycling. The advantages of concrete crushing on-site are as follows:

Figure 9.20 Articulated vibro-roller compactor.

Figure 9.21 Articulated vibro-tandem roller compactor.

- Segregation of demolition materials – useful for salvage
- Reduced volume of materials for off-site haulage – smaller pieces will compact more densely than larger pieces
- Reuseable materials negate the need to buy in hardcore and fill aggregates.

There are several types of **crane** used on construction sites. The main ones are:

- **Mobile cranes** – used as and when required on sites where a static crane would be unsuitable due to cost or space. Figure 9.23 shows a mobile crane with the boom partly erect. Note the large stabilising bars (**outriggers**) at the front and rear of the machine. They provide a broader rigid base,

Figure 9.22   Mobile concrete / rock crusher.

Figure 9.23   Telescopic masted mobile crane setting up. Note the outriggers.

increasing the stability of the crane. The extra space required must therefore be taken into account when planning the site.

- **Static derricking jibbed cranes** – used where air space prevents the use of rigid jib tower cranes. This is a very common type of crane, especially on congested sites where multiple cranes are required (Figures 9.1 and 9.24).

Figure 9.24 Derricking jibbed tower cranes attached to the outside of the building during construction.

Note the counterbalance concrete slabs to the rear of the crane. Complex crane programmes are required to prevent cranes operating in the same air space. The cranes have laser alarms warning if another crane is within the contact area. As with most plant, there are many variations of tower cranes, too many to be included in this book. Commonly, tower cranes are bolted securely on concrete pads that eventually will be the base of the lift shafts. However, on some projects they may have temporary bases that are either covered over or removed after the cranes have been dismantled. There are climbing cranes that are attached to the main structure as the building progresses in height. To provide more rigidity the mast can be attached to the structure under construction. Figure 9.24 shows the mast bolted to the structure. When the project is complete a mobile crane will be used to dismantle the fixed crane.

- **Rigid jib tower cranes** – the jib is counterbalanced with large concrete slabs. The crane hook is suspended on a bogie that travels along the length of the jib, enabling the load to be raised and lowered at various

Figure 9.25 Crane with slewing mast and ground level counterbalance operated by remote control.

points according to the weight. The crane driver has markers indicating the maximum load that can be supported at points along the jib. Figure 9.1 shows a rigid tower crane. The driver is positioned in the cab next to the slewing ring. A banksman would be in telephone contact with the driver, providing instructions from the ground.

- **Mast cranes** – are static cranes counterbalanced by concrete weights at ground level. The lifting capacity is less than that of tower cranes so they are ideal for smaller sites. Figure 9.25 shows the counterbalance sections at ground level. The cranes are not bolted to the ground, but instead have adjustable pads to spread the load over a large area. In the background is a larger capacity mast crane. Smaller versions can be operated from the ground by the crane driver using remote controls.

Telescopic fork lifts/handlers used on site have pneumatic tyres. They require road tax and insurance as they are commonly used on the public highway. They differ from conventional forklifts in that they can operate on relatively rough terrain. Larger machines can lift parcels of bricks or mortar tubs onto scaffolding to a height of up to 9.5 m (Figure 9.26).

Multi-purpose excavators are basically a tractor that has been adapted (Figure 9.27). The front loading shovel is ideal for moving volume materials such as sand and spoil. The shovel can be split to form a blade suitable for light scraping of, say, topsoil. A bar on top of the shovel has two adjustable forks that enable wooden pallets to be moved or offloaded from a delivery vehicle. At the rear of the vehicle a backactor arm is used for trench excavation or attaching demolition points, etc. A lifting hook enables heavy steel plates to

Figure 9.26   Small telescopic fork lift for use on or off the public highway. It must be licensed as a road vehicle.

Figure 9.27   Multi-purpose excavator for use on or off site. It must be licensed as a road vehicle.

be manoeuvred into place as shown in Figure 9.4. The diesel engine doubles as a compressor to power pneumatic tools such as road breakers.

9.5   Temporary water removal

Where excavations fill with rainwater or the formation level (the level at the base of the excavation) is below the water table, submersible pumps are required. They can be used in a sump as shown in Figure 9.28 where a section deeper than the formation level allows the water to drain. The pump normally is a diesel fuelled machine that continually sucks up the water through a metal gauze that prevents stones and other debris from damaging the pump (Figure 9.29). On some sites backup pumps are used to allow for maintenance

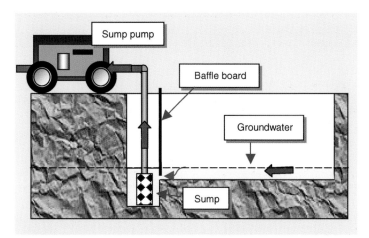

Figure 9.28   Sump pump used for temporary site dewatering.

Figure 9.29   Pump in continuous use dewatering an excavation.

or possible breakdown as the removal of water must be continual. Where concrete is to be placed during wet periods, a baffle board with a 10 mm gap at the bottom will allow the water to pass under the baffle board to drain the trench. When the concrete is to be placed by the board the pump nozzle is removed and the concrete is placed. Obviously some fines from the concrete will pass through the gap; however, the main bulk will be held back. When the concrete has set, the baffle board is removed and the sump/well is backfilled and compacted.

Chapter 10

Construction and the environment

Global warming and climate change are high on the agendas of government policy making and the national press. This chapter looks at some of the terms used and reasons behind the issues, starting with the three main features of the natural environment and moving on to the causes of pollution. The topics highlighted cover several parts of the Edexcel unit, Construction and the environment. The chapter concludes with discussion of various constructional techniques and their impact on the environment.

10.1 Natural environment

The three main components of the natural environment are:

- soil
- fresh water
- air.

Without them it would be impossible to exist.

Soil

Soil is a mixture of rock fragments and decomposing tissue from plants and animals. The oldest soils in the UK were formed about 10 000 years ago, after the last ice age. As the bedrocks changed from igneous to sedimentary and metamorphic rock, various minerals became deposited on the surface. Water within the rocks can transport minerals and deposit them many kilometres from their origin. Plant and tree seeds are blown across the barren rocks and they begin to take root and spread the flora. The process is similar to grass and larger plants growing in guttering and on top of roofs. As the plants form roots they catch any small particles blown or washed by and eventually build up soil that retains water for continual growth.

Storms and winter rain wash deposits of clay minerals and fine particles rich in minerals from high ground into rivers, spreading onto the lower grounds or out to sea. Seeds from plants can be transported in the storm water, deposited in bird droppings, or blown by the wind. The seeds germinate and grow producing seeds of their own continuing the spread of flora. Some plants die off as the seasons change and decompose, forming a layer of humus we call topsoil, which contains bacteria and fungi, and hosts countless insects so is considered to be alive. The layer can be a few millimetres thick to 200–300 mm thick; however, for estimating purposes we consider 150 mm as an average. Beneath the topsoil is a sterile layer of subsoil where, apart from tree roots seeking water and anchorage, most things will not grow. Topsoil is unsuitable to build on as it is likely to contain roots and seeds, so it is removed prior to building.

The quality of the soil depends on its origin. Soil formed by decaying heather, bracken and reeds will be acidic, and water passing through the soil will also become acidic. Acidic water will dissolve softer limestones, changing the chemical content which becomes alkaline.

Bracken surrounded by moss and reeds in boggy conditions will eventually die off each winter and become submerged, only partially decomposing due to the lack of oxygen, to form peat. Peat can be many metres thick, as in parts of East Anglia where historically it has been a source of fuel when dried. The waterlogged part of the east coast of the UK has provided generations of people with local fuel; during the middle ages peat pits were allowed to flood, forming man made rivers we now call the Norfolk and Suffolk Broads. The marsh lands have been drained, leaving very fertile soil which is ideal for grazing cattle and grain crops. In recent years the land has become more difficult to work as the fields are too small for modern high performance farm machinery. The shape and sizes of the fields have been changed; more dwellings have been erected, more waste produced and high intensity farming is now taking place. The crop yield and types of crop have also changed and require more fertiliser with high nitrogen and phosphorus content. Fungicides and insecticides reduce the loss of crop to fungi and insects, and weedkillers ensure the crop can take full advantage of the soil.

During the summer, short heavy bursts of rain wash some of the topsoil from the fields into the broads, taking with it the fertiliser, etc.; this is known as **runoff**. The intensified nitrogen and phosphorus levels in the water change the chemical balance, enabling algae to flourish and form **algal bloom**. As the blue-green algae spread they prevent aquatic plants photosynthesising and oxygenating the water, with the consequence that plant life (fish, snails, etc.) suffocate and die. Eventually the water becomes lifeless and affects the wildlife food chain.

Another water pollutant is effluent. In some areas untreated human effluent is disposed of directly into rivers. Sewage treatment works occasionally receive too much mixed surface water and soilwater during flash floods in summer or more prolonged flooding in autumn and winter. The treatment works are designed to accept a specific amount of mixed water and have extra holding tanks to allow for occasional stormwater. However, if the capacity is exceeded, the sewage treatment processes cannot be carried out and the effluent is dropped into the river or sea untreated. The consequence is an increase in ammonia, effluent and solid matter, including toilet paper, sanitary products and condoms, much of which is not biodegradable and will lay as waste on shore lines and river banks.

Farmyard effluent is another increasing source of nitrogen, phosphorus and ammonia. Factory farmed animals kept in sheds produce large quantities of urine and effluent. If the liquor flows through the ground or over the ground to rivers, the chemical balance is thrown out of control and aquatic life is affected. In contrast, animals that are allowed to graze naturally over large areas of land will fertilise the pasture, encouraging plant growth. In countries where concentrated farming techniques are used, such as The Netherlands, animal effluent is a major pollutant of rivers that flow into the North Sea and of great concern.

Fresh water

Fresh water accounts for about 4% of the total water on the planet. Only about 1% is accessible, with 3% stored in the ground or frozen as snow and ice. There are very few plants that can be used to feed animals or humans that survive in salt water, so as the population of the world grows, more demands are made on the available fresh water. The effects of pollution and climate change on fresh water supply are major concerns. There are two types of pollution:

1. **Natural pollution** such as tidal waters, storm induced floods and rising sea levels mixing with fresh water or covering low lying fertile land and rivers.
2. **Man-made pollution** – waste materials dumped under or on the land or discharged into the water and air.

There is little that man can do to change the natural pollution in the short term other than to understand the problem. Traditionally settlements grew up near fresh water on lower more fertile ground and are therefore more susceptible to flooding. The Netherlands and Bangladesh are just two of many countries worldwide that have highly populated land below or at sea level. Global warming is causing higher tides and sea levels to rise globally.

Global warming is a popular term that has been taken to the front of politicians' agendas and makes headlines for the press. It is a natural cyclical process that has occurred throughout the 4.6 billion years of the Earth's existence. The causes of global warming are open to debate, but the effects are both historic and current. In the early 1960s the British press published headlines like, 'The world is heading for another ice age', and then a decade or so later, 'The northern hemisphere will enjoy a Mediterranean climate that will allow us to holiday in the sun at home in the UK'. Historically, the global climate has changed, gradually becoming warmer; however, regionally large areas have experienced extreme temperatures in contrast to the global temperature increase. In the 1800s the weather in Europe became so cold that the River Thames froze over thickly enough for oxen to be roasted on the ice. The 'frost fairs' marked a period of extreme cold where even the summers were barely above freezing. There were massive shortages of food, animals could not feed, and many people and animals perished from the cold. In contrast, at the time of the Great Fire of London in 1666 the summer had been abnormally hot and dry and London was like a tinderbox.

Air

Air is a mixture of gases: 78% nitrogen and 21% oxygen with the balance 0.4% carbon dioxide and inert gases such as helium and argon. However, there are pollutants in air which include the following:

- carbon monoxide (CO)
- oxides of nitrogen (NO_x)
- volatile organic compounds (VOCs)

- particulate matter – normally referred to by size such as PM_{10} which are particles 10 micrometres in diameter
- nitrogen dioxide (NO_2)
- sulfur dioxide (SO_2)
- lead (Pb)
- ozone (O_3)
- toxic organic micro-pollutants, e.g. polyaromatic hydrocarbons (PAHs), polychlorinated biphenyls (PCBs), dioxins, furans.

In addition, there are many airborne natural pollutants such as sand and salt. If you visit the beach you may have noticed that the air tastes different. Salt cannot evaporate like water; however, it can become airborne as sea spray from wave action. Sand from the Sahara Desert has been deposited with rain in England after travelling thousands of miles high above the Earth's surface. Volcanoes erupt, sending millions of tonnes of naturally occurring pollutants into the atmosphere. On 18 May 1980 Mount St Helens erupted and sent an estimated 4 bn tonnes of matter into the atmosphere. The SO_2 content ranged from 500 to 3400 tonnes per day for more than six months. Two years later, El Chichón erupted sending an exceptionally acidic plume into the stratosphere that was estimated to contain 20 bn tonnes of SO_2 (see p. 389, What makes acid rain?).

To see what the air quality is where you live, look on the website http://www.airquality.co.uk.

10.2　What causes pollution?

Natural pollution from spectacular volcanic activity is limited to a few incidents per decade; however, sulfur dioxide is being vented continually from many places across the world. Methane levels are said to have risen due to increasing number of pigs and dairy herds being kept to supply the world's population with milk and meat. Paddy fields have increased in number to supply the increasing population with rice: the fields give off large volumes of methane during the flooding and maturing periods, but how much impact they have is debatable. Methane must have been produced before man killed off the massive herds of free roaming animals and drained the natural swamps. Volcanic activity is perhaps less now than thousands of years ago, so sulfur dioxide and carbon dioxide levels from these sources may have actually dropped. The main cause for concern is the change in the types of pollution, those made by man.

10.3　What is thought to cause global warming?

On Earth we cannot feel the heat from the Sun. Heat cannot travel the 150 million kilometres through the vacuum of space: we do experience the radiation energy though. The radiation travels in the form of very powerful short

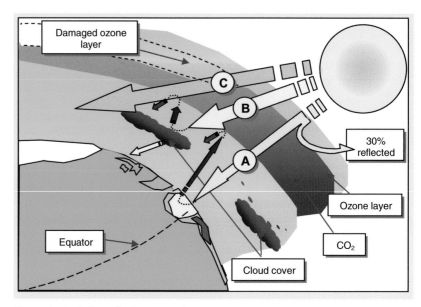

Figure 10.1 Solar radiation on Earth.

waves that cause whatever they collide with to vibrate submicroscopically and produce heat energy. Darker matt colours tend to absorb more radiation and therefore produce more heat than light, shiny surfaces. The thickness and the density of the material plus the surrounding materials or gases also have an effect. For example, the uninterrupted radiation from the Sun will produce a lot more heat energy on a clear day in the Sahara Desert (Figure 10.1 A) than say on a very cloudy day in the north of England (Figure 10.1 B). The short wave energy will bounce off the moisture forming the clouds, and the amount actually passing through will be very small in comparison to the clear sky above the desert.

Surrounding the planet Earth are several layers of gases known as spheres. The atmosphere acts as a shield repelling about 30% of the radiation back into space; however, that means that 70% gets through the layers, which is equivalent to about 10 000 times the amount of energy we currently use by burning fuel (Figure 10.1 A). When solar radiation collides with liquids, solids and gases on Earth it is either absorbed or changes to longer wavelength radiation and bounces off. Long waves cannot pass through gases as easily as short waves and therefore bounce back towards the Earth, heating it up. This is known as the greenhouse effect. Some gases, for example ozone, are more effective at stopping short wave radiation than others.

10.4   Ozone and other greenhouse gases

The layer of ozone high above the Earth's surface in the stratosphere provides an effective filter to exclude damaging ultraviolet (UV) radiation

(Figure 10.1 A). Each molecule of ozone (O_3) comprises three atoms of oxygen. In nature, oxygen atoms like to go around in pairs as oxygen gas: our respiratory system needs oxygen gas which makes up 21% of the air. However, when lightning occurs some of the oxygen molecules split apart and each atom joins another pair, forming ozone. Atmospheric ozone is also formed by intense radiation, such as UVB and UVC from the sun. Unlike oxygen, ozone is a poisonous oxidising agent that can be used to sterilise water.

Carbon dioxide (CO_2) has been identified as the major greenhouse gas. As the volume of CO_2 in the atmosphere increases, it effectively insulates the Earth, preventing the reflected long wave radiation passing to outer space and the planet increases in temperature (Figure 10.1 B). Other gases such as methane (CH_4) are more effective insulators that eventually break down to become CO_2. Trees and green plants photosynthesise during full daylight, absorbing CO_2 and producing the maximum amount of oxygen. Younger trees are more efficient than older trees at conversion. At dawn and dusk the exchange rate balances out as the tree absorbs CO_2 and gives out a similar amount of CO_2; at night trees give off CO_2 (Figure 10.2).

The gases CO and CO_2 are given off as combustion products of most fuels. When plants and animals die they decompose and produce CH_4; animals and humans also produce CH_4 as a waste product of their digestion processes. This all increases the amount of greenhouse gas in the atmosphere. Tonne for tonne, CH_4 is about 100 times more damaging than CO_2 in the atmosphere, and some scientists have suggested that CH_4 accounts for about one third of global warming.

Water vapour is another of the greenhouse gases. As the global climate warms, the rate at which water vapour is sent into the atmosphere increases; consequently the amount of rain increases together with the incidence of flooding worldwide. The cloud cover acts like a blanket, preventing heat from escaping into space.

Global warming is thought to be causing climate change. As the oceans warm, the number and strength of hurricanes, typhoons and tornados worldwide increase. Marine life migrates or becomes less fertile and reduced in numbers. Levels of phytoplankton and zooplankton have increased in some areas, resulting in more radiation being absorbed into the oceans, thus increasing their temperature and causing stronger and more frequent winds.

The winds above the polar icecaps form vortices, gathering ice crystals and chlorine molecules high above the Earth's surface which react with ozone. The resulting hole in the ozone layer allows unfiltered UV radiation access to the snow covering (Figure 10.1 C).

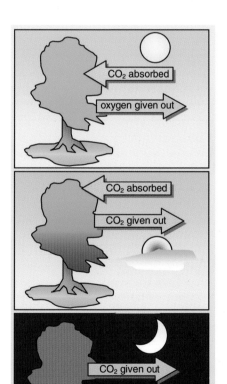

Figure 10.2 Oxygenation from trees.

The declining snowfall at the poles has resulted in a lower reflectance factor for solar radiation, and consequently more is absorbed by the ice. The ice pack cracks, allowing massive icebergs to break free and float away to melt in the warmer oceans. A similar change is happening to snow peaked mountains and remaining glaciers, resulting in an increase in sea level.

Frozen land masses such as the tundra in Siberia have begun to thaw, exposing millions of hectares of partially rotted plant life from the time before the last ice age. As the vegetation is exposed to the atmosphere, and in particular oxygen, decay can once more take place, giving off huge volumes of CH_4 to add to the increasing quantities of CO_2.

In Hawaii, the levels of CO_2 in the atmosphere have been measured and recorded since 1958. Data prior to 1958 has been obtained from ice cores from the South Pole. The levels of CO_2 have risen steadily from about 260 ppm (parts per million) in 1854 to 375 ppm in 2003 and are still rising. Anthropogenic (man-made) CO_2 has increased significantly since the 1950s, mainly as the result of electricity generation. An interesting website is http://www.metoffice.gov.uk/research/hadleycentre/pubs/brochures/2005/clim_green/slide11.pdf.

10.5   The Kyoto Protocol

In 1997, 168 countries (including European Union member states) signed and ratified (meaning that the official national government agreed) the Kyoto Protocol to work towards reducing greenhouse gas emissions to at least 5% below 1990 levels for the period 2008–2012. The targets cover emissions of the six main greenhouse gases:

- carbon dioxide CO_2
- methane (CH_4)
- nitrous oxide (N_2O)
- hydrofluorocarbons (HFCs)
- perfluorocarbons (PFCs)
- sulfur hexafluoride (SF_6).

Who monitors the levels though? Are they providing honest answers? Independent monitors suggest that the official data presented by the governments is incorrect by anything up to or more than 100%.

In November 2006, 6000 participants (including 100 ministers and two heads of state) at the United Nations climate change conference held in Nairobi discussed how they could support developing countries to reduce greenhouse gas emissions. In contrast to governments trying to reduce emissions, individuals are blatantly destroying rainforest and increasing CO_2 levels in catastrophic proportions. For example, according to Greenpeace International, Blairo Maggi, the largest individual producer in the world annually destroys 26 130 km^2 of rainforest in Brazil to grow soyabeans for animal feed; this is equivalent to about six football pitches per minute.

In Indonesia, the rainforests have been also been destroyed: in 1997–1998 over 2 000 000 hectares of forest were clear cut or burned, and the CO_2 released from the fires in that year is estimated to be equal to that from all the automobiles in Europe over the same period. Clear cutting means that literally every tree is cut down, leaving the stumps: environmentally it is a disaster. The trees would have converted the CO_2 into oxygen; their roots held the soil in place during the rainy seasons; without the roots massive mud slides occur as seen in the news. The shelter provided by the tree canopy held a humid layer beneath an umbrella of leaves; without the canopy more fresh water is required for irrigation of the new plantations. The rainforest was home to several endangered species, such as the great apes, orang-utans, Sumatran tigers and Asian elephants. Various native tribes have lived within the forests for countless generations with all their needs provided by the natural environment. Who allowed the devastation to take place? The government had laws in place preventing clear cutting without a permit, and burning was banned. It was subsequently found that corrupt government officials, including the ex-president Suharto, had made fortunes through the environmental destruction.

Where is the market? Things can only be sold if there is a buyer. In Europe and the USA, products have been promoted under the guise of being from a sustainable source. In most cases this was probably true; however, many came from plantations or shallow lakes that were previously rainforest or mango swamps. There is great demand for palm oils, rubberwood and farmed shellfish, so there is good incentive to provide them. Labels that claim an item comes from a sustainable source do not include a statement that the land was previously forest and had been cleared to allow the sustainable plantation to be planted. Shellfish have been grown in ponds where previously mango swamps had been before they were drained and commercialised. The goods are cheap and available; there is a label stating something about sustainability, so even people who are concerned about where the resource originates will have a clear conscience. The Japanese, however, look at the situation from a different perspective and just buy the hardwoods, being the biggest importer of hardwoods in the world.

10.6   Chlorofluorocarbons (CFCs)

Chlorofluorocarbons (CFCs) are man-made non-toxic, non-flammable non-carcinogenic gases. They contain chlorine, fluorine and carbon atoms, and were designed as propellants for aerosols, foam producers, solvent cleaners and refrigerants. When CFCs escape into the atmosphere they photodissociate and the resulting chlorine molecules destroy ozone, damaging the ozone layer. In 1987 the Montreal Protocol on Substances that Damage the Ozone Layer was signed by 24 countries. Since 1995 no new CFC gases have been produced in developed countries; however, there are still many appliances containing the gases. When these appliances, such as refrigerators and air-conditioning units, break down or need replacing, specialist equipment is required to remove the gases safely. In many countries, disposal is not controlled and the gases are allowed to vent into the atmosphere where they will cause damage for the next 200 years or so. In the 60 years or so of industrial use of chlorine, levels in the atmosphere have risen by 600%; chlorine is still a much used gas in the manufacture of plastics worldwide.

10.7 Pollution from fossil fuels

As far back as 1273, the use of coal was prohibited in London as it was considered 'prejudicial to health'. Later, in 1306, a Royal Proclamation prohibited craftsmen from using sea-coal in their furnaces because of the smoke emitted. With the introduction of mass coal burning in industrialised towns and cities in the early to mid 1700s, air pollution increased. During the 1800s several Acts of Parliament came into force regulating smoke emissions. Before the Industrial Revolution most people in Britain either worked on farms or from their homes in cottage industries. In London, there were a large number of small glass-works, tinkers and smiths. Each specialist joined a guild which looked after the interests of their members: fishmongers, masons, carpenters, lace makers and so on. (A visit to the various Guild Halls around the country will show the importance of the guilds of workers.) To supply the workers in the cities with food and drink, farmworkers produced crops, meat, ales and wines. The conditions in which city dwellers lived would be considered appalling today: the stench of the rivers, heavy sulfurous smoke, poor sanitation and lack of access to safe drinking water. To live into your thirties would be rare; to make it to five years old would be an achievement. The infant mortality rate was very high. Those who lived in the countryside may have had a cleaner environment and fresh food, but oppressive landowners kept wages low; most workers lived in tied accommodation but lived longer.

The Industrial Revolution has been blamed for climate change. Since the invention of machinery powered by burning coal, then burning oil and gas, pollution rates have soared. Scientists and clergy were concerned about pollution as far back as the early 1700s and tried to persuade the world not to pollute the land and seas. However, scientists do not make the policies; politicians do. They are elected and laws are made not by the poorest but the wealthy – those who 'have'. Rarely do people who 'have' openly show concern for those who 'have not'. Exceptions include George Cadbury, son of the founder John Cadbury of Cadbury chocolate fame. George tried to change the attitude of those in power; the mill and factory owners who employed the mass population. He wanted better living conditions for his workers and set up the town of Bournville as an example. George Cadbury junior started teaching on Sundays, enabling education for the workers in their time off.

Ebenezer Howard was another notable figure; however, unlike the wealthy Cadbury family he had spent time in America in the wake of the great fire of Chicago. He observed how the town was being rebuilt, and on return to England published a book in 1898 entitled *To-Morrow: A Peaceful Path to Real Reform*. His persuasion of like minded people gave birth to the association that built the first garden city in Letchworth. Both the Cadbury family and the wealthy Lever family provided the power and influence to further the idea of providing a much healthier environment for the general population.

William Lever, son of the wealthy soap manufacturing Lever family, bought land in the Wirrel to build Port Sunlight. He built new villages and factories, and provided amenities for the workers: meeting halls, recreation areas and places to worship. These developments were far from the hellish environments of 'back to back' houses in long terraces, with no gardens, poor sanitation, and no recreational buildings or open spaces. Typical living areas were 13–16 m^2 in damp, cold and filthy conditions. Unfortunately, there is not enough space in this book to allow further reflection on the history of the Industrial Revolution. It was not an isolated part of middle England, most of Europe had been developing in a similar way. Industrialisation spread like a plague across Asia and to the Americas. Revolutions or civil wars followed in many civilised countries.

10.8   Green belt

In response to instructions from national government, town, borough and parish councils in the UK designated areas around heavily built up industrialised centres that could not be built on. The land formed a belt that prevented spreading industrialisation consuming more green field sites. The owners of the land could continue to farm and erect farm-associated buildings, but were not allowed to build factories, etc. Today's government is suggesting that green belt is no longer required and the land should be made available for building.

10.9   Air pollution

Historically, all major industrialised towns and cities in the UK were divided into the poorer inhabitants living to the east and north east, managerial and professional workers in the north west, very wealthy owners of business in the south west and the less wealthy in the south east. The reason for the divide was to do with air pollution. As the predominant wind direction is from the south west, if your dwelling was to the east or north east you would be living continually in a sea of sulfuric smoke from the many coal fired power houses; consequently your health would have suffered and your life expectancy would have been very much reduced. If you find an old graveyard near an industrialised town, look at the tombstones. If they are limestone, the acidic air and rain will have probably eroded all of the text; however the inscriptions on some granite tombstones may still be legible.

The government tried to reduce smoke emissions from industrial sources with the Smoke Abatement Act 1926. Further measures came into force after the killer smogs of the 1950s. It is thought that in December 1952 there were about 4000 smog related deaths. In 1956 the government brought in the Clean

Air Act, making it illegal to emit smoke from homes in designated smokeless zones; the Clean Air Act 1968 regulated the height of chimneys and the content of emissions.

Air pollution today

So far we have looked at the historical aspects of air pollution, but what about today in the 21st century? Air pollution is still a major cause for concern. The levels of sulfur dioxide that caused the high level of deaths in London are still a problem.

Today, coal burning power stations tend to use Canadian and Australian coal with a lower sulfur content than British coal. The coal is cheaper to extract from thick seams close to the surface where opencast mining allows gigantic plant to be used. The coal is then transported to Britain by ship, again keeping the cost per tonne below that of British deep shaft coal.

British coal fired power stations are becoming less polluting. For example, the massive Drax power station consumes about 10 million tonnes of coal per annum. Limestone scrubbers filter the pollutants, removing above 90% of the SO_2 by converting it to flue desulfurisation gypsum, the main material used in plasterboard (Figure 10.3).

Air pollution can be categorised as follows:

- local
- regional
- global.

By permission of Drax Power Ltd

Figure 10.3   An aerial view of the Drax power plant.

Figure 10.4 An aerial view of a cement works.

Local air pollution

Local air pollution could be caused by a factory burning fossil fuel to make cement (Figure 10.4). Cement manufacture is the fifth greatest cause of pollution on Earth. Fuel to provide heat during the process is commonly coal, gas or oil, all three of which produce large volumes of CO_2 when burnt, together with several other polluting gases. Modern plants have filters that help reduce the pollution; however, they do not stop it entirely and CO_2 is still emitted. Castle Cement has been trialling supplementing normal fuel with old rubber tyres. The large number of old tyres that go to landfill can be significantly reduced by burning them in a controlled manner: rubber burns at a very high temperature; therefore this is an ideal way of disposing of the tyres which also have a much lower nitrogen content than coal, so far less NO_2 is emitted.

Some coal fired power stations also burn alternative fuels and waste products under trial conditions. Eventually, scrubbing devices and CO_2 sinks (to sequester carbon) will be fitted to minimise the effects of atmospheric pollution, and enable power generation and cement making to be clean operations.

At present, most waste from domestic or commercial sources is sent to landfill where CH_4 will be vented into the atmosphere as it decomposes. Modern landfill sites could incorporate impervious gas layers to cap the fill so that as the gases form they can be piped and stored as fuel for space heating or electricity generation. When waste is incinerated, plumes of polluting gases are sent high into the sky to fall onto the surrounding countryside as dust or in the rain.

Probably the most dangerous local air pollution is from transport in city and town centres. Catalytic converters on car exhausts are designed to reduce the emissions of CO and hydrocarbons (unburned fuel) significantly. These were originally designed for engines running on unleaded petrol, but versions for diesel fuelled cars and lightweight vans have been developed; however, neither version reduces CO_2 emissions. Catalytic converters only work when they reach their optimum temperature, which takes about 7 km or 10 minutes' running; before the optimum temperature is reached, the gas emissions are more toxic than a car without a catalytic converter. Studies show that about 58% of car journeys at rush hour, including the school run, are less than 8 km of uninterrupted driving and therefore create higher levels of pollution.

Vehicles in town and cities rarely achieve the correct speed for optimum efficiency and therefore consume more fuel per kilometre than cars on fast roads. Although diesel fuelled vehicles are proven to achieve better fuel economy (more kilometres per litre of fuel) than petrol engined vehicles, they also emit more volatile organic compounds, particulates (PM_{10} down to $PM_{2.5}$) and nitrogen oxides, which are the main causes of asthma and respiratory problems. Black sooty carbon PM_{10} particles are known to be carcinogenic. Recently, studies have shown that $PM_{2.5}$ may be more dangerous as they can be inhaled further into the lungs. Particulate matter is so small it can remain floating in the air for hours, days or even weeks.

Another gas emitted during the combustion of petrol and diesel is SO_2. Many oil companies market sulfur reduced fuels under various trade names. In cities, during hot days with high air pressure, sunlight converts some of the greenish brown smog into low level ozone (ozone smog) which is very damaging to the respiratory system.

Car manufacturers have produced models that are able to run on biofuels such as methane gas and petrol, petrol and electricity, or alcohol from distilling sugar and other plants (Figure 10.5). These are termed **hybrid vehicles** by the United Nations. At present, the technology for running a car on electricity whilst in town is very limiting: in theory the petrol engine charges batteries in the car whilst it is in conventional mode and then the car can be powered at low speeds in congested towns by an electric motor. Feedback from owners of such cars has revealed that petrol consumption is higher and distance travelled on electricity far lower than claimed by the manufacturers; however, new improved versions are being developed. In The Netherlands, a commercially produced car is available that can run on hydrogen fuel cells; it is very expensive, both in terms of initial capital cost and fuel, but as production increases the capital costs should fall.

Design

Environmentally, a great deal can be accomplished by design. In the past, accommodation was frequently designed to be in close proximity to places of

Figure 10.5 A hybrid car – petrol and electric powered.

work, reducing the amount of distance travelled to work. Many European countries have kept to the design of apartments over the shops and offices so keeping the need for personal transport to a minimum. However, over the few past decades domestic housing has been reduced in towns and shopping parks have been built on the outskirts. Lifestyle has changed from small quantity shopping as required on a daily basis, to monthly shopping plus top up shopping, requiring vehicles to transport the goods. Several larger chains of shops have returned to offering delivery services, something that smaller shops had been offering in the past.

Solutions

London Docklands is an example where new housing/apartments have been built close to the shops and offices (Figure 10.6). London's Barbican is home to several thousand people. It has been criticised as being a concrete jungle, but its tranquil water gardens and colourful balconies provide terraces and towers minutes from a good public transport system and a major city centre.

The city transport system is much hampered by illegal users of bus lanes and the poor attitude of many drivers who stop in restricted areas. In an effort to reduce unnecessary traffic a congestion toll has been introduced. It is debatable whether it has reduced the amount of congestion or whether it is a levy on the people travelling in the congestion zone. By analysing the availability of public transport against the number of new offices built in London, it is easy to see that the planning authorities have allowed an excessive increase in office

Figure 10.6   Canary Wharf offices, shops and housing – a modern local community.

accommodation without increasing the infrastructure, such as new roads or railway lines. Virtually all of the overground railway lines and stations were built in Victorian times, with the exception of the London Docklands Light Railway. At the time of its construction, local people were very opposed to any development as it entailed the demolition of thousands of homes and businesses. If London wanted to build more railway lines and stations there would be major problems in providing the necessary land.

Efforts to reduce pollution and increase the number of places on public transport have been tested in Croydon, South London, with multi-car electric powered trams (Figure 10.7). Electric powered trams and trolley buses were previously used in the period between the horse drawn omnibus and the petrol engined 'old bill' buses of the early 20th century and the diesel engined buses of the 1950s. Electric trams are almost silent in operation, fast, and can transport large numbers of people in and around town centres. They are non-polluting at the point of the vehicle, although the method of producing the electricity may be a source of pollution. They are restricted by the power supply whether it is rails or overhead catenaries. If a fault develops, the system is blocked until the bus or tram can be removed. Tram rails are similar to railway lines; however, the tops of the rails need to be flush with the road surface to enable other vehicles to pass over without hindrance. Cyclists and motorcyclists have to be wary in case their tyres become caught between the rails, especially at junctions. The dedicated tramways in Croydon ensure that motorists cannot illegally park in or use the routes; where road surfaces and tramways cross, the rails are flush with the road surface.

A limitation of trams is rigidity of route. In most cases, any maintenance of the utilities that pass under the rails will require the route to be closed. Unlike other vehicles, trams cannot follow diversions or drive over steel plates bridging the utility trenches whilst they are open.

Figure 10.7 Electric powered modern tram.

Trolleybuses with overhead catenaries do not require rails. Essentially the vehicle is a double deck bus powered by mains electricity. The power input is via cables suspended from poles. The trolley bus has two arms with strong springs pushing the connecting arms against the cables. In contrast to the rigidity of trams on rails, the pivoted arms on the roof of a trolleybus allow some flexibility of movement either side of the power cables. Both trolleybuses and trams are currently being considered as solutions to London's traffic pollution problems.

Hydrobuses are currently being tested in several major cities around the world. These single deck buses use hydrogen fuel cells to produce electricity to power the motor (Figure 10.8), as mentioned in Chapter 11, Section 11.3. The exhaust is purely water vapour formed by hydrogen combining with oxygen. However, energy is required to produce hydrogen and this is expensive. European countries obtain hydrogen from methane gas and steam. Canada is developing ways of using hydroelectricity and electrolysis to produce hydrogen, and other countries are looking at solar methods of producing the gas. At present, hydrogen buses cost about ten times more per mile to run than a conventional diesel powered bus, but in terms of the cost of pollution there is no contest, especially in city centres.

Regional air pollution

When large areas create high levels of pollution, it no longer affects only local people. The intensity of the pollution enables it to be carried on the wind for hundreds of miles without much dilution. Coal fired power stations, steelworks, cement works and brickworks generate enormous volumes of CO_2 and

The only emission is water vapour

Figure 10.8 Hydrogen powered bus in service producing only water vapour.

SO_2. In compliance with the Clean Air Act 1968, chimneys were increased in height (see Section 10.9) and consequently the pollution travelled further on the prevailing wind to Norway and Sweden where it was deposited as acid rain. Rain is naturally slightly acidic with a pH of about 6. However, with large regular discharges of SO_2 from industrial sources and NO_x and SO_2 from vehicles, the rain becomes a weak mixture of sulfuric acid and nitric acid of pH 4, or in severe cases pH 2.

When acid rain falls on tree foliage, it removes the waxy film from the leaves and interferes with photosynthesis. The trees in Norway and Sweden are mainly coniferous and have needle-like leaves (Figure 10.9 A) that turn brown before dropping off and decaying over a very long timescale. Coniferous forests have reduced vegetation at ground level (Figure 10.9) due to the naturally acidic decomposing leaves. In contrast, deciduous trees (generally broad leaf types; Figure 10.9 B) drop their leaves once a year. As their leaves rot they form a rich humus promoting a diverse range of flora and an ideal habitat for wildlife.

What makes acid rain?

Sulfur (S) produced naturally or by the combustion of fossil fuels oxidises to form SO_2. As it enters the troposphere it bonds with single oxygen molecules

Figure 10.9 A coniferous forest and examples of coniferous (A) and deciduous (B) leaves.

to form sulfur trioxide (SO$_3$) which in turn bonds with water vapour (H$_2$O) to form sulfuric acid (H$_2$SO$_4$; Figure 10.10).

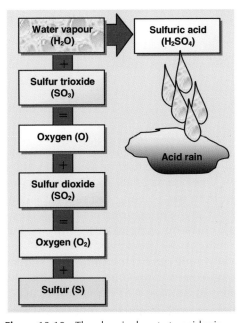

Figure 10.10 The chemical route to acid rain.

Nitrogen oxides (NO$_x$) are given off during the combustion of fossil fuels. Nitric oxide (NO) is oxidised by ozone, forming nitrogen dioxide (NO$_2$) which in turn reacts with hydroxyl radicals (OH$^{\bullet}$) to form nitric acid (HNO$_3$) that falls back to earth with the rain. The consequence of acid rain can be seen on many ancient limestone and marble buildings, from the ancient Greek and Roman temples to the cathedrals of northern Europe (Figure 10.11) and on round the globe to the Taj Mahal in India.

Air pollution does not stop at the frontiers of each country so is termed transboundary air pollution. Wealthy countries like the former Western Germany have introduced legislation to reduce pollution; however, the situation was very different in poorer post-communist East Germany which, like many poorer countries, had old inefficient manufacturing and energy production facilities. Burning of the local coal, lignite, which has a high sulfur content, produced extremely high

Figure 10.11   Effects of acid rain on a limestone building.

levels of air pollution. Neighbouring Czech Republic and Poland are also relatively poor, and also contributed to a regional air pollution problem known as the 'black triangle'.

Air pollution can be attributed to old technology and a lack of finance. We may want to travel in a car powered by hydrogen but can only afford an old-technology polluting diesel car. Likewise, large companies in poorer countries may not be able to afford low sulfur coal and a carbon sequestration plant, and continue to use highly polluting plant.

Global air pollution

As the heading suggests, global air pollution affects us all. A wealthy country can afford to ban certain industrial techniques, either by legislation or by heavy taxation. Cement is an example. One country may reduce its local air pollution, yet import tonnes of cement from less wealthy countries that cannot afford to forego the export income. The manufacturing country perhaps produces more pollution per tonne of cement by using older technology, but continues production as part of the economy. As previously mentioned, the pollution adds to the CO_2 levels associated with global warming. In a similar way, the production and use of CFCs have been banned in countries like the USA and most of Europe; however, there are still less wealthy countries using them or dismantling old machinery and venting CFCs to the atmosphere with consequent damage to the global environment.

10.10 Environmental audit

An environmental audit usually takes stock of what is there, what is planned for the future and the cost implications to the environment. For example, companies employ consultants to prepare environmental audits as part of their business plans. Even councils and public authorities carry out environmental audits. So what would an environmental audit look at?

1 Running costs – both financially and environmentally.
2 Waste disposal – how can it be minimised?
3 End of life – what happens to the buildings after they are no longer required?
4 What effect will the project have on the natural environment?

Running costs

Energy is a costly item. Changing the method of space heating from, say, oil fired boilers to natural gas fired boilers will probably reduce carbon emissions and allow cheaper fuel to be used. Installing monitoring meters enables running costs of isolated plant to be recorded. One major hospital spent about £250,000 per year on electricity just to power the lifts.

Electronically monitoring the use of lighting to turn it off when there are no workers such as at night and over the weekends can reduce the amount of energy used and related pollution, and also reduce light pollution at night.

Installing thermostatic controls for individual areas that enable adjustments to be made when one part of the building is heated by the Sun whilst the remainder is in shade reduces energy demand and associated pollution. Buildings currently monitor air temperature during the air-conditioning process, something we will look at again in Section 10.11. It is better to use less energy to heat than more energy to cool. Upgrading electronic valves can make significant savings in energy demand, especially in commercial and industrial buildings.

Increasingly, ventilation of buildings is achieved by air-conditioning systems. Sick building syndrome is regularly associated with air-conditioned or air-regulated buildings. It is estimated that we currently use more energy cooling our buildings than heating them, a problem that will dramatically increase with climate change.

Waste disposal

Waste is very costly. Has too much of a material been ordered or delivered? If so, what will happen to the surplus? It is costly to dispose of waste. Since the advent of computers and photocopiers, the amount of paper used and wasted has increased dramatically. Not only is this wasted paper, but wasted energy and pollutant chemicals such as inks, bleaches and clay washes.

What about the packaging? Can it be recycled? What about the transport and associated pollution to bring and dispose of the waste? Environmentally this is damaging. Paper can be recycled several times, but fibre quality tends to deteriorate and the process to improve the quality increases costs, making it more economical to buy new paper each time. Have you seen recycled photocopy paper? Some companies are publicly advertising that they use recycled paper, but how much is recycled and how much is new? There is no excuse that using recycled paper will clog machines or give inferior copies. Even down to the loo paper: is it 100% recycled in your building? One use of loo paper and it is disposed of, so why do most people want new paper?

Printer cartridges and photocopier cassettes are very costly items environmentally. Some companies will refill them with ink and recycle the plastic containers. However, manufacturers commonly state that any non-genuine printer cartridges will invalidate their warranty. Waste is costly and often harms the environment: it may go to landfill, be incinerated, go into the sewage system, or just be dumped illegally.

End of life

In the context of buildings there are several important factors (see also Section 10.11). Building materials commonly require maintenance. It may be the occasional wash, such as for the windows, new paint, new coverings to the floors and walls, and so on. However, there comes a point when it may be too expensive to maintain a building or the running costs are too high. The size of the building may no longer suit its use or there is no longer a requirement for that type of structure. The building must go. Or can it be modified? A good example would be the old dock buildings of East London. Massive buildings that once housed cargos from around the globe stood empty for decades. Developers bought the empty warehouses and converted them into luxury waterfront apartments. Buildings sometimes can be reused to the benefit of the environment.

Some buildings do have to be demolished. Environmentally many contain hazardous materials such as asbestos, which would be highlighted in the environmental audit. What will happen to the many different materials used in the old structure? Can some be salvaged and reused? What will happen to most of the debris? Landfill is often the final destination; however, the following European and national legislation is having an effect on legal disposal:

- Control of Pollution Act 1974
- Environmental Protection Act 1990
- Pollution Prevention and Control Act 1999
- Pollution Prevention and Control Regulations 2000 with amendments
- Directive on the Landfill of Waste (1999/31/EC) – a tax on waste per tonne was levied
- Directive on Hazardous Waste (91/689/EEC, as amended)
- Hazardous Waste Regulations 2005

- Lists of Wastes (England) Regulations 2005 – this regulation replaced the Special Waste Regulations 1996
- Directive Concerning Integrated Pollution Prevention and Control (IPPC) (96/61/EC)
- Landfill (England and Wales) Regulations 2002 – the regulations categorise waste into twenty main groups using about 900 codes and require a brief coded statement of waste being transferred on the public highway and in the event of spillage and final disposal, similar to road tankers having a code identifying what material is being hauled
- Landfill (England and Wales) (Amendment) Regulations 2004
- Directive on Waste Incineration (2000/76/EC) – the principle is to limit air pollution, possible soil pollution and water pollution.

A very useful website for further study into waste is www.wasteonline.org.uk.

What effect will the project have on the natural environment?

Disposal of a building can have a detrimental effect on the local environment. If the building is of a traditional construction, some of the materials may be salvageable such as:

- roof tiles
- roof timbers if the roof is of a cut design and not gang nailed trusses
- timber joists and floorboards
- doors and windows
- steel joists
- stone cladding
- bricks – with older buildings that have lime mortar it may be possible to salvage the bricks which are very sought after for extending or altering other old buildings of the same brick type.

Services in traditional housing can be removed after the floorboards have been removed. Cabling and water pipes buried behind the plaster may be reclaimed from the debris after the walls have been pulled over. Larger more exposed materials are removed prior to demolition, such as steel radiators, cast iron or pressed steel baths, copper hot water cylinders, steel water cisterns: stainless steel sink units, unless virtually new, are usually sent for scrap. The value is measured per unit of weight set by the scrap metal market. Ceramic products such as WC pans and close coupled cisterns, shower trays, bidets, etc., will have their metal fittings removed. The common metals used are plated brass and, depending on the age of the item, some lead and brass water traps. If the ceramic is undamaged it will be reclaimed; however, it should ideally be removed to prevent it contaminating the demolition masonry of the walls. Door frames and linings and window frames, if not salvageable, should also be removed.

10.11   Is timber frame so environmentally friendly?

It is estimated that 800 000 tonnes of construction and demolition timbers are not salvaged. About half goes to landfill, over a third is burned on site, with the remainder burned off site or just dumped. It is estimated that currently about a quarter of all construction and demolition waste is timber. How much will that increase when the timber frame houses erected over the past three decades come to the end of their useful life in the next half century?

What is sustainable construction?

Before considering different forms of 'green' construction techniques, we should look at the wider question; what is sustainable development?

Sustainable is a relative concept. There is no time factor. Sustainable suggests that something should last a long time, but how long? Politicians like to talk about sustainable development and sustainable construction, and perhaps one of the most often quoted definitions of all came from the World Commission on Environmental Development report 1987 entitled *Our Common Future*.

> Sustainable development is development that meets the needs of the present without compromising the ability of future generations to meet their own needs.

As a principle it cannot be faulted; however, there are factors that prevent its progress.

Sustainable construction in the UK is considered in terms of selecting materials that can be replaced. Trees, for example, can be farmed for use as timber and fuel. To an extent they are a sustainable resource that can be recycled or disposed of at the end of their usefulness. Timber frame housing in theory should be sustainable; however, the common practice of nailing sheet materials such as orientated stranded board (OSB) to the timber frame as sheathing presents a problem. Many of the buildings have a brick veneer attached to the sheathing by wall ties. The insulation is sandwiched between the OSB and the plasterboard wall lining (see Chapter 8, Figure 8.4). The electrical cables and service pipes are built in, threaded through the forest of studs and noggins (see Chapter 8, Figure 8.6). When the buildings are to be demolished the contractors will have to remove the services, cables, pipes, etc., and separate the materials. At present, demolition companies 'ball' the outer skin of brickwork (a large metal ball is suspended on a cable and swung against the brickwork) and, where possible, remove any salvage materials before depositing in landfill sites or open fire burning. Timber frame panels will be difficult to dismantle and dispose of. Machine nailed sheathing to the timber framing and the plasterboard on the other side will leave metal fixings that will prevent chipping for use as fuel. The contaminates such as plasterboard will require time consuming removal, especially where plasterboard screws have been

Figure 10.12 Demolition ready for recycling the majority of the building materials.

used as opposed to nails. The insulation should be easily removed by contractors wearing full appropriate personal protective equipment (PPE) to prevent breathing in the hazardous dust from glassfibre insulation. Will the contractors erect specialist removal screens in a similar way to those used for asbestos removal?

Contrast timber frame with traditional brick and block, and skeletal frame structures. The norm is for the demolition contractor to remove as much of the contaminates as possible, including asbestos. In domestic buildings the process starts with removing all salvageable materials as mentioned previously. Plasterboard ceilings may be taken down and placed in disposal skips, then the walls are demolished, either by a ball suspended from a cable or a large hydraulic armed machine (Figure 10.12). The main difference between modern timber frame and traditional construction is the sheet material sheathing. Traditional construction rarely has a problem. Traditional construction materials, other than gang-nailed roof trusses and timber studwork partitions, can be salvaged or recycled, and therefore landfill is minimal. In contrast, timber framed housing has fewer recyclable materials and therefore more disposal problems.

If cement works can be fuelled by alternative means and the pollutant gases scrubbed before carbon sequestration, the fifth major source of global pollution would be significantly reduced. Concrete, glass and steel can be recycled, and last for generations. They all require large amounts of energy in their manufacture and transport, but with the correct design should outlast many 'green' materials that require more maintenance.

It is inconceivable that cement, concrete and steel will be replaced, so the manufacturing processes must be cleaned up. The Japanese have produced steel in greater volumes and with far less pollution than steelworks in other countries. They have invested in very sophisticated and expensive technology, proving that improvements can be achieved. In contrast, the poorer countries of Europe and Asia are still using highly polluting methods of production, because the financial investment cannot be recouped. Pollution is a product both of the wasteful rich and the poor countries of the world.

Chapter 11

Energy

Our use of energy is perhaps one of the most important factors in today's world. To understand what energy is, where it comes from and how can we harness it, this chapter sets the scene by starting with the basic science of energy. This is followed by energy calculations to enable comparisons to be made between various sources of available power. The main body of the chapter discusses the range of power sources, together with calculations to show efficiency and comparative costs in terms of fuel use and pollution.

11.1 The science of energy

What is energy? The dictionary definition is:

> The capacity of matter or radiation to do work. The means of doing work by utilising matter or radiation.

Scientists tell us that energy cannot be created or destroyed, it just changes its form. Most of the energy arrives from the Sun as light and radiation varying in wavelength across the **electromagnetic spectrum**. On Earth, we cannot feel any of the heat from the Sun, just the light and radiation. The other main source of energy is from the decay of radioactive rocks within the planet, again radiation. So what is radiation?

The dictionary definition is:

> The emission of energy as electromagnetic waves or as moving particles.

Without going too deeply into the world of physics and chemistry, **chemical elements** are made up of **atoms**. If we look at an atom it comprises three types of particle:

- proton – positively charged (+1)
- neutron – no charge
- electron – negatively charged (−1).

Every atom contains a **nucleus** of positively charged protons and, usually, the same number of neutrons (Figure 11.1). Surrounding the nucleus are negatively charged electrons orbiting in set layers termed **shells** which can each hold a limited number of electrons. The negative charges of the electrons balance out the positive charges of the protons, rendering the atom neutral. If, however, another electron joins the atom, it becomes negatively charged and is known as a **negative ion**; for example, chlorine ion, written as Cl^-. If an electron breaks away from the atom it becomes a **positive ion**; for example, sodium ion, written as Na^+. Although the nucleus usually comprises an equal number of protons and neutrons, sometimes additional neutrons can be added without altering the overall electric charge of the atom, giving **isotopes**.

Atoms have a specific mass based almost entirely on the mass of the protons and neutrons in the nucleus (the mass of the electrons in insignificant by comparison). Over 100 known elements are listed in the **Periodic Table**, sorted according to their atomic mass. Hydrogen is the smallest element so has been given atomic number 1.

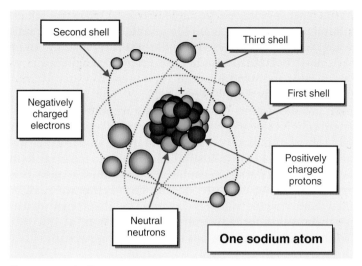

Figure 11.1 Sodium atom.

Atoms can be attracted to or repelled by other atoms. Where two atoms are attracted to each other they may **bond** and share some electrons; the resulting entity is called a **compound**. For example, one atom of hydrogen bonds with two atoms of oxygen to form water (H_2O). The ratios of elements making up a compound are constant.

In contrast, where elements or compounds are blended together they form a **mixture**. They can be in any proportions and each element or compound keeps its own properties. For example, salt crystals and water can be blended together relatively easily but the atoms that make up the salt do not bond with the water, so the water can evaporate leaving the salt. It is this process that takes place with efflorescence.

In summary, substances are made up of:

- elements
- compounds
- mixtures.

Why, you may ask, do we need to know about atoms to understand energy? Everything is made up of atoms. Although the ancient Greeks came up with the theory over 2000 years ago, it is only in recent decades that we have begun to understand the properties of matter more fully.

Energy exists: we cannot make it or destroy it. The Sun provides light in the form of radiation and virtually everything around us depends upon it to exist. Trees and green plants use sunlight to drive photosynthesis which changes carbon dioxide and water into oxygen and sugar. Without light the process ceases. Plants also require water, without water in some form, life cannot exist.

As we humans evolve, we are consuming more natural resources in our quest for energy. Initially, we used fire to keep warm and eventually to cook food. Then demands for energy increased as wood was burned to provide heat to produce metals from ores (earth-like rocks). During the hours of darkness and to illuminate the insides of buildings, oils, mainly from plants and animals including whales, were burned. Candles made from tallow (animal fat) provided solid, more portable sources of light. Energy from muscle power, water power, wind pumps and windmills was used to move water for irrigation and grind dry foods such as corn. However, things changed with the discovery of coal which produced more heat energy than wood or oil and took longer to burn. Water boiling vessels increased in volume, allowing steam to be produced in usable quantities. Larger quantities of metal, in particular iron, could be produced; this lead to the Industrial Revolution in the early to mid 18th century. The invention of the steam engine allowed valuable metal ores such as tin and copper to be mined at greater depths. The engines were originally designed to pull water out of the mines; however, as their efficiency increased they became the power source for industrial mass production.

With the introduction of machine power the population of the world significantly increased:

- 750 000 000 in 1750
- 1 200 000 000 in 1850
- 2 500 000 000 in 1950

Currently the world population is estimated at:

- 6 500 000 000 in 2006

and is set to exceed:

- 9 000 000 000 by 2032.

11.2   Energy calculations

As the world's population increases, the demand for energy expands, although not proportionally; wealthy countries obviously use more energy than poor ones. To compare energy demand a common unit or material has to be used. For many years coal dominated the production of energy, so statistics were compared as per tons of coal (the imperial ton). Britain dominated the world in the production of coal, iron and steel, so the use of the BTU (British thermal unit) became popular. Energy could be measured as units of heat produced from the combustion of a set amount of coal.

One BTU is the heat energy required to raise the temperature of 1 lb of water by one degree Fahrenheit. This equates to 1055 joules.

The BTU is an imperial measurement still used in the USA. Eventually, Britain adopted the European unit the **calorie**.

One calorie is the heat energy required to raise the temperature of 1 gram of water by 1 degree Celsius. This equates to 4.19 joules.

Oil became the dominant tradable commodity, so energy was compared to the amount of heat produced by a barrel of oil. The unit 'barrel' is based on the size of the barrel used to transport the oil in the 1860s which contained 42 US

Table 11.1 World Energy use for the year 2004.

	Use of energy expressed in billions of tonnes of oil equivalent (billions of toe)	Increase expressed as a percentage over the previous year (%)	Increase expressed in millions of tonnes of oil equivalent (millions of toe)
Rest of Asia	3.4	5.8	200
Europe and Eurasia	3.0	1.1	33
North America	2.8	0.3	8.4
South America	0.6	4.5	27
The Middle East	0.5	4.5	20
Africa	0.3	1.8	5.4

Source: BP Statistical Review of World Energy June 2006.

gallons (equal to 35 imperial (British) gallons or about 160 litres). The density of oil varies between the different types, from thick oil (heavy oil) down to thin oil (light oil), so a standardised unit of 7.33 barrels per tonne has been agreed upon. The energy contained in a barrel also varies; therefore for statistical reasons one barrel contains 5.71 GJ. The oil industry equates oil production in units of the **barrel**, i.e. millions of barrels per day (**Mbd**) whereas energy statistics can be quoted in barrels of oil equivalent (**boe**). World energy statistics are compared using the unit of tonnes of oil equivalent (**toe**). When quoting energy statistics every primary fuel can be converted to **toe** (Table 11.2).

How much energy does the world use? World use of primary energy has been calculated as 424EJ for the year 2000. When discussing enormous quantities of energy, specialised SI (Système Internationale) units are used. Based on the power of 10, Table 11.2 shows the more commonly used multiples. Using the prefix letters enables very large or very small numbers to be considered; for example, it is relatively easy to imagine a bottle full of sand but it is almost impossible to visualise how many particles of sand are contained in the bottle. Energy is measured in the unit **joule**. As previously mentioned, 1 calorie equals 4.19 joules, a very small quantity, and one barrel of oil contains 5.71 GJ. Using Table 11.1, write down the number of joules in a barrel of oil, showing all the noughts behind the number. Then write down the number in full for 424EJ. When calculating energy use and production/conversion, it is much easier to use prefix letters instead of all the noughts. Calculations technically should contain numbers; however, in this book units have been shown in the calculations for clarity.

Task
Calculate the world energy use for the year 2000 in **boe**.

Table 11.2 SI prefixes.

Symbol	Prefix	Multiply by the power of	Written number
E	Exa	10^{18}	1 000 000 000 000 000 000
P	Peta	10^{15}	1 000 000 000 000 000
T	Tera	10^{12}	1 000 000 000 000
G	Giga	10^{9}	1 000 000 000
M	Mega	10^{6}	1 000 000
k	kilo	10^{3}	1 000
h	hecto	10^{2}	100
da	deca	10	10
d	deci	10^{-1}	0.1
c	centi	10^{-2}	0.01
m	milli	10^{-3}	0.001
μ	micro	10^{-6}	0.000 001
n	nano	10^{-9}	0.000 000 001
p	pico	10^{-12}	0.000 000 000 001

Standard notation

To make calculations with large numbers easier, standard notation (SN) is used. This means that the number should be between 1 and 10 and be multiplied by 10 to the appropriate power. Therefore the biggest that the number before the decimal point can be is 9. For example:

- 10 would be written as 1×10^1
- 100 becomes 1×10^2
- 1000 becomes 1×10^3

The power of ten is termed the **exponent**. The number is shown before the decimal point followed by the decimal of one, if appropriate, so 7.75 is the decimalised way of showing $7\frac{3}{4}$. In SN

- 7750 would be written as 7.75×10^3
- 775 becomes 7.75×10^2
- 77.5 becomes 7.75×10^1
- 7.75 becomes 7.75×10^0

How to enter numbers in standard notation in a calculator

Enter the number between 1 and 10 (remember the largest number is 9). If using a Casio scientific calculator, enter the number and then press the button marked '×10^x' (for other makes of calculator consult the instructions booklet for 'exponential' notation). You should have '1×10' on the screen (Figure 11.2). Now enter the exponent. For example, to enter 2000 (2 × 10^3) in SN (Figure 11.2), enter the number 2 and then press '×10^x' followed by the number 3. To check that you have entered the numbers correctly, press the equals button. Does it now display 2000?

Try entering the following numbers in SN:

- 125
- 5 000 000
- 7580
- 8.675

Quantities smaller than one are expressed as decimals of one, meaning how many parts of one shown in tenths, hundredths, thousandths, etc. To enter smaller numbers in SN format into the calculator use the following procedure.

SN requires a number between 1 and 10. The number being considered is less than one; however, with SN we can adjust the magnitude (size) by using the exponent. Enter the first number behind the decimal point followed by the

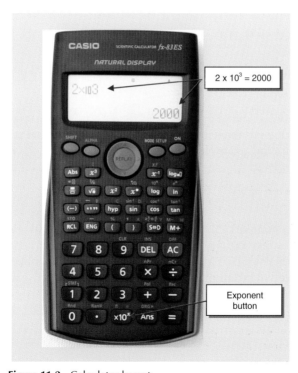

Figure 11.2 Calculator layout.

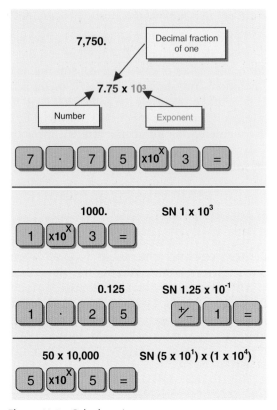

Figure 11.3 Calculator input.

decimal numbers (Figure 11.3).

> Q1. How would 424 EJ be written?[1]
> Q2. How would 5.71 GJ be written?[1]

To multiply the numbers in SN, multiply the numbers and add the exponents as in the following example:

$$50 \times 10\,000 = 500\,000$$

If this is written in SN it would be

$$(5 \times 10^1) \times (1 \times 10^4) = 5 \times 10^5$$

Multiply the numbers (5×1) then add the exponents $10^1 + 10^4 = 10^5$. Therefore $5 \times 10^5 = 500\,000$.

To divide the numbers in SN, divide the numbers and subtract the exponents as in the following example:

$$500\,000 \div 50 = 10\,000$$

If written in SN it would be

$$(5 \times 10^5) \div (5 \times 10^1) = 10\,000$$

---

[1] Q1: 4.24×10^{20}; Q2: 5.71×10^9.

Divide the numbers $(5 \div 5)$ then subtract the exponents $10^5 - 10^1 = 10^4$. Therefore $1 \times 10^4 = 10\,000$.

Some calculations require comparisons of energy used in a specific time unit. For example, 1 joule of energy used in 1 second is given the unit 1 watt, named after the scientist James Watt. Electricity can be measured in watts with the unit prefixes of kilo for 1000 watts, mega for 1 000 000 watts, giga for 1 000 000 000 watts, and so on (Table 11.1).

Task

Write 1 kW, 1 MW and 1 GW in SN (note that the prefix kilo is shown in lower case; upper case K represents kelvin which is a unit of heat. The *unit* (watt) is normally shown in lower case with the *symbol* (W) in upper case as it is the first letter of the scientist's surname).

If 1 watt is the unit for 1 joule per second $(1\,\text{W} = 1\,\text{J/s or } 1\,\text{J s}^{-1})$ how many joules would be used in an hour? The answer is:

$$1 \times 60\,\text{seconds} \times 60\,\text{minutes} = 3600\,\text{J per hour}$$

This would be written as 1 kWh.

$$3600\,\text{J} \times 1000\,(\text{kilo}) = 3\,600\,000 \text{ could also be expressed as } 3.6\,\text{MJ}.$$

We can now calculate how many kWh there would be in 1 toe if 1 toe equals 7.33 barrels and there are 5.71 GJ per barrel:

$$1\,\text{toe} = 7.33 \times 5.71 = 42\,\text{GJ}$$

If 1 kWh $= 3.6$ MJ, then 1 kWh also equals 0.0036 GJ or 3.6×10^{-3} GJ. Therefore

$$42 \div 3.6 \times 10^{-3} = 11667\,\text{kWh}$$

As values are not precise when considering energy on a large scale, it is common to round up to the nearest whole number which gives 12 000 kWh.

$$\text{Therefore 1 toe} = 12000\,\text{kWh}.$$

In the UK energy is converted to a usable form from a range of sources:

- wind
- wave
- tidal
- hydroelectric
- current
- solar
- geothermal
- fossil fuel combustion
- nuclear
- incineration and waste.

Much of the energy is converted into electricity which is convenient, clean and instant at the point of use. However, we shall compare some of the underlying factors.

Generally, electricity is an energy associated with wealth. Richer countries use a disproportionate amount of the world's energy. In 2000, the average North American used about 6.4 toe, whereas the average European used slightly over 3 toe (similar to that of the former Soviet Union which has significantly reduced consumption since the 1990s) and the rest of the world about 0.75 toe. These figures are rapidly changing, with China and India projected to be using ever increasing amounts of energy over the mid 2000s.

Task

Convert the total world primary energy use of 424 EJ to watts.

Start by calculating the number of seconds in a year:

Days in a year × hours in a day × minutes in an hour × seconds in a minute

$$365 \times 24 \times 60 \times 60 = 31\,536\,000\,\text{s}$$

Divide 424 EJ by 31 536 000 seconds

$$4.24 \times 10^{20} \div 3.1536 \times 10^7 = 1.344 \times 10^{13}\,\text{W} = 13.4\,\text{TW}.$$

If there are 6.1 billion people on the planet what is the average annual use of energy?

$$1.34 \times 10^{13} \div 6.1 \times 10^9 = 2196\,\text{W per person}$$

If we round that figure to 2.2 kW per person, then multiply it by the 24 hours in a day we obtain a consumption of 53 kWh which can be converted to joules as follows:

$$1\,\text{kWh} = 3.6\,\text{MJ}$$

Therefore energy consumption is $3.6 \times 53 = 190\,\text{MJ}$.

A **boe**, as previously mentioned, contains 5.71 GJ of energy per 160 litres.

$$5.71 \times 10^9 \div 160 = 3.56875 \times 10^7 \text{ or } 36\,\text{MJ per litre}$$

$$190 \div 36 = 5.28$$

Therefore the average energy consumption per capita (per person) expressed in litres of oil is slightly over 5 litres per day, or globally 3.172×10^{10} litres.

Energy consumption for one year is given by the equation:

Litres of oil per day × days in a year ÷ litres in a barrel ÷ barrels in a tonne

$$3.172 \times 10^{10} \times 365 \div 160 \div 7.33 = 9\,871\,930\,423\,\text{toe}$$

or, as is more usually quoted, 10 100 Mtoe per year.

The energy requirements of the world are not increasing in line with the population growth of the planet. As countries try to survive, they become more dependent on energy. This is an issue more fully discussed in Chapter 10.

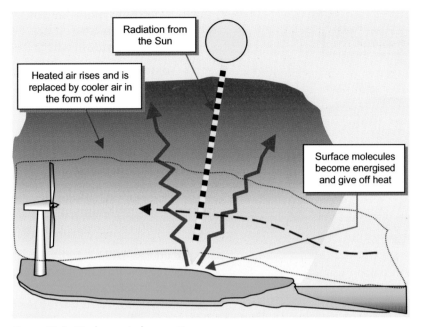

Figure 11.4 Onshore wind generation.

11.3 How is energy converted into electricity?

Wind power

As a group of islands surrounded by seas and an ocean, the UK has probably the greatest potential for using wind power. **Wind turbines** can be used:

- onshore
- offshore.

Note that they are *not* windmills: their function is to generate electricity via a turbine. The principle is based on warm air rising and cooler air taking its place – wind (Figure 11.4). Wind turbines should be placed where the wind regularly blows; therefore much research must take place before choosing the location. The wind turbine blades are angled (**pitched**) to resist the force of the wind, causing the blades to be pushed aside. As the blades are fixed to a central shaft the forces acting on all three blades are transferred into the rotational forces required to turn the turbine. The area that the rotating blades cover is termed the **sweep** (Figure 11.5).

We can calculate the amount of energy and express it in watts based on the following data:

- Air density at normal pressure is 1.29 kg m^{-3}
- Wind speed is 15 metres per second (m s^{-1}) = velocity (v)
- Length of the turbine blades is 10 m.

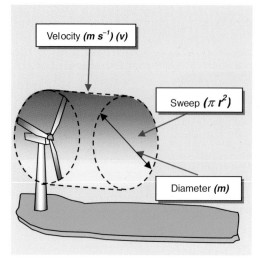

Velocity *(m s⁻¹) (v)*

Sweep *(π r²)*

Diameter *(m)*

Figure 11.5 Wind turbine energy input and potential output calculation.

Consider how much air will pass through the blades in one second. It can be calculated as a cylinder, a volume of air:

Volume = Area of the sweep

$\times$ distance travelled in 1 second

$= \pi r^2 \times \text{length}$

$= \pi 10^2 \times 15 \, \text{metres}$

$= 4712 \, \text{m}^3 \, \text{of air}$

To find the total mass of the air multiply the density by the volume of air:

Mass $= 1.29 \times 4712$

$= 6079 \, \text{kg}$

The energy will be derived from the mass of air pushing the turbine blades every second; it is termed **kinetic energy**.

Now, using the following formula the data can be converted to output in watts.

Output $= \frac{1}{2} \, m \, v^2$

$= \frac{1}{2} \times 6079 \times 15^2$

$= 683\,888 \, \text{W}$

$= 684 \, \text{kW}$

Q3. If the wind speed dropped to, say, 10 m s⁻¹, what would be the total output?[2] A large amount of energy is given off as waste heat caused by friction and the number of revolutions the blades can physically make.

Wind turbines can only work when the wind speed is fast enough to rotate the blades and below a speed above which it becomes too dangerous. These are known as the **cut-in** speed and **cut-out** speeds (Figure 11.6). Some of the smaller turbines do not have cut-out speeds.

Vestas are one of the biggest manufacturers of wind turbines in the world. They currently produce 3 MW turbines that cut in at 4 m s⁻¹ and cut out at 25 m s⁻¹. 25 m s⁻¹ is about 56 miles per hour. Therefore potentially there are periods that the turbines cannot produce electricity. Average annual offshore coastal wind speeds range between 9 and 10.5 m s⁻¹, compared with onshore sites ranging between 6.5 and 8 m s⁻¹. Note that these are average annual figures so calm days and strong winds have also been averaged out.

There are other considerations with onshore wind turbines (Figure 11.7):

- Aesthetics – some people object to onshore wind turbines on the basis that they are unsightly. Perhaps the communication masts, the rows of electricity pylons and the industrial processes that enable the built environment are also unsightly; however, without them could we exist?

---

[2] Q3. 304 kW.

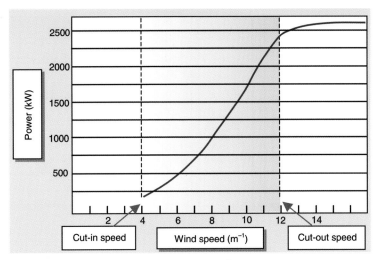

Figure 11.6 Wind turbine energy output graph.

The American saying NIMBY (not in my back yard) can frequently be used when debating technology. We have looked at the subject in Chapter 10.

- There are claims that they are noisy – that is true of some of the early models where the turbines were connected to a generator at ground level. The new models generate within the head of the turbine, significantly reducing the noise. Typically they produce about 100 – 105 dBA at the head. If the turbines are sited correctly they are less noisy than road traffic. When grouped together in valleys they are very noisy as the sound reverberates between the valley walls. In California, USA, the noise from the hundreds of wind turbines near Palm Springs can be heard miles away; however, no dwellings are close enough for this to be a problem.

- Wind turbines are a very expensive way to generate carbon free electricity – it is true that the unit cost per kilowatt is significantly higher than fossil fuel power stations. Also, the materials and transport costs in carbon emissions should be taken into account (copper for the windings, steel and concrete required to form the towers, extra runnage of less efficient cabling to transfer electricity to the user) together with higher maintenance costs and the conditions under which the engineers will have to work.

- The efficient use of resources and economy of scale. An argument against the use of many wind turbines based on the amount of copper and aluminium required per unit output of electricity is valid. Turbines used in large scale power stations use less copper windings than the equivalent power output from wind turbines. The wattage output is higher from power station turbines (more accurately the generators); therefore power losses to transmission are less. Less power cabling is required from a power station than the number of wind turbine arrays capable of generating the same amount of electricity.

97 m tall
1.5 MW max output
Blade length 33 m
Cost > £1 m

ecotricity

Figure 11.7 Wind turbine with an observation platform.

- Wind turbines cannot reliably generate electricity at a constant rate – again this is true when compared with conventional power stations. See Chapter 10 for arguments for the use of wind technology.

Offshore wind turbines can overcome some of the problems of continuity of wind, as wind speed is almost constant; also, noise becomes less of an issue. However, other problems arise:

- It is more difficult to attach the turbine tower to the sea bed. Caissons of concrete and rocks or reinforced concrete piles/shafts are required on sand beds in the North Sea. It is more difficult to bury the cables when siting turbines in the North Atlantic off the west coast of Scotland because of the rock sea bed.
- Maintenance can be a problem in stormy weather, and the barge cranes for heavy lifting require local harbourage.

- Fishing using long drag net trawling cannot take place between the turbine towers, having an effect on the fishing industry. (It can also be seen as a benefit as the fish stocks can rejuvenate.)
- The location of the wind turbines requires many kilometres of cabling to link up with the main grids and therefore significant transmission consumption.
- Location must be relatively near an onshore link if the power is to be run on the National Grid.

Wind turbine technology is progressing rapidly, with several major companies investing in very expensive production lines which have the effect of reducing the unit cost per turbine. Countries such as Germany, Spain, Denmark, France and North America are very committed to wind power energy. In the UK progress is somewhat slower due to objections and regulations. The UK land mass and population density, plus the social and governmental approach of the people, are different from the above named countries.

Wave power and sea pressures

Wave power generation is still in its early stages with the Limpet, sea snake and several small scale experimental systems. The main difference between wave power and sea pressures are that waves are surface movements mainly caused by changes in wind pressures. The gravitational pull of the moon also causes waves from tidal effects. In contrast, sea pressures termed **swells** are below the water surface. The swells can also result from gravitational effects of the moon and weather conditions, such as atmospheric conditions of high and low pressures, but can still be present below the water surface and not reliant on localised winds.

Sea snakes require constant wave motion. Each sausage-shaped float has a generating device attached to the articulated coupling which produces electricity during every motion. The greater the wave activity, the more electricity the sea snake can produce.

The advantages are:

- No fuel is required so there are no carbon emissions
- Carbon-free electricity is generated from water power
- Waves are naturally generated and should be inexhaustible.

The disadvantages are as follows:

- Equipment can be damaged by fishing processes – although the areas are identified by marker buoys
- Equipment can be damaged by vessels during storm conditions
- There is a requirement for strong anchorage to prevent the snake floats being taken out to sea
- Generation is reliant on wave generation and therefore subject to seasonal changes and weather conditions
- Power cables must be run over the sea bed and protected from anchor drag damage from vessels in stormy weather

Picture courtesy of Wavegen

Figure 11.8   Limpet electricity generator.

- Maintenance is restricted to periods of relative calm and availability of seaworthy craft.

The **Limpet**, in contrast, is a land based rigid power plant. The locations require relatively deep water adjacent to the land to utilise the swell of the sea with a minimum depth of 5m; therefore they tend to be in isolated locations such as northern Scotland (Figure 11.8). The Limpet is not reliant on surface conditions such as waves. (Swells are caused by pressures that could be kilometres away pushing downward on the water or in some case upward from the ocean bed. The water cannot compress therefore the pressure radiates from the source causing the swell effect beneath the surface throughout the full water depth in a pumping action.) The ocean floor must be deep enough for there to be little to no effect by friction or drag, so about 5m depth is ideal for harnessing the energy (Figure 11.9).

   The concrete sidewalls direct the flow of water into a narrowing chamber, pushing the trapped air upward and through the turbine causing it to rotate. When the full energy of the positive pressure has been exhausted, the water rushes back causing a negative pressure sucking air into the chamber (Figure 11.10). The **Wells turbine,** named after its inventor Professor Alan Wells, can operate in either direction, unlike most turbines. The blades are symmetrical and look similar to an elongated tear drop without a pitch. As the air pushes against the fatter surface of the blade, the force causes a lift. The blades cannot move in the direction of the force and are therefore pushed aside causing the hub and shaft to rotate (Figure 11.11). A heavy flywheel is

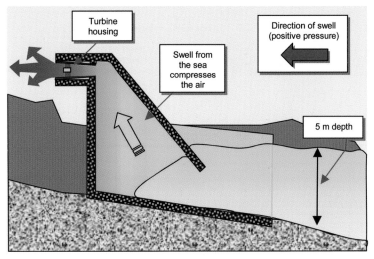

Figure 11.9 Diagrammatic section of the Limpet.

connected to the turbine shaft continuing the rotational force during the short periods of time that the direction of pressure changes.

The advantages of the Limpet are as follows:

- It is ideal for remote communities requiring constant electricity
- Power cable runs on land are short so there is reduced power loss, and ease of access for maintenance purposes
- Power cables cannot be damaged by fishing processes or anchor drag
- Virtually guaranteed constant swell, unlike tidal systems
- Ease of access for maintenance of the turbine in any weather condition without the need for barge cranes or seaworthy craft
- No fuel is required
- Can operate day and night, even in snowy conditions
- Can be linked to the National Grid system 11kV AC
- Power generation schemes are less obtrusive as they can be designed to blend into the local terrain and require relatively low turbine tower heights.

There are the following disadvantages:

- relatively small turbine sizes at present
- locations tend to be away from the mass population.

The Japanese have recently installed shore-mounted oscillating water column (OWC) power plants. Incorporating the Wells turbine, they use a tower with the turbine mounted horizontally to generate electricity. Other projects include breakwater barriers that both reduce the erosive power of the oceans on the coastline and have integral OWC power units to produce electricity. The development of water powered devices has tended to be slower than wind powered

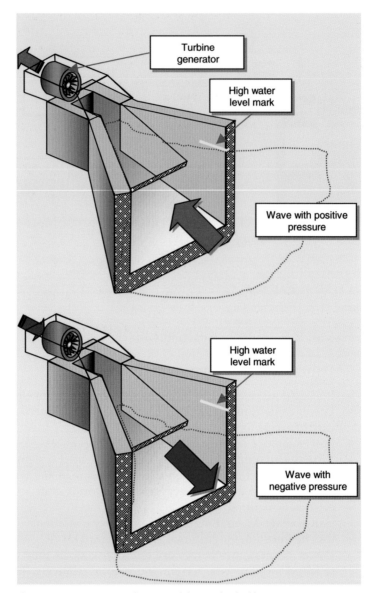

Figure 11.10 Isometric drawing of the method of harnessing water power.

ones due to market size. OWC power plants need to be sited along the coastlines of active oceans to harness enough power.

Tidal power

Tidal power generation has been under consideration for several decades. The guaranteed regularity of two tides per day has been considered by the British government for both the River Thames and the River Severn. The principle is to construct a barrier across a strong tidal river and direct the water through

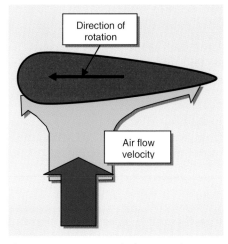

Figure 11.11 Sectional drawing showing the aerodynamics of the turbine blade.

turbine generators. The tidal water flows upriver through the turbines generating electricity. At high tide the barrage is closed off, trapping the tidal water and forming a reservoir. When the water level of the receding tide is low enough the entrapped water in the reservoir is allowed back to the sea via the turbines, generating more electricity (Figure 11.12). Perhaps the most famous barrage is the La Rance barrage, finished in 1967 in northern France. Although it is still producing electricity, the French government and power generating companies have not planned any further barrages. The scale of the proposed Severn Barrage would dwarf La Rance several times over, so based on the French experience there are many objections to the proposal.

The following main factors have prevented the schemes going ahead:
- colossal cost and lack of finance
- the large amount of concrete required to form the barrages

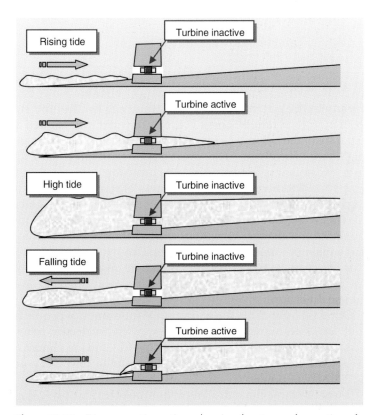

Figure 11.12 Diagrammatic sections showing the stages of operation of a water barrage.

- extent of the natural environment that would be changed forever
- barrage size.

Other smaller schemes have been considered such as the Mersey Barrage. The latest information is dated 1989 when the UK government announced in the House of Commons that they were in negotiation to see if it was commercially viable.

- La Rance is 740 m long with an output of 240 MW
- the Mersey Barrage would be 2 km long with a potential 700 MW output
- the Severn Barrage would be 15.9 km long with a potential 8640 MW output.

The issues relating to tidal barrages include:

- Initial cost and final costings – who will pay if the project becomes out of control? Historically, large projects always go well over budget, the Channel Tunnel being a good local example.
- Period of construction time
- Environmental upset
- Output is only over a set timescale dictated by nature
- Output is not constant. The power input is based on the head of water and the speed of the tide – there are designs to store tidal water in reservoirs to spread the output time.

Hydroelectric power

In 2000 about 2.3% of the world's energy needs were met by hydroelectricity. Hydroelectric power is ideal where water can be captured at high levels such as in mountain ranges. Large volumes of water are required and, with the geography of the UK, can be found in Wales, parts of northern England and in the highlands of Scotland. Those areas tend to be of low population density; therefore transmission factors have to be considered. As an alternative, damming rivers and flooding valleys is a good way of storing water in theory is in low lying land nearer to the mass population; however, there is potential for massive disaster if the dam fails. This can be due to natural ground movement caused by significant changes to water courses or the extra mass of stored water. Water has a mass of 1 tonne per cubic metre. Diverting natural water courses can, in extreme cases, even dry up seas to become deserts. For example, the Aral Sea, the fourth largest lake in the world, has almost dried up as a result of water course diversion.

Other issues include the relocation of thousands of people from their homes and farms to allow the world's largest hydroelectric schemes to be built. The Three Gorges hydroelectric scheme in China is almost complete. It harnesses the power from the Yangtze River and in 2008 will potentially generate 18 200 MW of power.

There are three main designs of hydroelectricity power stations:

- high head

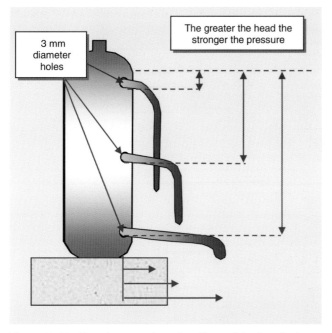

Figure 11.13 Experiment to show the effects of the 'head' of water.

- medium head
- low head.

The energy is derived from the gravitational pull on the water. Water is subject to atmospheric pressure on the surface; however, the deeper the water the greater the pressure from the mass of water above.

To visually prove the theory use an empty 2 litre fizzy drink bottle. Drill three holes and cover them with sticky tape (Figure 11.13). Fill the bottle with water and remove the sticky tape from all three holes, noting how far the water is spurting from the bottle. The pressure is related to the amount of water above the hole which is termed the **head**. Hydroelectric installations with a head greater than 100 m will produce pressures in excess of 11 atmospheres. Atmospheric pressure is 100 000 pascals (Pa) in SI units (14 lb per square inch in imperial units); therefore high head will be at least 1 100 000 Pa, medium heads range between 10 and 100 m providing 200 000–1 100 0000 Pa and low heads range between 100 000 and 200 000 Pa (Figure 11.14). The turbine blades are driven by water funnelled through nozzles concentrating the kinetic energy from the water to rotate the turbine. There are a range of turbine types such as the Francis turbine, the Pelton, Kaplan, Turgo, crossflow and fixed pitched propeller. The turbine type is matched against the head of water and the efficiency required. The pipes that transfer the water from the reservoir to the turbines are termed **penstocks**. The Hoover dam, for example, has sixteen 4 m diameter penstocks fed from the dammed Colorado River reservoir named Lake Mead holding 30 billion cubic metres of water.

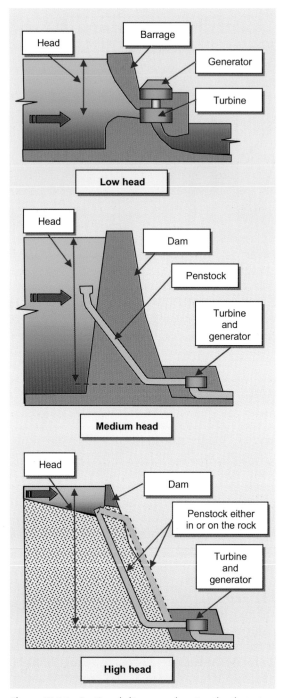

Figure 11.14 Sectional diagrams showing the three types of hydroelectric generation.

The energy within water currents can be harnessed using submersible turbines. In principle they are similar to wind turbines with only two horizontally opposed blades submerged in the path of the current. The turbines are attached on a sliding mechanism to steel masts fixed to the sea bed. To enable maintenance the turbines can be raised out of the water. Unlike wind power, water currents are relatively constant and unaffected by weather on the surface of the water. The density of water is significantly greater than that of air so the ratio of revolution to torque is far better. This means the current turbine output is about four times greater than the same size wind turbine. Potentially large tidal rivers could be harnessed; however, the turbine blades would require modification to facilitate the changes in flow direction. In the Thames Estuary a new offshore wind farm has recently been completed. At present there are planning difficulties with the onshore connections; however, it would be ideal to site a farm of water current turbines within the forest of wind turbines. The electrical connections to the grid are or will be in place very soon and the strong tidal current twice per day would significantly increase the potential power generation. There are issues of ships colliding with masts either submerged or above the water line, but locating both forms of renewable energy in one area should reduce the risks.

The benefits of current devices are:

- Single seabed fixing of the mast – installed using barge platform drilling rigs
- Minimal concrete required when compared with barrages and dams
- Minimal environmental disturbance
- Unaffected by the weather conditions
- Maintenance can be carried out just above waterline
- Less obtrusive than most power plants, although this is only relevant to NIMBY objectors
- All construction materials can be recycled safely
- No fuel is required and there is no pollution to air, land or water.

At the time of writing, the SeaFlow project offshore from Lynmouth, Devon, is currently producing electricity, although it is not connected to the National Grid. Phase 2, named SeaGen, comprises the installation of a turbine with a 1.2 MW output off the coast of Northern Ireland in 2007 (Figure 11.15).

The method for calculating the potential energy from the current is similar to that used for the wind turbine. The volume of water displaced can be calculated as a cylinder. The example is based on a turbine sweep of 10 m and a current velocity of 2.5 m s^{-1}. The density of water is 1000 kg m^3 (Figure 11.16).

$$\text{Volume} = \text{Area of sweep} \times \text{distance travelled in 1 second}$$
$$= \pi r^2 \times \text{length}$$
$$= \pi 5^2 \times 2.5 \text{ metres}$$
$$= 196 \text{ m}^3 \text{ of water}$$

© Marine Current Turbines Ltd

Figure 11.15 Water current submersible turbines generating electricity.

To find the total mass of the water multiply the density by the volume of water:

$$\text{Mass} = 1000 \times 196$$
$$= 196\,000\,\text{kg}$$

The energy will be derived from the mass of water pushing the turbine blades every second.

Now, using the following formula the data can be converted to output in watts.

$$\text{Output} = \tfrac{1}{2}\,m\,v^2$$
$$= \tfrac{1}{2} \times 196\,000 \times 2.5^2$$
$$= 612\,500\,\text{watts}$$
$$= 613\,\text{kW}$$

The potential energy is greater than the actual output; however, the velocity of the water current is regular and predictable.

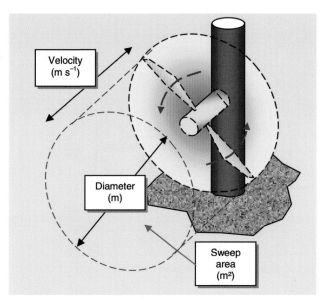

Figure 11.16 Submersible turbine energy calculations.

Solar power

Solar power technology can be classified into two groups:

- heating units
- direct electricity units.

Heating units collect radiation from the Sun and produce steam to drive electricity generators. They can be on an enormous scale, such as the parabolic solar power stations in America and Spain. Highly polished mirrored surfaces reflect the solar radiation and concentrate it either onto a feed pipe or solar store (Figure 11.17). In a similar way to concentrating the Sun's rays using a magnifying glass, the solar power stations concentrate the radiation, raising the temperature in the feed pipe liquid to between 200 and 400°C (water boils at 100°C). Solar stores can achieve temperatures up to 3800°C.

As the energy is a direct result of solar radiation, the output is limited to the day time; however, recent developments have enabled some storage of energy using molten salt for up to two hours. The efficiency of solar power stations is less than 25%. The feed pipe systems take up considerable space and require clear skies and strong radiation. Areas in Australian and American desert regions are most suited. The solar stores can be small scale dishes with the solar store positioned in the front of the dish (ideal for isolated villages) or solar towers with arrays of mirrors reflecting the radiation with a 10 MW output.

Evacuated tubes

These are the most efficient of the small unit collectors (Figure 11.18). In contrast to the previous solar collectors, evacuated tubes are used to heat water and not

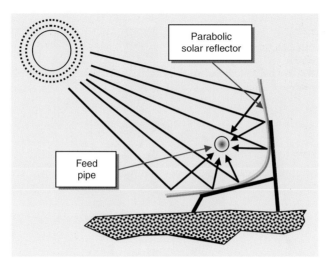

Figure 11.17 Parabolic solar collectors.

Figure 11.18 Evacuated tube solar water heaters.

generate electricity. The efficiency is excellent; however, it is proportional to scale and location. In the summer months the ability to heat water for showers and body washing is very good. In contrast, during the periods of thick cloud, rain and snow, the output becomes very low. The glass cover plate must remain clean to optimise the radiation onto the tubes. The method of collection is via a line of clear tubes made with thin glass walls (Figure 11.19). The air has been removed (evacuated) producing a vacuum similar to a Thermos flask. Inside the tube is a black coated pipe containing a long black coated strip of copper about 50–100 mm wide. Inside the inner pipe a liquid with a low boiling point surrounds the copper strip. The radiation passes through the outer tube of glass through the vacuum and is absorbed into the black surface of the copper pipe. Most of the radiation is absorbed into the metal, generating heat. The hot

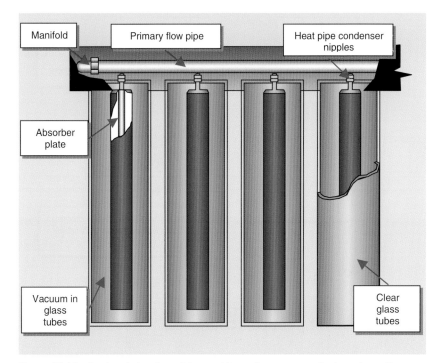

Figure 11.19 Sectional drawing of the workings of an evacuated tube solar collector.

metal conducts the heat energy into the metal strip inside the tube which in turn causes the liquid to boil. Boiling liquid becomes a gas that rises within the metal tube where it comes into contact with the heat exchanger containing the primary flow water. As the heat is transferred, the gas condenses to a liquid and falls to the bottom of the tube where it will continue in the heat cycle. A thermostat stops the water moving in the primary feed tube whilst it absorbs the heat. At the required temperature the thermostat opens and the circulator pushes the heated water through the heat exchanger in the hot water cylinder (Figure 11.20). The primary feed pipe should be well insulated (**lagged**); however, antifreeze similar to that used in a car radiator will prevent freezing. The circulator requires an electrical input which could be powered using a photovoltaic cell. Where used in the UK, another method will be required to heat water in the cooler months. Also the heat build-up takes several hours and therefore limits the amount of hot water that can be used at one time.

Other flat plate solar collectors are less efficient than evacuated tubes. If the amount of energy needed to produce and deliver the materials is compared with the amount of energy saved with its use, it will take many years to even balance out. In areas where cloud cover is less, the equation becomes very different. The heat gain is via solar radiation, light, and has nothing to do with air temperature; therefore it is akin to comparing, say, a mountain village in winter in Switzerland on a clear day with north west England. The ambient temperature may be 10°C higher in England, but the cloud cover will prevent efficient working.

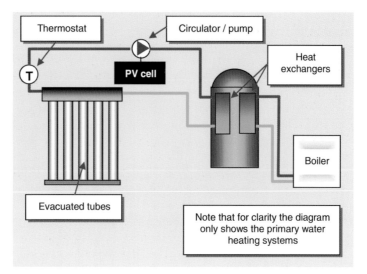

Figure 11.20 Diagrammatic layout for evacuated tube solar water heaters.

Photovoltaic cells

Photovoltaic (PV) technology has developed more rapidly over the past decade than any time since its inception in 1839 by Edmund Becquerel. The process is based on the movement of electrons. Two dissimilar semiconducting materials, one negative and one positive, are layered together. Although commonly silicon, other materials can be used. To modify the silicon to become n-type (negative), phosphorus is added containing a surplus of free electrons. The other layer is coated with boron, causing the surface to have a deficit of electrons, and becomes p-type (positive). The process of exchanging electrons between the layers is caused by sunlight. As previously mentioned, sunlight is radiation and arrives in bursts of energy known as **photons**. The stronger the light, the more energy is converted, up to a maximum of about 0.5 V at 2.5 A which equates to 1.25 W, typically from an area of 100 cm^2.

As you can see, it requires a large number of PV cells and direct sunlight to produce a relatively small amount of electricity. Where several PV cells are placed together, they are termed an **array**. Machines requiring a low electric current, such as remote telephones and illuminated road signs, can be powered by batteries trickle charged throughout the day (Figure 11.21). The PV cells produce direct current (DC); therefore connecting them in series will provide usable voltages of 12, 24 or 36 V DC.

Orientation has a significant effect on the amount of direct sunlight received by the array. Remote emergency telephones in France have timed tracking motors attached that rotate the array to maximise the charging time. In contrast, UK road signs like that shown in Figure 11.21 have fixed arrays and therefore only receive the maximum light input for a few hours per day per year.

The efficiency of the PV cells is greatly reduced when clouds mask the sunlight, so fluctuations in output occur. The most common approach to overcome

Figure 11.21 Photovoltaic cell powered street sign.

the fluctuations is to use PV cells with storage batteries and a secondary backup system. In countries where the strong sunlight is virtually uninterrupted by clouds, the output can be very high. The angle of receipt changes with the seasons; therefore countries on the same latitude as Spain or below will benefit the most. Other countries further north, such as Germany and The Netherlands have invested heavily in PV technology. At best the highest efficiency conversion to electricity is 16%.

The energy payback has become a prominent factor showing how long the PV cells will have to operate at zero carbon emissions to balance out the energy and carbon release during their manufacture. At present, the best estimates are about 1.7 years based on the most efficient products. The estimated life of a PV cell ranges between 25 and 30 years; however, where used in countries with very high irradiation levels the useful life is reduced. The technology over the past decade has increased the efficiency whilst reducing the overall costs. Earlier examples from the 1990s are being renewed with the higher performance versions.

PV cells are ideal for low voltage lighting and light load machines such as domestic freezers and refrigerators. Computers and similar communications equipment require relatively low power DC consumption. However, heavy loading machinery, such as washing machines, dishwashers, air-conditioning units, vacuum cleaners and the like, would require an enormous number of PV cells for minimum use, so they are therefore impractical for this application.

Example

If a PV cell maximum output is 1.25 W per 100 cm^2, calculate the area that would be required to produce 2.5 kW to power a washing machine (make no allowance for energy losses due to inversion – converting DC to AC, etc.).

$$\text{Power} \div \text{area} = 1.25 \div 100$$
$$= 0.0125 \, \text{W/cm}^2$$
$$\text{Energy required} = 2.5 \, \text{kW}$$
$$= 2500 \, \text{W}$$
$$2500 \div 0.0125 = 200\,000 \, \text{cm}^2$$

Area conversion 1 m = 100 cm

$$\therefore 1 \, \text{m}^2 = 100 \times 100 \, \text{cm} = 10\,000 \, \text{cm}^2$$
$$200\,000 \div 10\,000 = 20 \, \text{m}^2$$

Therefore at least 20 m^2 of PV cells would be required, without any allowance for energy loss due to converting the DC output to AC or transformer losses converting the low voltage to 230 V AC. The output would fluctuate significantly with cloud masking and be greatly reduced during the winter months if used in the northern hemisphere above, say, 49° latitude.

Where energy figures are quoted they are frequently based on 1000 W/m² at 25°C. This figure is approximately equal to that experienced on a horizontal surface in Saudi Arabia at noon in June, and somewhat different from a rainy day in October in the north of England.

Geothermal energy

There are many places where the Earth's crust has been significantly fractured or is relatively thin, or radioactive decomposition is close to the surface. Volcanoes mark weak points within the Earth's crust where the enormous energy trapped within the planet escapes.

The rock formation around Southampton is basin shaped, creating a zone where hot rocks are relatively close to the surface. The extinct submarine volcanoes still provide heat to the underground waters of Bath and Southampton. The brine water percolates through the hot Devonian beds of rock at 76°C about 1.7 km below the surface. It pushes upward through the well pipe under its own pressure to about 100 m below the surface dropping in temperature by only 2°C before it is pumped into the power station. The heated water goes through heat exchangers to transfer as much heat as possible into the secondary circuit. Originally the heated water was used for hot water and space heating; however, the secondary circuit has recently been modified by the addition of a gas fuelled combined heat and power (CHP) plant with a water chiller for space cooling. The 5.7 MW electricity generator supplies electricity to the system and to Powergen into the National Grid. The gas fuelled turbine is used to top up the heated water in the system. The naturally heated brine provides about 15% of the energy for the local scheme and when exhausted of potential energy it is pumped into the Solent (Figure 11.22).

The process utilises water that flows through cracks in the hot rocks at lower levels which is then brought to the surface under pressure from a pumping station before it is pumped back down into the hot rocks again. The amount of heat transferred is governed by the speed with which the intrusion can reheat the water; therefore there is a finite amount of energy that can be claimed. In regions where volcanic activity is still present, such as northern Italy and Iceland, power stations generating electricity have been set up. The superheated water from the hot rock is kept under high pressure in pipes at temperatures several times that of boiling water. The water is then allowed to convert to steam by reducing the pressure to atmospheric. The steam rotates the generating turbines and produces electricity for mains distribution. There are versions that reuse the steam in a secondary circuit utilising as much energy as possible before returning the water back into the ground.

Fossil fuel combustion

The main fuels are:

- coal

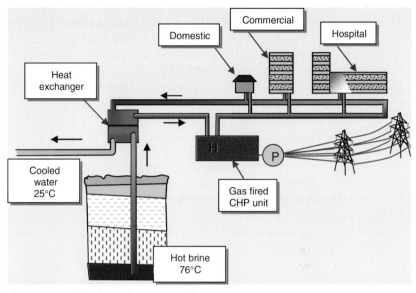

Figure 11.22 Geothermal energy supply.

- oil
- natural gas.

The method of claiming the energy from the fuel is by combustion. The process either produces steam to drive turbines or uses a higher pressure gas exhaust in a similar way to a jet engine aimed at turbine blades. Efficiency varies depending on the energy within the fuel and the process used; however, in general fossil fuel power stations rarely exceed 36% overall efficiency. The greatest losses are through waste heat.

Steam engines developed in the early 18th century used large quantities of wood or coal as fuel. The main objective was to produce motion up and down to power pumps. The technology evolved and the efficiency increased as the Industrial Revolution took hold. Actual examples of the early engines are on display in the Science Museum in London and are worth viewing.

In 1824 Sadi Carnot, a captain in the French army, published his theory which still stands today. He stated, 'It is impossible to have a perfect heat engine', which we now consider to be the second law of thermodynamics. He concentrated on the efficiency of the heat engine: the heat output from the boiler compared with the heat from the exhaust. The higher the inlet temperature and lower the exhaust temperature, the more efficiently the heat energy is being used.

$$\text{Efficiency} = 1 - \frac{T_2}{T_1} = \frac{T_1 - T_2}{T_1}$$

where T_1 = temperature from the boiler and T_2 = exhaust temperature from the engine.

For the formula to work, the temperature scale needs to have a starting point of zero. A Belfast born Irishman, William Thompson, who later became Lord Kelvin, produced a theory that when atoms stop vibrating, they must be at the point of absolute zero – no heat. The Kelvin scale uses temperature increments of the same size as those used in the Celsius scale (another French idea) in which the temperature of the conversion points of water to ice and to steam were divided by 100. Absolute zero (0 K) is the same as $-273°C$; therefore water will freeze at 273 K and become steam at 373 K. Note that the temperature scale uses the capital K and there is no degree sign.

Example

What would be the efficiency of a heat engine if T_1 is $100°C$ (boiling point of water at atmospheric pressure) and the exhaust temperature T_2 is $75°C$?

$$\text{Efficiency} = \frac{T_1 - T_2}{T_1}$$

$$\text{Efficiency} = \frac{100 - 75}{100} \times 100 = 25\%$$

The heat engine would be 25% efficient; however that would be incorrect.
    If we use the Kelvin scale the temperatures would be as follows:
T_1 becomes $100°C + 273 = 373$ K
T_2 becomes $75 + 273 = 348$ K

$$\text{Efficiency} = \frac{373 - 348}{373} \times 100 = 7\%$$

The correct efficiency would be 7%.
    What would be the increase in efficiency if the water was pressurised and the steam temperature increased to $110°C$?

$$\text{Efficiency} = \frac{383 - 348}{383} \times 100 = 9\%$$

The efficiency of the heat engine would be increased to 9%.
    In commercial power stations, the water is heated to very high temperatures in pressurised systems. When released, the steam forces the blades of the turbine round, causing the turbine shaft to rotate. The shaft is connected to a generator which produces electricity. To maximise the difference in heat input to exhaust and therefore efficiency, the steam is condensed back into water. Steam produced by the boilers is at very high pressure before driving the turbines; when the steam leaves the high pressure turbines, it still contains sufficient energy to power a second turbine before going to the condenser (Figure 11.23). The condensed steam becomes water once more and is pumped back into the boiler to continue the cycle. The fuel to heat the boiler can be coal, oil or gas. Some power stations can run on alternative fuels, utilising the fluctuations in market price and availability, such as oil or natural gas.
    As previously mentioned, different fuels contain different amounts of energy. Coal, for example, can be classified on the basis of energy content, with lignite

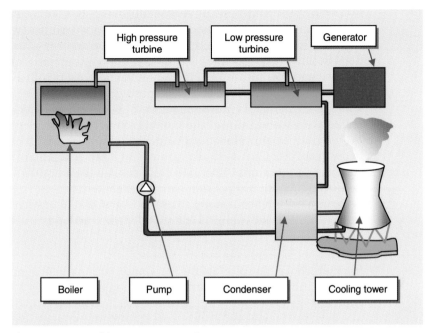

Figure 11.23   Coal burning power station.

or brown coal having the least, and black coal (anthracite) containing the most energy.

What is coal?

Coal comprises mainly prehistoric trees and plant life that lived in hot swamp-lands which became covered by sedimentary rocks before being highly compressed and heated over millions of years. In Britain, during the period of coal formation, the land mass would have been close to the equator. Mammals had not evolved and this was the period of the great reptiles. Plant life during the carboniferous period was nothing like that around us today. The Variscan Orogeny had started; the landscape folded, burying the decaying swamps at various angles and thicknesses, and forming what we now term **coal seams**. Many of the coal seams in England require deep mining from beneath the surface covered with sandstones and limestones. Where coastal waters and rivers eroded the rocks and overburden, some of the seams became closer to the surface, so the coal could be extracted from **open cast pits**. During the erosion process, water washed away lumps of coal, allowing easy collection; however, the large volumes of coal required in the Industrial Age meant that many men, women, children and animals spent most of their relatively short lives working in the pits. The UK became the world leader in coal production which increased from about 10 000 000 tons in 1800 to more than 200 000 000 tons per year by 1900. Battersea Power Station required tens of thousands of tons of coal per week, every week of the year which was brought in by ship and offloaded onto a wharf before being crushed and sent to the boilers. In contrast, modern coal fired power stations pulverise coal into dust before blowing it into the boilers.

By ensuring the particle size is very small, combustion is almost immediate. The gas contained in the pulverised fuel burns with extreme heat when mixed with oxygen. The combustion produces exhaust gases and a residue termed **furnace bottom ash** (FBA) or **clinker**. The clinker is a waste material and conveniently the raw material for clinker blocks like the now extinct 'Breeze' block. There are some manufacturers using clinker as aggregate. Pulverised fuel ash (PFA), which is lighter, is the main aggregate of some aircrete blocks such as Thermalite, Celcon and Tarmac. The subject is covered in more depth in Chapter 10.

Coal burning power stations worldwide burnt at least 2 100 000 000 tonnes in the year 2000. Coal is mainly carbon, but it can also contain sulfur, water, oil and several other impurities. We can calculate the amount of carbon dioxide produced by burning coal from its carbon content.

Carbon has a relative atomic mass of 12 and that of oxygen is 16; 1 tonne of coal = 1000 kg. The chemical equation for the reaction is:

$$C + O_2 \rightarrow CO_2$$

from which we can calculate the amount of carbon dioxide produced by 12 kg of carbon (the relative atomic mass of carbon multiplied by 1 kg). The mass of added oxygen is $16 \times 2 = 32$ kg, so the yield of carbon dioxide is

$$12 + 32 = 44 \, \text{kg}$$

Therefore 1 kg of coal will produce 1/12 of this amount of $CO_2 = 44/12 = 3.67$ kg of CO_2.

Knowing that 2.1 billion tonnes of coal were used to fuel coal burning power stations worldwide in one year, it is easy to calculate that this put 7.8 billion tonnes of carbon dioxide into the atmosphere.

Coal also contains sulfur and nitrogen; these combine with oxygen during combustion to produce sulfur dioxide (SO_2) leading to acid rain, and polluting oxides of nitrogen (NO_x).

The various types of coal and other solid fuels, such as wood and peat, have different energy contents. The moisture contained in the fuel will not combust, and this must be taken into account when calculating emissions. Some fuels contain volatile matter, such as resins and trapped gases, that are released during combustion; therefore there are many factors to be taken into account when comparing fuel use. If we take three comparisons:

- air dried wood contains about 16% carbon
- lignite contains about 30% carbon
- anthracite contains about 90% carbon

Carbon burns hot in oxygen. Coke, for example, is virtually all carbon and will produce a very high temperature output per tonne when compared with wood. Let us compare the energy output per tonne of the three materials:

- air dried wood contains about 15 GJ/tonne
- lignite contains about 19 GJ/tonne
- anthracite contains about 30 GJ/tonne

1 gigajoule (GJ) = 1 000 000 000J = 1×10^9J

Specific heat capacity (SHC) is the heat energy required to raise the temperature of 1 kg of any substance by 1°C. Water has a SHC of 4200J kg^{-1} K^{-1} (where a unit is followed by $^{-1}$ it means 'per', or 'for every'. e.g. per kg, per kelvin; 1K = 1°C).

Example

A boiler has water entering at a temperature of 20°C and heats it to a temperature of 85°C. How much energy is required per 1000 litres of water heated, assuming that at 20°C 1 litre of water has a mass of 1 kg and the SHC of water is 4200 J kg^{-1} K^{-1}?

The energy required to heat water is given by the following equation:

$$\text{Energy (J)} = \text{Mass of water (kg)} \times \text{SHC} \times \text{temperature difference (°C)}$$
$$= 1000 \times 4200 \times (85 - 20)$$
$$= 273\,000\,000\,\text{J}$$
$$= 273\,\text{MJ}$$

If the fuel was **dry wood** then 1 tonne ≡ 15 GJ or 1 kg ≡ 15 MJ of energy, so the amount required would be:

$$273/15 = 18\,\text{kg}$$

If the fuel was **lignite** then 1 tonne ≡ 19 GJ or 1 kg ≡ 19 MJ of energy, so the amount required would be:

$$273/19\,000 = 14\,\text{kg}$$

If the fuel was **anthracite** then 1 tonne ≡ 30 GJ or 1 kg ≡ 30 MJ, so the amount required would be:

$$273/30\,000 = 9\,\text{kg}$$

However, the efficiency of transferring all the heat energy from the fuel to the water is very poor, so the real answer will be several times that calculated; perhaps 45 kg of dry timber would be required in an enclosed stove.

Based on, say, 45 kg of **dry timber** of which 16% is carbon, how much CO_2 will be produced when it is burnt?

$$\text{Mass of carbon} = 45 \times 16\% = 7.2\,\text{kg}$$

As 1 kg of carbon produces 3.67 kg of CO_2, the amount from 7.2 kg will be

$$7.2 \times 3.67 = 26\,\text{kg of } CO_2.$$

If the fuel had been **lignite**, it would be used in a boiler and therefore combustion would be more efficient. Assuming that the efficiency of combustion is 70% (very unlikely), the amount of lignite actually required would be

$$14 + (14 \times 30\%) = 18\,\text{kg}$$

Lignite comprises 30% carbon, so 18 kg lignite contains

$$18 \times 30\% = 5.4\,\text{kg of carbon.}$$

As 1 kg of carbon produces 3.67 kg of CO_2, the amount from 3.67 kg will be

$$5.4 \times 3.67 = 20 \, \text{kg of } CO_2$$

If the fuel had been **anthracite,** it would be used in a boiler and therefore more efficiently combusted. Assuming that the efficiency of combustion was 70% (very unlikely), the actual amount of anthracite required would be

$$9 + (9 \times 30\%) = 11.7 \, \text{kg, say } 12 \, \text{kg}$$

Anthracite comprises 90% carbon, so 12 kg contains

$$12 \times 90\% = 10.8 \, \text{kg of carbon.}$$

As 1 kg of carbon produces 3.67 kg of CO_2 the amount from 10.8 kg will be

$$10.8 \times 3.67 = 40 \, \text{kg of } CO_2$$

Using the three examples of fuel, the approximate indications are that more dry wood would be consumed than either lignite or anthracite to heat the water. However, when pollution is considered, although less tonnage of anthracite is consumed, the amount of CO_2 produced is far more than from lignite. Dry wood in stick form would be less efficiently burned, but if it had been chipped or pelleted and used in a high efficiency boiler, pollution would be significantly reduced.

British coal tends to have a relatively high sulfur content, so UK coal fired power stations tend to use coal from as far away as Canada and Australia where surface mines (we term them open cast pits) allow massive machinery to excavate, crush and move colossal volumes of coal more cheaply than the deep shaft mines in the UK.

Coal

Coal varies in its energy content and heat output. There are five ranks of coal:

1   **Lignite** or **soft brown coal** can be found in many parts of eastern Europe – mainly Germany, where there is estimated to be about 55 billon tonnes in the Cologne area. Canada, USA and Greece also have massive reserves. Closer to home, lignite has been found in Northern Ireland. Lignite is more like peat than black coal and contains woody materials. As decomposition of the plants and trees took place, gases within the fibres produced volatile matter (VM) as the lignite formed. When heated the volatile matter becomes gaseous and easily ignites, producing more heat. Typically lignite contains about 50% volatiles.
2   **Sub-bituminous coal** or **hard brown coal** contains about 5% less volatile matter than lignite; however, the carbon content increases to 75%. The oxygen content also drops to about 20%.
3   **Bituminous high VM coal** has more carbon and slightly less volatile matter than sub-bituminous coal.
4   **Bituminous low VM coal** has a lot more carbon and less volatile matter than bituminous high VM coal. It also has virtually no moisture and therefore higher combustion rates and less ash.

5 **Anthracite** has the highest rank rating with a 90% carbon content and 10% VM. The combustibility is high, producing similar quantities of ash to that of dried wood.

Volatile matter enables easier combustion but produces more smoke and particulates; coals with low VM, including anthracite, are virtually smoke-free, although all coals produce varying amounts of pollutants. If the coals are heated in ovens known as **retorts**, the VM can be taken off as coal gases and then condensed as various coal by-products, leaving a residue known as **coke** which is almost all carbon and can be used for industrial processes including carbonising iron and steel.

Coal power stations tend to burn coals with higher VM as they combust more easily even though they produce less heat energy per tonne (Figure 11.24).

Coal fired power stations have generated most of the UK's electricity for almost a century. Starting as privately owned power stations, they generated electricity for lighting, supplying direct current (DC) electricity at various voltages. Distribution was via bare cables fixed to the outside of buildings en-route to the customers. Only wealthy people could afford to use the clean form of energy. In 1917 there were 70 independent companies with 50 different systems and 20 different voltages operating in London alone. Enormous power stations were built, perhaps the most famous being Battersea Power Station on the riverside of South London. At the time it was the largest brick built structure in Europe and on being commissioned in 1934 produced 400 MW of power purely from burning coal. In 1946 new turbines were added increasing the output to 509 MW, making it the third largest power station in the UK.

Figure 11.24 Aerial view of a coal burning power station with the coal stockyard to the rear.

In 1926 the UK national government set up the Central Electricity Board to oversee the production of electricity. In 1947 the industry was nationalised and it was decided that only 50 cycles alternating current (AC) should be used and 240 V. Power stations originally were set up to supply clean efficient lighting, and compete directly with the dirty and dangerous coal gas lighting. However, technology exploded and the new clean energy was used as a source of industrial power. On the domestic side, washing machines, vacuum cleaners and other labour saving machines became available to the mass population. Coal fired power stations billowed enormous plumes of carbonised smoke into the sky, increasing industrialised pollution.

Since the Industrial Revolution, Britain had been burning coal in open fires for space heating and industry had been burning coal to power steam engines, leading to very high levels of smoke pollution. However, the scale and intensity of pollution from the giant power stations and steel industry produced killer smogs. In London over 3000 people died of respiratory disease in one winter alone. Giant power stations were built near coal mines and large rivers to reduce transportation costs, and produced a band of pollution across England. The high levels of sulfur emissions from the smoke stacks blew across the North Sea to Scandinavia, causing acid rain which was blamed for killing their rivers and forests. The UK had to significantly reduce its pollution levels.

The introduction of oil burning power stations in the 1950s significantly reduced the pollution levels and compared to coal, oil was relatively cheap and easier to handle. An enormous oil fired power station was built on the Isle of Grain in north Kent, specifically to use cheaper cleaner oil (Figure 11.25). Being next to the River Thames, the oil tankers could easily offload with the minimum workforce. However, in 1973 oil prices soared when the Organisation of Petroleum Exporting Countries (OPEC) blocked exports to the West after the Arab–Israeli war. Commentators at the time softened the political statement by reporting that the increase was due to oil being undervalued and therefore wasted. The oil demanding West (now referred to as the north) plunged into recession. Countries such as Denmark were almost completely dependent on oil-fired power generation, so the Danish government changed policy to being as self sufficient as possible by using wind power and local natural gas from neighbouring countries.

Oil fired power stations became expensive to run, so were only brought on line when required. Many oil fired boilers have been converted to burn natural gas (Figure 11.26).

Oil

Oil, when it was first commercially found, was used as a fuel for lighting. The first commercial oil well was drilled in 1859 in Titusville, Pennsylvania, USA. The locals had found the oil seeping to the surface for centuries and when the European settlers arrived they used the oil for lighting oil lamps. In the same period Carl Von Reichenbach developed a process of manufacturing paraffin from beechwood tar. It could be used in lamps for lighting or in solid form as candles. Scot James Young developed a way of refining oil from shale found in Scotland, again for lighting.

Figure 11.25 Oil fired power station with oil holding tanks.

Figure 11.26 Aerial view of a gas and oil burning power station with the oil holding tanks in the foreground.

We take artificial lighting for granted today; however, before the various fuels became available, artificial light was expensive, dangerous and of relatively poor quality. Wealthy people could afford whale oil and candles made from animal fat (tallow) and other waxes. Others in Europe used plant oils to burn. The new oils found in the USA were converted to kerosene for lamps and as a new fuel for heating ovens for cooking. Oil was also used for lubrication of machinery. The remainder of the oil was discarded as being too volatile and therefore dangerous to handle. However, in the late 1880s the German born Gottlieb Daimler developed the first motor car that could be run on the waste product from oil – gasoline.

There are several types of oil found naturally, ranging from thick oil that can be cut with a spade through to thin liquid oils that gush to the surface under pressure. Oil can also be found in rocks such as shale, and in gas form similar to petroleum spirit.

Oil burning power stations use **fuel oil** which is pre-heated to lower its viscosity and enable it to flow and burn more easily. The process is the same as in other fossil fuel power stations: to heat water to produce steam and power turbines. In contrast to coal, oil can be moved relatively easily by pipe.

In the UK, political decisions were made to change power stations from expensive oil burning boilers to locally found natural gas in a period labelled 'the dash for gas'. Ironically, the oil found off the coast of Scotland had been of high quality and too good to be used as a fuel oil. Much of the oil was sold on the open market and brought in high tax revenues for the UK government. The oil fields of the North Sea lay below impervious rock capping enormous volumes of methane (Figure 11.27). Initially the methane had been considered as waste and was vented into the atmosphere; however, as with oil, the technology

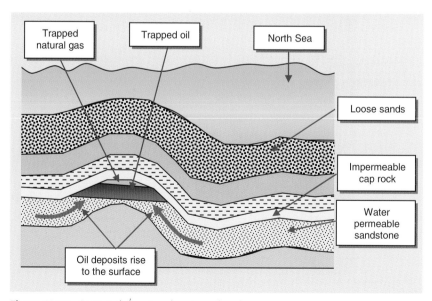

Figure 11.27 Sectional drawing showing oil and gas reserves.

found a use for the waste North Sea gas. Shared with the Norwegians, massive oil and gas rigs were floated out to sea to tap the oil and gas reserves.

Oil originates from the decomposition of marine and aquatic plants and creatures buried by mud and silt. Successive layers of sands further compressed the decomposing layers in anaerobic conditions (in the absence of oxygen). It is likely that the aquatic and marine life flourished in shallow warm waters similar to those near the equator, and a major catastrophe over the land caused a layer of mud and silt to be deposited over very large areas of shallow seas. As the layer of sands settled they became cemented together by the minerals in the shallow seawater, forming a capping stratum impervious to water. The water below the cap percolates between the porous sand grains carrying the decomposing matter. Tectonic action caused the stratum to fold or fault, creating pockets. The density of the decomposing matter is less than water so the deposits gathered at the highest point. The gases filtered through the oily decomposing mass to form gas pockets. Where the capping rocks lay beneath shallow seas, the naturally cemented sandstones and limestones trapped the gas beneath them (Figure 11.27). In areas of seismic activity the gas may have escaped through the fractured capping rocks, or in arid conditions percolated through the porous sandstone escaping into the air. There are biblical references to rocks in the Middle East having everlasting flames: these are likely to have been fuelled by the oil gas, methane, escaping slowly (Figure 11.28). Most of the world's oil and gas reserves are in the Middle East.

Oil deposits that remain within the dry rocks are commonly shale. The black oily deposits are trapped in the pores of the rock, making it expensive to extract. Expense is relative though: when any material is plentiful, it is valued less and is cheaper. That was the argument of OPEC. As easily accessible oil runs out or demand increases, as it has done from China and India, the price of oil will rise. Then the cost of extracting oil from oil impregnated rock becomes viable. Canada has vast reserves of oil as impregnated rock.

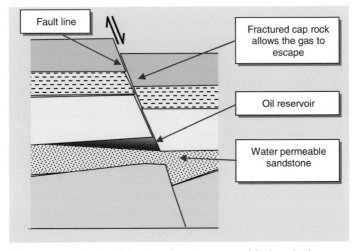

Figure 11.28 Sectional drawing showing a natural fault and oil reserves.

The reserves of oil left on Earth could last another 40 years according to expert predictions. However, the same thing was said 30 years ago. Most experts agree though, that the peak of oil supplies is about now in the early part of the 21st century and that the supplies will decline. There are vast areas of oil contained in rock that will become viable to claim as the price of oil increases; therefore it is possible that the 40 year period of decline may be extended by several decades.

Natural gas – including carbon sequestration

The gas taken off from the oil fields can be piped relatively easily directly to shore. The natural gas from some of the fields has a very high carbon dioxide content, making it difficult to burn. Statoil is a private Norwegian company with about 70% of its shares owned by the Norwegian government that has pioneered removal of CO_2 from natural gas taken from the North Sea by a process of **carbon sequestration**. The methane is processed to remove the high levels of CO_2 which are then pumped back into a saline aquifer about 800 m below the sea bed. About one million tonnes of CO_2 is pumped into the aquifer per year. The technology of carbon sequestration can be adapted to remove CO_2 from other sources, including power stations and heavily polluting industrial manufacture of steel and cement.

Nuclear

Nuclear energy has been naturally occurring for at least a billion years. As mentioned in Section 11.1, atoms of a given element will always have the same number of protons in the nucleus, but may have different numbers of neutrons: these **isotopes** are of different atomic mass and some are unstable. Those that decay spontaneously are said to be **radioactive**. Most radioactive isotopes with atomic number greater than 83 (bismuth) decay by emitting alpha particles (helium nuclei) which can be stopped by paper or even about 50 mm of air, so alpha radiation is the least dangerous form of radioactivity. However, alpha particles should not be inhaled or ingested as contaminated food, when they can become a health hazard.

Naturally occurring uranium has three isotopes: ^{238}U is the main one, ^{235}U occurs as one atom in 140, and ^{234}U exists in very small quantities. The **half life** (time taken for initial radioactivity to fall by half) of ^{238}U to decay to thorium is 4 500 000 000 years.

It was found that upon bombarding ^{235}U with neutrons, the atom absorbed an extra neutron and became ^{236}U. Although the extra neutron added to the mass of the atom, the number of electrons remained the same at 92. ^{236}U is very unstable and disintegrates into two smaller nuclei, two or three neutrons, and a large amount of heat energy in a process known as **nuclear fission**. In a controlled environment, the process takes place in a reactor where the heat energy is removed by circulating carbon dioxide; the hot gas is then used to heat water to produce steam to drive turbines which are connected to gener- ators. To regulate fission, rods containing cadmium or boron can be lowered into the reactor to absorb the neutrons and slow the process. To control the amount of reaction and therefore the heat output, the rods can be lowered into **moderators** of graphite (most UK reactors use this system), ordinary water or heavy water. Unlike the control rods, the moderators do not absorb neutrons but are designed to slow them down. The fission process gives off radiation in the form of gamma rays and beta rays which are more dangerous than alpha particles. Beta radiation can be stopped by thin sheet aluminium or other metals. Gamma radiation is the most powerful of the three types. It contains the same level of energy as X-rays and is therefore hazardous to living things. To contain gamma rays, thick layers of lead, steel or dense concrete must be used. Radiation from X-ray machines in hospitals, though less intense than emissions from a nuclear power station, must be similarly contained.

'Nuclear power stations are dangerous and should be banned!'

Why? Yes, there have been accidents involving the escape of radioactive gases and radiation. In 1979, at Three Mile Island in Pennsylvania, a technician refused to acknowledge that there was a faulty valve, and decided that the fault was with the electrical monitoring equipment. He didn't want to close the system down until the electronics had been checked out. In consequence the coolant system burst and a partial meltdown occurred. The failure was contained and the pressure vessel remained intact. The Chernobyl disaster in April 1986 was the result of testing how long the turbines would produce power following a loss of main electrical supply. The automatic safety system was switched to manual operation by the plant management to carry out tests for increasing output. The reactor became too hot and distorted; it then ruptured, lifting the cover plate to the reactor and releasing fission products into the atmosphere. The official list of fatalities shown in the UN reports totalled 56 up to 2004. The UN report in 2000 stated there was no scientific evidence of any significant radiation related health effects to most exposed people. The subsequent UN report in 2005 confirmed the study.

If put in the context of hazards to health, there are more fatalities per year, every year from smoking in England alone. Looking at the benefits of nuclear energy, there are low carbon related emissions, other than those associated with producing the materials to build the plants; therefore arguments for nuclear energy to help reduce global warming are very strong. For further discussion see Chapter 10.

Incineration and waste

The issues of what to do with waste are becoming of major importance. Old tyres will not rot away, and theoretical mountains are building. Burning old rubber tyres has a two-fold use: to obtain the energy contained in the rubber tyres and to dispose of the waste tyre mountains. With the increasing number of vehicles in the UK and changes in legislation, mountains of used tyres have to be disposed of. If they are used as landfill they are unlikely to decompose for thousands of years.

Traditionally, household waste has been dumped in old quarries and pits. Commercial wastes containing toxic chemicals have been dumped, forming soups or cocktails that will be inherited by future generations. Incineration will dispose of large quantities of waste. The heat given off during the process can be used to produce steam and drive turbines and generate electricity; the process is similar to burning any other fossil fuel and will produce pollution in the form of gases such as CO_2, CO, NO_x, SO_2 and so on. The very small particles (**particulates**) that are measured in micrometres (thousandths of a millimetre) are also given off by all fossil fuel powered transport, including large volumes from aircraft. However, when incineration plants are proposed objectors strongly oppose their being built. There is technology to sort the waste before incineration and reduce the toxicity of the exhaust emissions; filters known as **scrubbers** can be used to remove much of the polluting gases and particulates.

Biomass as fuel

All living things will produce gas as part of their digestive or degrading process. Farms keep large numbers of animals in captivity, such as pigs, cows and chickens, which produce enormous amounts of effluent, disposal of which can be costly. However, countries such as Denmark collect the effluent using road tankers and deposit it in biodigesters. Anaerobic bacteria digest the effluent, speeding up the degradation and producing methane gas which can be compressed for use as fuel for transport and substitute heating gas; it is known as **biogas**. The residual effluent is inert and is dried ready for returning as humus to the fields or growing materials for garden centres. All the harmful chemicals, such as ammonia have been decomposed by the bacteria. Household or small volume commercial waste from food preparation can also be added to the digesters.

Purpose grown crops can be used as fuel. Sugarcane, for example, grows easily in warmer countries and the sugar can be distilled into combustible spirit. Rum is such a spirit; however, a less refined liquid is an ideal transport fuel and sold as ethanol. Brazilians run more than four million vehicles on pure ethanol and the other nine million run on a 20% ethanol blend. Other plants such as maize and sorghum can also be used to produce alcohol for direct combustion or blending.

Waste and biofuels can also be used for direct heating. Commonly used on a larger scale and termed district heating, the waste materials are incinerated in combustion chambers with water-filled heat exchangers. The heated water is then distributed throughout a network of underground pipes to estates of buildings such as dwellings, schools, and offices. Denmark has pioneered district heating and further developed the schemes to heat store or district cooling. The geothermal district heating scheme in Southampton also has an attached district cooling scheme.

Hydrogen as fuel

Unlike most other fuels that produce CO_2 as a product of combustion, hydrogen and oxygen can produce electricity and emit only water vapour. The inventor

Sir William Grove designed a fuel cell in 1839 which, in principle, worked as the reverse of electrolysis, combining hydrogen and oxygen to produce electricity and some waste heat. Fuel cells have two **electrodes,** a negative **anode** and a positive **cathode**, immersed in electrolyte. Both electrodes are porous and allow the gases to pass through into the electrolyte where the hydrogen splits into hydrogen ions (protons) and free electrons. The free electrons flow from the anode along a cable to provide useful electricity and on to the cathode. The protons flow through the electrolyte to the cathode where they combine with the oxygen to produce H_2O. To improve the efficiency of the original fuel cell both electrodes are coated with a **catalyst**, such as platinum, palladium or ruthenium.

There are several different versions of the fuel cell, with efficiencies ranging from 30 to 60%. Modern techniques enable the hydrogen to be highly compressed, making it a usable source of fuel for transport (see Chapter 10, Figure 10.8).

Chapter 12

Utilities

12.1 Introduction

The main services required are:

- water
- gas
- electricity
- sewage
- drainage
- telecommunications.

They are referred to as **utilities**. Utilities are the Cinderella of the construction programme and are often ignored until road works make you late for work. This final chapter looks at the main utilities and how to identify them.

Perhaps the most important utility of all is drinking water, for without fresh water we cannot exist. We look at where water comes from, how it is delivered to all premises throughout the UK, the reasons for leakage, control of the network, and take the subject right through to the taps in the buildings and maintenance issues. Services, including hot water and wet heating systems, are grouped with utilities, emphasising how they are linked. We have already examined the subject from a global aspect in Chapter 10, questioning the political and social ownership of the world's fresh water supply and here explore future trends, such as the use of desalination plants.

Following on from the subject of water supply, we examine how gas is distributed via mains, where it originates and general maintenance issues for supply to the consumer.

We live in an electronic age, with electricity perhaps the most used source of energy in the UK, but how is electricity delivered? We look at methods of supply from power generation through to end-of-line use. This section is not intended to enable the reader to become an electrician, but to understand the route each unit of electricity has to take from generator to consumer.

Another most important utility we take for granted is mains drainage. We are very fortunate living in a country that has a fine network of sewers. Different types of sewer are compared, maintenance issues are discussed, and the terms and regulations from the water closet (WC) or bath to sewer outfall are introduced. To complete the subject, black, white and grey water are defined.

The laying of communications cabling, the most recent of the utilities, has resulted in miles of scarred footpaths and carriageways; identification of such installations is described. The chapter concludes by discussing how contractors can find and identify the complex network of subterranean utilities.

Although most buildings have all the utilities, they may not all be **mains** connected. For example in the countryside where only four or five dwellings exist in a large area, it would be far too expensive to run mains gas, drainage or sewage pipes to them. Electricity could be supplied using overhead power lines as these are less expensive than running cables below ground. Telecommunications cables could also be run overhead for the same reason.

Potable water is the most important utility; therefore there is a statutory requirement for all dwellings to have access to a water main. The statutory requirements for water supply are given in the Water Resources Act 1991 and the Water Industry Act 1991 as amended by the Water Acts 2003 and 2006 in response to recent changes in climate and water demand. The Environment Agency, a government department, oversees the use of water in the UK. Other government departments, such as the Department for Environment, Food and Rural Affairs (DEFRA) and the Drinking Water Inspectorate (DWI), also work with the Environment Agency to monitor and advise on water usage.

12.2   Water

Water supply is perhaps the service most taken for granted. As long as water is available from the tap, people rarely even consider it. Unlike many countries of the world, the UK has excellent water services: the water is potable (drinkable) and available virtually everywhere. For over a century major cities such as London, Manchester, Birmingham and Liverpool have benefited from networks of pipes allowing water to be supplied from taps. Originally, the **mains pipes** were made of cast iron and covered in pitch and hessian to prevent corrosion. The joints were flanged and held together with nuts and bolts to keep the pipe ends tightly butted up. To supply water to each building smaller **service pipes** were needed that had to allow for ground movement; therefore lead pipes with a '**goose neck**' fitting were used. (The 'goose neck' is a bend in the pipe shaped like a goose's neck that allows the ground (and therefore the pipe) to move without straining the pipe joint.) The service pipe then required a valve known as a **service cock** to enable the supply company to regulate the water flow to private ground: the service cock is normally either under the public footpath or in the road, and is covered by a service flap (Figure 12.1). The service pipe continues onto the private ground and is then the responsibility of the property owner. Some properties have an outside valve in the front garden also covered by service flap, or the valve may be inside the property, in which case it is likely to be below the stairs, under the sink, by the front door or in the room with the WC: the position of the valve depends on how old the property is.

Why is so much water lost to leakage from the water mains?

Any pipework that is up to a century old has fulfilled its useful life; however, there are many miles of original or near-original pipes still in active service. One of the causes of pipe failure is the significant ground movement that has occurred in recent years due to:

- water table levels changing
- larger and heavier structures being built close to the mains
- heavier transport and more regular transport passing over
- heavy goods vehicles driving off the strengthened road onto footpaths and verges directly over the water mains

It is not surprising that older water mains fracture, either as a leak or more spectacularly as a full burst. It is costly to close off large areas of city centres

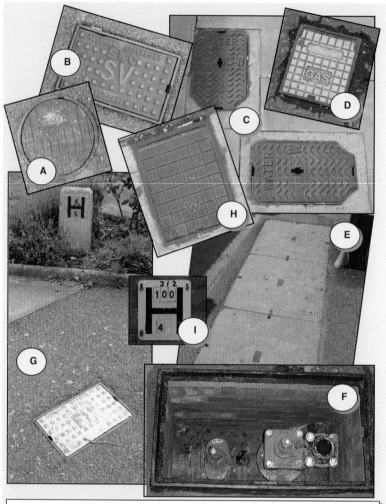

A – water meter cover plate in black polythene
B – water sluice valve cover plate in cast iron
C – two water control valves at junction cover plates in cast iron
D – gas service valve cover flap in yellow plastic
E – telephone cables underground junction chamber covers in reinforced precast concrete
F – a brand new fire hydrant with control valves in a brick chamber – note the new cast iron frame
G – fire hydrant cover plate in cast iron painted yellow for easy recognition and marker post in precast concrete. The numbers on the 'H' show it is an imperial 4" main about 5' 0" in front of the marker. Particularly useful when leaves or snow are covering the cover plate
H – water company's service valve cover in black polythene
I – fire hydrant plate in metric. The top number shows a 100 mm diameter main about 4 m from the front of the plate

Figure 12.1 Utilities cover plates and markers.

to replace the larger water mains. The pipes often run across roads; therefore re-routing already congested traffic is a major problem and the consequence is water leakage.

Detecting water leakage is another costly problem. A water company knows how much water it has pumped into the system and by using water meters at strategic points can carry out secondary metering; however, it is difficult to determine where along the miles of pipe, much of which is under concrete, the leak is situated. Very sophisticated sensor devices, either hand held or vehicle mounted, are available. They send radio waves into the ground and monitor the rebounds to detect pipes and cables. Specialist equipment can detect fluid movement or the presence of electricity. The water company can shut off sections of the water grid and search for water movement along the pipeline to find the area of possible leakage. The company then has to consider whether:

- the pipework is old and fragile and likely to leak elsewhere
- the pipework will be disturbed by major works in the near future
- other utilities are being increased or renewed
- the diameter of the main is sufficient for future demands

Water company planners have to audit their areas to enable money to be allocated according to the priorities. Local Authorities have to produce a 25 year plan specifying the use of land over that period; water companies have to respond by providing new water mains to areas where, say, farmland has been reclassified for change of use to housing. If the existing water main is old and leaking it is financially more efficient to leave the leaks until the whole main can be renewed or re-routed.

Water falls as rain so why should we pay for it?

This is a familiar statement, especially when water bills increase. Most of the water we use is 'winter water' falling as rain or snow. The rate of evaporation during the winter months is slower than the summer; therefore the rain can soak into the ground and restock the aquifers. Summer rain frequently results from thunderstorms which cause flash flooding; most water will run off into rivers or the drains and not into the ground and aquifers.

The hydrological cycle has not been included in this book as there are many excellent geography texts covering the subject. We will look at water as a utility.

The water source is normally a reservoir, freshwater river, or well. Originally, settlements would have been close to freshwater rivers or wells, so subsequently water companies would have built their purifying systems near towns to reduce distribution costs. London has many wells that still retain the original pump heads used by the local population. However, as the demand for water grew, pumping stations were built to enable buildings to have piped water on tap (Figure 12.2). There are still several private water companies in major towns extracting water from ancient aquifers. Some breweries have their own water supply, enabling them to produce beer with a specific flavour, something that would be lost if the water came off the mains supply.

Figure 12.2 Historic city pump.

How much water does the average person use per day?

To calculate this is not as simple as counting how many drinks, washes and visits to the loo. Virtually everything in the built environment has water use somewhere in its process; therefore perhaps finding out how much water is purified and how many people live in the area may give the answer, but will it be accurate? What about the hundreds of thousands of commuters who travel long distances from the countryside to work or visit cities? As you can see, it is far from easy to balance the demand for water and the supply. If there are one or two relatively dry winters with low rainfall and no snow, the rivers cannot refill the reservoirs. If the summers are then hot and dry, large volumes of water evaporate before they can be purified. Therefore water shortages, hosepipe bans and water emergency procedures have to be enforced.

Is it practical to sink more efficient wells and remove the water from deeper aquifers?

It can take years before water falling as rain can be removed from an aquifer. Dense strata that are fractured, such as granite or dense limestone, or permeable limestone or sandstone can cover the seabed; in this situation only the fresh water held in the aquifer prevents the seawater from entering it (Figure 12.3).

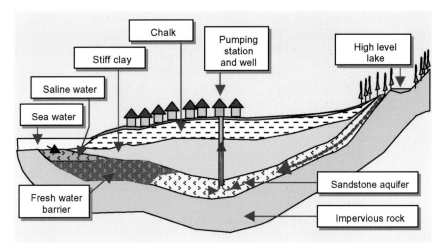

Figure 12.3 Fresh water aquifer.

In this example the rain falls on the hills forming rivers and a high level lake beneath which the rock is impermeable. As the lake fills the water will run off into the end of the permeable sandstone bed forming an aquifer. The rock below the sandstone is impermeable so the water cannot drain away. Above the sandstone is a thick bed of stiff clay that prevents any water permeating its way down into the aquifer. The benefits are that the water cannot become contaminated from the village waste or chemicals used on farmland or from factories. Also, during hot weather the water cannot evaporate, whereas rivers and lake reservoirs could lose tens of thousands of litres of water every day. However, if the fresh water is removed from the aquifer more quickly than the rain can recharge it from other openings, seawater which is saline may enter further. The eventual contamination of the aquifer will make the well become unusable, probably for ever.

Desalination plants

What about desalination plants? London is considering a new desalination plant to provide fresh water for the Thames Water Company. Several decades ago the same idea was heralded as the means to supply fresh water for crops and irrigate the desert in California. We now know the consequence. Vast areas of land are now salt desert which is so saline that virtually nothing can grow. Originally all the salt generated was deposited in the Bay of Mexico and this killed off almost all the fish stocks as the seawater became saturated with salt. Major international pressure stopped the dumping, so an alternative had to be found: the salt was dumped on land. The area is now so polluted that rangers have to shoot all birds to prevent them breeding in the salt marshes: those that do survive are often born badly deformed and die painfully because they are unable to fly or feed. The areas are now completely dead and will remain so, probably for ever. Removing salts from water requires a large energy input and therefore causes more indirect pollution.

Figure 12.4 London rubbish barge.

London has a long history of dumping its waste on its neighbours' land or in the North Sea: every day barges of waste from London are taken down the River Thames to Essex to be dumped in old sand pits and quarries (Figure 12.4). Where will the salt and waste materials from the desalination plant go?

Potable water supply

Potable water supplied by the water companies is distributed via a network of pipes, pumping stations and covered treated reservoirs and water towers. New water mains can be run in large diameter spun concrete pipes, smaller diameter steel pipes and polyethylene pipes coloured blue for easy recognition. To service an area higher than the pumping station, **water towers** are built high above the ground to act as reservoirs of treated water: the water is pumped in and gravity provides a constant water pressure in the area network. Using gravity to pressurise the water saves having pumps continually cutting in and out when users run cold water; however, some areas do use this system despite its being more costly.

It would be impractical to provide potable water from one long pipe like a hose, as the user at the end of the pipe might or might not receive any water, or the pressure would be so low that the water would only trickle out of the tap. Supplying commercial, industrial and domestic users with potable water requires networks of **grids** where large diameter pipes feed into smaller branches (Figure 12.5). The large diameter **ring main** branches off into a smaller diameter pipe that feeds a specific number of buildings before returning to the ring main. Valves are located at each end of the ring to enable the water company to isolate a ring or run of main. Figure 12.5 shows how shutting down **sluice valves** 3 and 4 would isolate the burst main near the house, leaving the remainder of the system unaffected. In addition to the sluice valves there are **air valves** that allow any trapped air to be released when refilling the pipes:

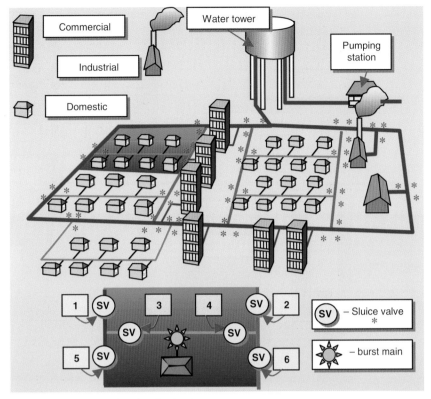

Figure 12.5 Water utility distribution layout.

this is known as **purging** the system. Air valves have not been shown on the drawing.

Connecting to the mains supply

Water pipes in the UK are buried for three main reasons:

- to protect them against freezing and bursting – rarely do we have cold enough weather to affect the mains water supply; moving water does not freeze easily
- to protect the pipes against impact damage
- to protect the pipes from heating during hot spells.

Branches from the mains are connected by contractors on behalf of the water company. The main building contractor would trench out to the required depth of at least 850 mm from finished ground level and bury a blue convoluted pipe from the site boundary to the rising main intake. If deep strip foundations have been used, a short length of pipe can be cast in to act as a sleeve for the convoluted pipe (Figure 12.6). All new property should have an in-line water meter fitted, either behind a glass panel to enable it to be read without disturbing the occupants, or on the user's side of the company's service valve with

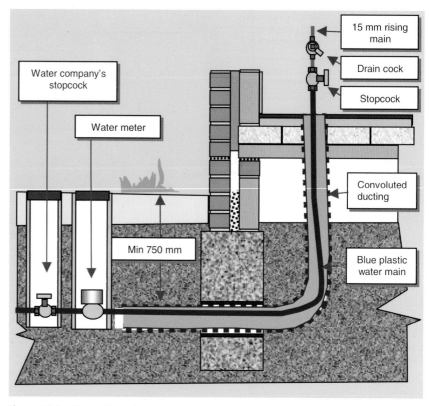

Figure 12.6 Domestic mains water supply.

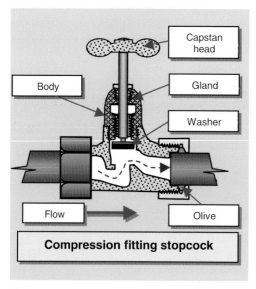

Compression fitting stopcock

Figure 12.7 Section through a water stop-cock.

a larger service flap cover (Figure 12.1). The pipe continues in blue coloured polyethylene and connects to copper pipe near or below the kitchen sink. The pipe will be vertical and is referred to as the **rising main**. The valve is termed a **stopcock** and when operated should stop any water entering the property (Figure 12.7). In the event of an emergency such as a burst pipe the stopcock should be turned down (off) to prevent any more water flowing into the property. However, another valve should be positioned directly above the stopcock: this is the **drain cock**. It has a **spigot** to which a hose pipe can be attached; when the square-headed tap is turned anticlockwise all the water above the drain cock should drain via the hosepipe to the outside, thus emptying the system. To help the flow, all cold water taps should be turned on, especially those upstairs in a house. Apart from draining the system down more quickly, this will also prevent airlocks holding water in some parts of the pipe.

From the rising main the first **take off** pipe normally goes to the kitchen sink tap. If salt type water softeners are used, a branch pipe should go to the kitchen sink to give a supply of salt-free water for drinking, etc., before it goes to the water softener. For the same reason it is important that the bathroom washbasin tap is also connected to a water pipe coming directly from the mains supply. All potable water in the UK has to meet stringent quality control standards and in many cases is superior in quality and cleanliness to many bottled waters.

Exercise

Collect the labels from as many bottled water containers, compare the impurities and enter the data on a spreadsheet. You can also contact your local water company or visit it online and compare the impurities in their piped water. Look at in particular at the salt (sodium chloride) levels.

Plumbing – hot and cold water supply

There are two types of cold water system:

- direct
- indirect.

In a direct water system the cold water pipes are fed directly from the rising main. The advantages are that:

- all tap water is of drinking quality
- there are no storage containers (tanks or cisterns)
- water is always at mains pressure, an advantage when it is required for showers.

Disadvantages are that:

- there is no stored water if connections are temporarily disrupted
- water pressure may fluctuate as demand varies.

In contrast, the indirect water system, after tapping off to the kitchen sink, continues in the rising main to water storage cisterns, commonly in the roof space. The cistern, valves and pipes should be adequately thermally insulated (**lagged**) to prevent freezing. There are purpose-made insulation packs for valves; however, a plastic carrier bag with a small quantity of quilt insulation can be loosely tied around the valve, identifying where the valve is and allowing easy access when required.

The water cistern should have a **warning pipe** also known as the **overflow pipe** to warn if the water valve is leaking or has been stuck open by an obstruction. To regulate the amount of water entering the cistern a **float valve** is fitted (Figure 12.8). The valve is opened and shut by an arm attached to a plastic or metal float that, as the name indicates, floats on the surface of the water. As the water is drawn from the cistern the water level drops and the float descends. The other end of the arm pivots from the valve body, rotating the cam away from the valve pin or plunger. The water pressure pushes the neoprene diaphragm washer away from the nozzle allowing the water to fill the cistern.

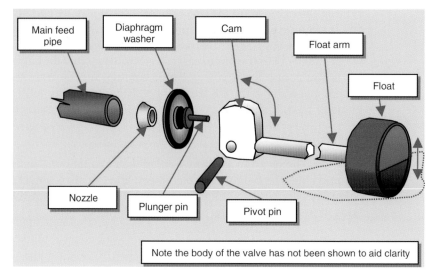

Note the body of the valve has not been shown to aid clarity

Figure 12.8 Exploded view of a float valve.

As the water level rises again the cam rotates, pushing the pin or plunger onto the diaphragm washer to stop the flow of water from the nozzle. To adjust the **shut-off level** of the valve (the depth of water in the cistern), manufacturers use threaded adjustment rods on the float and the plunger end. If the water level exceeds the shut-off level, the warning pipe should let the water flow out of the building, signalling the problem. The overflow or warning pipe should have an insect-proof mesh, preferably at the outermost end, to prevent spiders from laying their eggs and producing spider's webs, thus blocking the pipe.

To enable maintenance, a **service valve** should be fitted close to the float valve, WC feed and in line with all taps and appliances, such as washing machines and dishwashers (Figure 12.9). Service valves are also available with a short coloured plastic coated arm. When the arm is in line with the pipe, it indicates the pipe is 'live' and the water can flow. If the handle is at 90° to the pipe the valve is closed. By colour coding the valve handle, the water can be identified more easily: blue indicates drinking water and red indicates hot water.

The cold water is piped from the cistern to the bath, basin, bidet, WC and hot water cylinder. The draw-off pipe is positioned about 50 mm up from the bottom of the cistern to prevent any grit from leaving the cistern. During the cleaning of the water sand filters are used and very occasionally some of the particles of sand or grit enter the water system. Over several years the particles will form a thin layer of silt on the bottom of the cistern in a similar way to a settlement tank. There is no health risk. Indeed, if water cisterns are made from stainless steel and sealed, the water would still be of drinking quality. However, it is unwise to drink water from other water cisterns for the following reasons:

- the plastics used are not of food container quality
- the cisterns are not sealed so bacteria could gain access
- spiders and other small insects can access through any small gaps and fall into the water

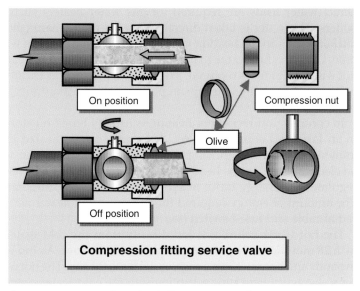

On position

Compression nut

Olive

Off position

Compression fitting service valve

Figure 12.9   Section through a ball type service valve.

- some larger water cisterns on high rise buildings do not have tight fitting lids and creatures such as pigeons and mice can either fall into the water or at least use it as a toilet.

The pipe from the cistern is termed the **distribution pipe** which is commonly either 22 mm or 28 mm diameter to optimise the flow of water. The distribution pipe feeds all water-fed appliances such as the bath, basin, WC and bidet. The diameter of the distribution pipe is reduced by reducing fittings: 28 mm is reduced to 22 mm in the bathroom and feeds the bath cold water tap; the 22mm pipe is reduced to 15 mm diameter feed to the basin, bidet and WC. The most common fittings are reducing tees:

- 22 mm and leaves as two 15 mm diameter pipes
- 22 mm and leaves as one 15 mm and one 22 mm diameter pipe

or a straight reducing fitting – in at 22 mm and leaves as a 15 mm diameter pipe.

Kitchen appliances are normally run directly from the mains supply. For example, many automatic washing machines and dishwashers are 'cold feed' only and require mains pressure to fill.

Hot water can be provided by several different methods:

- gas boiler – water heater
- immersion heater – mains electric water heating
- in-line electric water heater – including electrically heated showers
- other fuel water heaters – coal, wood, straw and oil.

There are two methods of supplying water for heating:

- direct
- indirect.

Direct water feeds are required for gas water heaters such as combination boilers. The water is taken directly from the rising main (should be after the kitchen tap) into the heating vessel and through to the hot water distribution pipe. Only the hot water that is immediately required is heated; therefore no hot water cylinder is required. The same applies to electrically heated shower units and local hot water taps in places such as restaurants and some public buildings.

In contrast, indirect water heating is usually based on storage of hot water in an insulated copper cylinder. The water can be heated electrically by an immersion heater comprising two heating elements similar to the element in an electric kettle. In the head of the immersion heater a bimetallic thermostat regulates the energy input when the desired water temperature is achieved. The amount of energy required for an immersion heater necessitates use of a radial cable and fused switch (see Section 12.4).

The hot water cylinder is fed directly from the cold water storage cistern by a 28 mm diameter pipe entering near the bottom. As the water is heated it expands and becomes less dense, so it rises to the top. The hot water distribution pipe also has a vent pipe returning into the cold water cistern to enable free expansion of the water plus a possible outlet if a malfunction of the thermostat occurs and the water overheats (Figure 12.10).

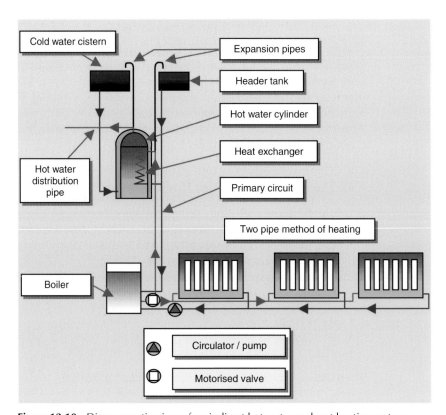

Figure 12.10 Diagrammatic view of an indirect hot water and wet heating system.

Where a wet system of heating has been designed, it is common to heat the water for the radiators and hot water supply using a boiler. Technically, the water should never boil; however, the term 'boiler' has been used for several decades. Although, as previously mentioned, the fuel can be coal, straw or oil, that most commonly used today is natural gas. The Building Regulations have been revised with a view to reducing pollution and obtaining the best output against energy input; therefore virtually all new heating systems will have condensing boilers.

An indirect water system usually has its own water cistern and supply independent of the main cold water storage. The cistern is referred to as the **header tank** and is required to provide water and top up for the hot water for heating *only*. The feed is taken as a branch from the rising main, usually as a tee from the feed pipe to the cold water cistern. A service valve and float valve would be fitted with a take-off pipe and warning pipe similar to those used in the cold water storage cistern. The take-off pipe feeds the boiler directly using gravity. The water in this part of the pipework is isolated from the hot water supply and is referred to as the **primary system**. The circuit is as follows:

1 Header tank to boiler, to ensure a constant supply of water.
2 Water is heated and monitored by the boiler stat (a thermostat to regulate the maximum heat of the water in the heating vessel).
3 The heated water goes to a motorised valve where it is directed to central heating radiators or to the heat exchanger in the hot water cylinder. The valve can also allow hot water to both areas. The motorised valve is basically two rubber or brass stoppers that block off the hot water from one or the other circuits or is left open to allow both circuits to operate.
4 The central heating circuit shown in Figure 12.10 comprises pressed steel radiators linked by the **two pipe system**. This means that the hot water travels to a tee piece where it can continue past the radiator or go through the radiator. A valve enables the water to be regulated manually, either by opening the valve fully to enable a full water flow or by closing it off to prevent any water flow. Alternatively, a thermostatic valve can be used where the temperature of the air near the valve is monitored by a bimetal spring that expands as it becomes heated thus closing the valve/water off. To regulate the speed of the water leaving and thus entering the radiator, another valve is fitted to the outward end of the radiator known as a **lockshield**. When the heating system is fitted the engineer should adjust the lockshields to allow the maximum flow at the last radiator in the circuit and the least flow in the first radiator. Using a thermometer the air temperature in each room can be measured and the radiator flow rate adjusted accordingly. This is known as **balancing the system**.
5 To push the water around the circuit a pump is positioned near to the end of the system to push the water back into the boiler. The pump or more correctly termed the **circulator** is basically an impellor blade on a small mains electric motor. An impellor blade comprises many slightly curved blades on a wheel in a metal housing. As the blade rotates it pushes the water along the pipe. It is the exact opposite operation to that of a water

wheel. Positioning the circulator to push the water into the boiler ensures the boiler always has water in the event of a leak in the system.

6　A room stat (thermostat) is commonly sited in the hallway or general area of the building to ensure an overall temperature. The room stat has a heat monitoring circuit that measures the air temperature around it. Therefore it is unwise to position it near an external doorway, near large windows or above a radiator. As the required air temperature is reached the room stat electrically switches the boiler to cease heating the water until the air temperature falls and switches it back on.

7　To heat the water the circuit starts at the motorised valve. As hot water is less dense than cold water it rises to the top of the pipe where it is connected to a heat exchanger inside the hot water cylinder. An expansion pipe goes over the header tank. If the heating system malfunctions, such as by a valve not operating or a thermostat failing, the overheated water will fall into the header tank mixing with the cold water thus cooling it down. In extreme cases very hot water will flow from the warning pipe out of the header tank, requiring that the heating system be closed off as quickly as possible.

8　In normal circumstances the hot water will flow into the heat exchanger where it is contained in a copper coiled pipe or small cylinder. The cold water in the main water cylinder is then heated by the hot water through the heat exchanger. Note the two systems of water do not mix and the system is known as an **indirect hot water system**. The primary system loses its heat through the heat exchanger, becoming cooler and therefore denser, and so falls to the bottom of the exchanger eventually to rejoin the cold water feed to the boiler (Figure 12.10). Note that the system is designed to ensure the boiler always has a water feed even if the system leaks.

9　A cylinder stat should be attached to the hot water cylinder; when the required water temperature is reached the stat electrically switches either the motorised valve and/or the boiler heating control. The hot water system can be operated independently of the heating system by the control of the motorised valve. This is important during the summer months when hot water is still required.

It should be noted that although the water heating system is fuelled by gas it requires mains electricity to operate. The water pipes, boiler, radiators, etc., are all linked by the water they contain, even if plastic pipes have been used; therefore it is a safety requirement of the Building Regulations Approved Document L and the IEE Wiring Regulations: BS7671 2001 incorporating amendments nos 1 & 2, published in 2004 by the Institute of Engineering and Technology, that the system is earth bonded to the main electrical earthing point (see also Chapter 11).

Other brief points include the following:

- There should be drain cocks in the pipework to enable the water from the lowest part of the central heating system to be completely drained for maintenance work.

- There should be a drain cock on the boiler to enable maintenance.
- There should be an air valve (manually operated or automatic) to collect any trapped air in the heating system. This is similar to bleed valves on the radiators. The water in the closed circuit will trap air when the system is first filled. Also, when the water is heated air will be driven off and will circulate with the water. In the cold water system, trapped air can reverberate at elbow joints causing a condition known as **water hammer**.
- Each radiator should have an air bleed valve to allow air to escape and be displaced by water filling the radiator.
- Hard water contains minerals that will leave the water and bond to any adjacent surface, especially during heating. The water in the primary circuit is a closed system; therefore in hard water areas it is likely that the pipes and valves may become covered with a deposit of calcite. People who live in a hard water area will notice a mineral coating on the heating element of the kettle and around the taps (especially hot taps) termed **scaling**. If allowed to form in the primary circuit, the deposit will reduce the size of the pipe and the flow of water by increasing the surface resistance within the pipe and valves. To overcome the problem an additive should be put in the header tank of the **hot water system only**. In a similar way to antifreeze in a car, the additive acts as a lubricant to the circulator, prevents calcite from leaving the water and forming scale, and prevents or at least reduces the formation of sludge in the radiators. Sludge is formed by impurities within the water. In certain water types an electrolytic reaction can occur between the metals in the system; for example, brass valve bodies, copper pipes, steel radiators, steel bosses in the radiators, cast iron heating chambers, cast aluminium circulator bodies and stainless steel heater vessels. The correct additive will prevent all the conditions listed above. Some additives will only solve part of the problem.

There are other systems for heating water using gas, such as the combination boiler ('combi') where the system is pressurised; however, these are outside the scope of this book.

12.3   Gas

Mains gas, also termed natural gas, is basically methane. Over the past few decades the UK has been self sufficient because of its vast reserves in the North Sea; however, as the resource is nearing its end large volumes of gas are now imported (see Chapter 11).

Natural gas supply

The gas at source contains impurities such as carbon dioxide that have to be removed before it can be sent to the wholesaler. The gas is pressurised to push it through the enormous network of pipes throughout the UK.

There are similarities with the water networks:

- Gas is piped through **grid networks** which are normally at least 600 mm below ground level and have valves to regulate or isolate sections of the grid.
- **Air valves** are fitted to enable any trapped air to be purged from the system, especially when re-commissioning the pipework with gas. The air has to be allowed to escape as the gas fills the pipes.
- The natural gas contains water vapour which will eventually condense to form **liquor**. To allow the liquor to flow back into the mains pipe all service pipes into consumers' premises should run at an incline from the mains. The gas supplier as part of maintenance will send a 'pig' (no, not the pink type: it looks more like a bowling ball) down the pipe under gas pressure to push the liquor out of the pipe.
- Each consumer will have a **service pipe** taken off the gas main with a valve owned and maintained by the gas supplier. From there the pipe passes through a **regulator** to maintain a constant gas pressure. The mains pressure fluctuates due to the irregularities of demand; therefore the regulator is essential. From the regulator the pipe passes through a meter that monitors how much gas has passed through the pipe and then through to the consumer's gas appliances. An in-line service valve enables the gas supply to be turned off before it reaches the regulator, allowing total isolation in the event of a leak or maintenance on the regulator and meter. All the gas appliances should have additional service valves, similar to those used on the water pipes but specially designed for use with gas.

Domestic gas supplies are now commonly housed in a meter cupboard that can be accessed from the outside of the property (Figure 12.11). The service pipe is now of yellow coloured polyethylene, either shrouded in a metal sheath or connected to a steel gas barrel to protect it from knocks or abrasions. Commercial premises have larger entries run either in larger diameter yellow plastic pipe or steel barrel with threaded joints in cast metal. Domestic feed is usually via yellow plastic convoluted ducting buried in a trench prepared by the builder. The gas supplier or contractor will insert the feed pipework through the ducting and make the connection to the gas main and service valve (Figure 12.12). The service valve is usually located outside the boundary of the premises under the footpath (Figure 12.1).

12.4 Electricity

Mains electricity supply

Mains electricity is transported via a network of overhead power cables suspended from steel pylons. There are two main networks, termed **grids**:

- the National Grid
- the Super Grid.

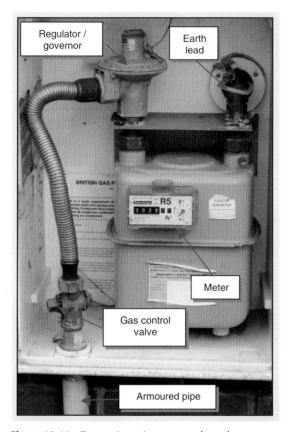

Figure 12.11 Domestic mains gas supply and meter.

Electricity is generated in various parts of the UK and also brought in from the Continent onto the main Super Grid at 400,000 V (Figure 12.13). The electricity voltage is reduced by transformers down to 275,000 V and 132,000 V to distribute the supply over the UK. Localised supplies are reduced to 33,000 V and 11,000 V for use by heavy industry. It is common for transformers to be close to an industrial site to reduce the power loss of transmission. Commercial and domestic users require the voltage to be further reduced to 415 V three phase and 230 V single phase, respectively.

Electricity is usually supplied to the consumer via armoured cables run below ground; these are coated in black plastic for the lower voltages and red plastic for the higher voltages such as 11,000 V. Commercial and multi-domestic premises such as flats have the supply brought into a service room where distribution can be arranged. The electricity cable contains a live line and neutral line known as a 'star' line.

Domestic supply

The service cable is protected by a **sealed fuse** of between 60 and 100 A rating before entering the **consumer's meter** where the power is monitored and

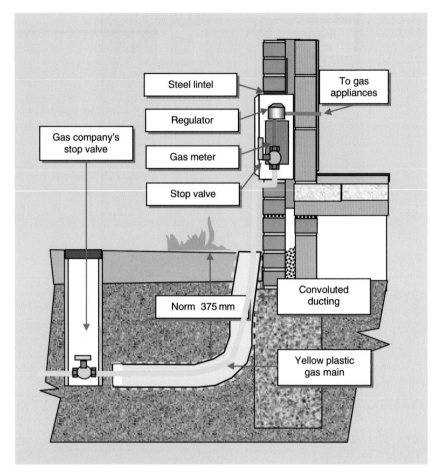

Figure 12.12 Domestic mains gas supply.

recorded for charging purposes (Figure 12.14). Two cables known as **tails** leave the meter and in new installations go directly into the **consumer unit** (Figures 12.14–12.16). There are still some old installations that have a fuse box.

Consumer units have a double pole switch that when thrown breaks both the live and the neutral lines, completely isolating the building. Many consumer units now have a **residual current device** (RCD) fitted within the unit. RCDs are particularly useful in monitoring the flow between the live and the neutral cables: they should continually balance. However, if more electricity is monitored going into the circuits than is leaving on the neutral, the imbalance immediately switches off the live line. The fault could be a lawnmower cable accidentally cut so the electricity is going directly to earth. Not all circuits are wired through the RCD; for example, a freezer. If the RCD is activated when, say, a 100 W security bulb blows, the sudden change could be enough to trigger the RCD. The mains would be switched off, including the freezer, and the food would defrost.

To protect the circuits within the building the main live cable from the RCD is connected to a continuous **live** brass bar onto which **miniature circuit breakers**

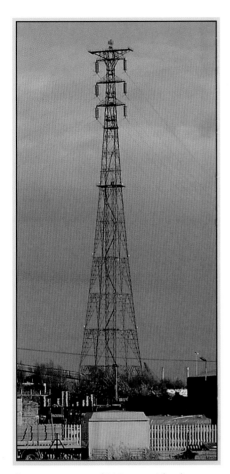

Figure 12.13 400 kV Super grid pylon.

(MCBs) are attached (Figure 12.17). MCBs are very sensitive switches which switch off if more than the rating of the switch is taken. An MCB is virtually the same as a fuse; however, the tolerances are far more accurate and when the fault has been rectified the MCB can be switched back on. There is no fuse wire or cartridge to replace. Each MCB has a switch to enable that particular circuit to be completed. New fixed cabling colour codes came into effect on 1st April 2004: brown insulation for live and blue for neutral replaced red for live and black for neutral. There is normally an MCB to protect each circuit.

The connection of the MCB is made by connecting it to the live bar and securing the brown live cable into the outlet port of the switch. The cable would then continue to all the sockets in the ring and return to the same screw port (Figure 12.17).

The neutral cable from the RCD is directly connected to a separate brass bar termed the **bus bar**. The blue neutral cable is secured onto the bus bar using the screw ports and continues to all the same sockets as the live cable before returning to the screw port on the bus bar. Note it has *not* gone through any switches as it is common to all circuits.

The third cable is the **earth line**. It is connected to the earth bar screw ports and continues to all the same sockets as the live and neutral before returning to the earth bar. The earth cable insulation is green and yellow striped.

Ring mains

- *Only* the live cable goes through a switch protected by an overload device called an MCB. The colour of the new fixed live cable is brown – old systems were red.
- The neutral is connected to the bus bar and does not go through the MCB. The new cable colour is blue replacing the old black.
- The earth cable is connected to the earth bar and does not go through any switches. The earth bar is connected to the earth on the supply cable.

Some lighting mains are wired as rings; however, due to their lower amperage they are more commonly wired radially. This means that the live wire is connected to the outlet side of the MCB but the cable does not return. This also applies to the neutral and the earth cables.

Where a heavy load rating is required, such as for an electric cooker and immersion heater, separate radial circuits are installed. The MCB rating will be higher and the thickness of the cables greater. External power would also have an independent MCB.

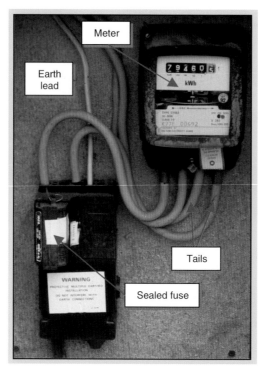

Figure 12.14 Domestic mains electricity supply and meter.

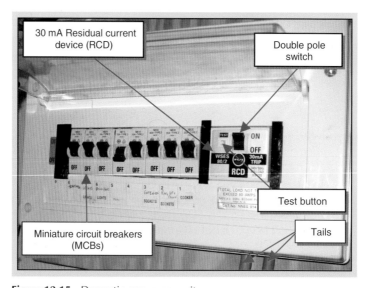

Figure 12.15 Domestic consumer unit.

A typical consumer unit for a house would have the following:

5 A   MCB upstairs lighting
5 A   MCB downstairs lighting

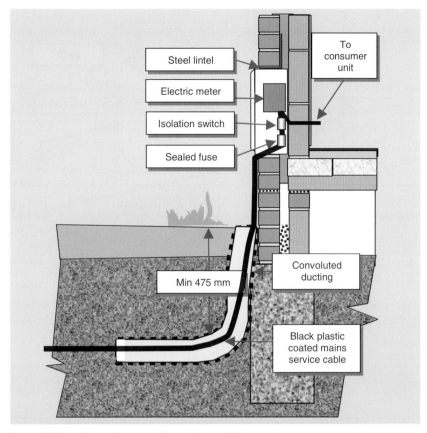

Figure 12.16 Domestic mains electricity supply.

30 A	MCB upstairs ring main
30 A	MCB downstairs ring main
35 or 40 A	MCB for the cooker
15 A	MCB for the immersion heater
15 A	MCB for the garage and garden.

The 100 A sealed fuse can carry the demand because it is unlikely that all the electrical appliances will be at their maximum rating. If they are, the main sealed fuse will break. For further information refer to the IEE Wiring Regulations: BS7671: 2001 incorporating amendments numbers 1 and 2, published by the Institute of Engineering and Technology in 2004.

12.5 Sewage

Mains sewage disposal in towns and cities has been available since the mid 1800s.

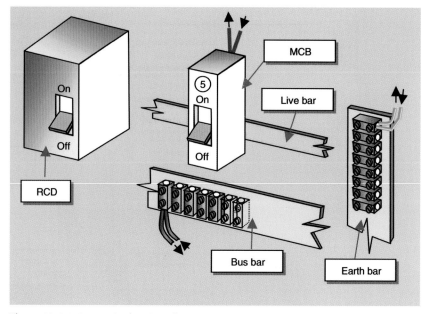

Figure 12.17 Isometric drawing of consumer unit components.

Domestic sewers

Sewage is referred to as **soil water** or **foul water** and requires treatment to prevent disease. Mains sewage systems enable the soil water to be removed from buildings via a network of pipes, mainly using gravity. As a rough guide, 100 mm diameter pipes should be laid at a fall of 1:40–1:60, meaning for every 40 metres of pipe run there should be a fall in height of 1 metre. Gradients greater than 1:40 or less than 1:60 will cause the fluid to run faster than the solids which thus block the pipe. Larger diameter pipes allow shallower gradients (see Approved Document H1 of the Building Regulations). Single dwellings are connected to the main sewer by 100 mm diameter pipe of plastic or vitrified clay (although in some circumstances grey iron pipe may be required) buried in the ground. The main sewer can be made of spun concrete pipe or be one of the original brick sewers built beneath the public highway in the 19th century.

For maintenance purposes inspection chambers should be located at:

- the start of the drain run adjacent to the building – the peak
- any change of direction or drop in height
- where other pipes join on – junctions or pipe size change
- intervals of no more than 45 m on straight runs (inspection chambers)
- intervals of no more than 90 m (manholes).

The difference between a manhole and an inspection chamber is access. A manhole is large enough to climb into for maintenance.

Inspection chambers are available in preformed concrete rings, plastic rings, or they can be built in semi or full engineering quality brick using English bond in 1:4 cement–sand mortar. The old trend of rendering the inside of the chamber is not recommended as the render can spall off and block the sewer pipe.

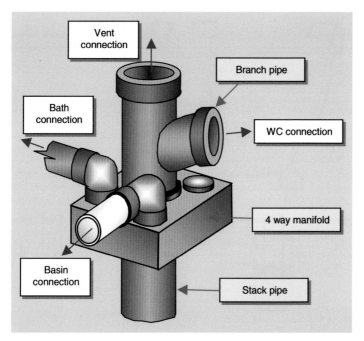

Figure 12.18 Isometric drawing of a soil pipe, branch and manifold junction.

For all works inside a building the word **sanitation** replaces the word **sewage**. Modern buildings usually have a main vertical plastic pipe either inside or attached to the outside wall of the building. This is known as the **stack pipe**. Most of the waste pipes are connected to the stack pipe via multi-point bosses, whereas the WC connection is via a branch pipe (Figure 12.18).

To prevent smells and gas from the sewer returning into the building, each waste pipe should have a water trap. Baths, basins, bidets, sinks and showers normally have 75 mm depth of seal whilst WCs require 50 mm. There are several different types of trap to suit various conditions; 'P', 'S', bottle and banjo (Figure 12.19). To ventilate the system and prevent birds falling into the stack pipe, a balloon basket should be attached to the top of the stack pipe.

The alternative to mains drainage is to store the effluent on site in a septic tank. The pipework is connected directly to a large plastic vessel buried near the building. Chemicals are added to reduce the gas build up and break down the solid matter into a soup like liquid. Periodically a tanker type lorry is required to suck the contents from the septic tank and deliver it to the sewage works for full treatment (see also Chapter 10, Section 10.1).

12.6 Drainage

The following terms are used to identify the liquids:

- Sewage water – **soil** or **foul water** – also termed **black water** – manhole identity red

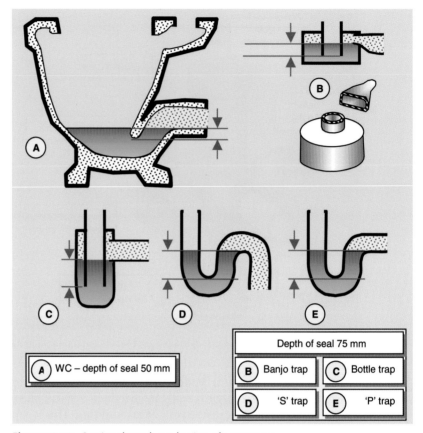

Depth of seal 75 mm

A	WC – depth of seal 50 mm
B	Banjo trap	C	Bottle trap
D	'S' trap	E	'P' trap

Figure 12.19 Section through a selection of water traps.

- Drinking water – **potable water** – also termed **white water**
- Drainage water – **surface water** – also termed **grey water** – manhole identity blue.

Rainwater falling onto the roof and surrounding hard areas is relatively clean and can be piped either into a **separate drainage system** or joined to the sewer system making it a **combined system**. In recent years the grey water has been collected for use, mainly to flush WCs, and therefore reducing the demand for drinking water. In some areas surface water must not be connected to the mains sewers; therefore a soakaway would be required comprising a hole filled with rocks or bricks positioned far enough away from other buildings where the rainwater can drain into the subsoils.

12.7   Telecommunications

In recent years the number of buildings having access to cable services buried underground has increased. Optical fibres transmitting light are replacing

Single phase electricity cables plus earth

Telephone wires

Figure 12.20 Overhead mains electricity and telephone wires.

Figure 12.21 A happy utilities technician showing a CAT radio detector.

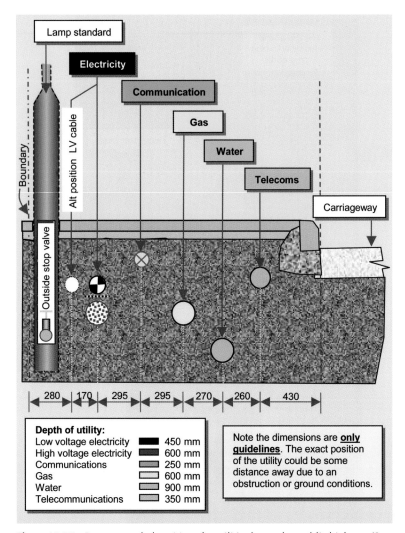

Figure 12.22 Recommended positions for utilities beneath a public highway (Source: National Joint Utilities Group).

traditional copper wires because they can transport far larger amounts of digital data more quickly. The service is provided on a grid originating from sender stations and exchanges. The builder would provide a green convoluted pipe from the boundary into the building in a similar way to the other utilities. After the building is finished a draw cord and mouse (small piece of wood to prevent the draw cord being removed) are run through the convoluted duct to enable the service cable to be pulled through (a draw cord would also be used for the other service entries). When the ducting has been run, the trench is then back filled with suitable spoil and compacted at 150 mm layers. The service provider will excavate the service main, attach their cable to the draw cord provided by the builder and pull the service cable into the

Figure 12.23 11 kV underground junction.

Figure 12.24 An excellent clean trench ready to receive a new gas main.

building. The radius of the bend in the convoluted duct can present difficulties so the connection to the draw cord must be very secure. At the connection point an access chamber is positioned and covered with a dark grey identity plate.

Table 12.1 Identification of utilities.

Utility	Duct	Pipe[a]	Cable	Colour of marking/ warning tape where used
Gas	Yellow	Yellow		Yellow with black legend
Water	Blue	Blue MDPE/ MOPVC/also blue coated ductile iron. Can be black in blue sheathing		Blue
Water pipes for special purposes, e.g. contaminated ground		Blue with brown stripes. Polyethylene/also blue coated ductile iron		
Sewerage		Black		
Grey water		Black with green stripes		
Electricity	Black		Black (red for some high voltage)	Yellow with black legend
Communications	Grey, green			White with blue legend. Green and/or yellow with identification showing co-axial or optical fibre cable

[a] MDPE, medium density polyethylene; MOPVC, molecular-oriented polyvinylchloride.
Source: National Joint Utilities Group, April 2003.

12.8 Street identification

When carrying out any building work it is essential to know where all the utilities are placed. Some will be above ground, although most are now below ground (Figure 12.20). The utility companies can provide charts in hard copy, or now more usually electronically, showing where they have their service. The charts are only a guide and should not be taken as being very accurate; therefore it is also necessary to look for the signage, cover plates, pillars, and scars on the highway (Figure 12.1). The use of service detection equipment

Table 12.2 Identification of utilities.

Highway Authority service	Duct	Cable	Tape
Street lighting England and Wales	Orange	Black	Yellow with black legend
Street lighting Scotland	Purple	Purple	Yellow with black legend
Traffic control	Orange	Orange	Yellow with black legend
Telecommunications	Light grey	Light grey or black	Yellow with black legend

Source: National Joint Utilities Group, April 2003.

such as a CAT (Figure 12.21) will register live services beneath the ground, even through concrete. The service provider will also mark out their utility using flags or spray paint. Most new utilities have placed a layer of marker tape over their service to provide prior warning during excavation (Figure 12.22). High voltage electricity cables should have concrete slabs about 200 mm × 450 mm with the name of the electricity supply company embossed into the top surface. Figure 12.23 shows 11 kV cabling in red going into a junction box and running under communications ducting/conduits in green and grey. When excavating by machine, the concrete marker slabs or tape should provide the warning to change over to a hand dig.

Utilities that have been run in the past decade or so will be colour coded according to the information in Tables 12.1 and 12.2.

Exercise

Using Tables 12.1 and 12.2, identify the various utilities shown in Figure 12.24. The contractor had located all the services using a CAT scanner and marked their positions on the road surface before excavating.

Index

I

J

K

L